SAMPLING METHODS FOR APPLIED RESEARCH

TEXT AND CASES

SAMPLING METHODS FOR APPLIED RESEARCH

TEXT AND CASES

PETER TRYFOS
York University

WILEY

JOHN WILEY & SONS, INC.

New York • Chichester • Brisbane
Toronto • Singapore

ACQUISITION EDITOR	Brad Wiley
MARKETING MANAGER	Debra Riegert
SENIOR PRODUCTION EDITOR	Tony VenGraitis
DESIGNER	Kenny Beck
MANUFACTURING MANAGER	Dorothy Sinclair
ILLUSTRATION COORDINATOR	Jaime Perea
PRODUCTION COORDINATOR	J. Carey Publishing Service

This book was set in 10/12 Times Roman by Eigentype Compositors,
and printed and bound by Donnelley/Harrisonburg.
The cover was printed by Lehigh Press.

Library of Congress Cataloging-in-Publication Data
Tryfos, Peter.
    Sampling methods for applied research : text and cases / by Peter Tryfos
        p.   cm.
    Includes bibliographical references and index.
    ISBN 0-471-04727-9 (clotch : alk. paper)
    1. Sampling (Statistics) I. Title.
 QA276.6.T78   1966
 519.5'2–dc20   95-35575
                                                                                    CIP

L.C. Call no.   Dewey Classification No.   L.C. Card No.
ISBN 0-471-04727-9

Printed in the United States of America

10 9 8 7 6 5 4 3 2 1

PREFACE

This book is intended primarily as a main text at the upper undergraduate or graduate level for a course in applied sampling for students of business, government or health administration, economics, political science, and other social sciences; as a main or supplementary text in service courses on sampling offered by departments of statistics or mathematics; and for self-study by professionals in marketing research, auditing, purchasing, production, opinion polling, and other areas of business and government.

The text accommodates at least two types of course design. The first (the traditional approach for a technical course) requires students to do exercises and problems and to write exams. The second approach (frequently used in business courses) requires students, perhaps working in groups, to read, report, and discuss cases using the text as a resource. They may also do a project and write a paper involving sampling or a critical appraisal of sampling. The text includes examples, exercises, problems, and small and large cases suitable for either approach or for a combination of them. An effort was made to make the text as flexible as possible, so that even the broader cases could be informative to students using the traditional approach, and some of the problems would be interesting to students using the case approach.

This is not a mathematical statistics text, but neither is it a handbook of formulas only. Formulas are indeed provided, and care has been taken to arrange them for easy reference, but a strong effort was made to give intuitive understanding of sampling concepts and confirm sampling formulas by means of simple numerical illustrations.

It is assumed that readers have taken an introductory course in statistics. Since such an introductory course is required by most academic programs, this prerequisite should not be restrictive. However, Appendix A reviews in some detail the essential prerequisite concepts so that the text can be understood in the event these

concepts are not well-remembered, or even (in the case of mature readers) in the absence of an introductory course. High-school mathematics is sufficient. An interest in quantitative methods is presumed.

An effort was made to proceed from need to solution, from the general to the particular, beginning with a view of the forest before dealing with the trees.

Thus, Chapter 1 attempts to demonstrate by means of short cases the need for and importance of sampling. Most of the cases are dealt with fully later in the text. Chapter 2 aims to establish first that the complexity of the problems suggested by the cases in Chapter 1 is more apparent than real. It draws attention to the common elements of these problems and to the fact that in the large majority of applied problems there is need for estimating only two types of population characteristics. Chapter 2 then describes how known auxiliary information suggests different methods for sample selection and estimation. The appropriateness of stratified and single-stage sampling and of ratio/regression estimators in some extreme but highly instructive situations is thus established early, for random as well as other types of samples.

Chapters 3 to 7 deal, respectively, with simple, stratified, two-stage random sampling, ratio and regression estimators, and such special topics as estimation of ratios of population totals and subpopulation means.

Of particular importance in this sequence is Chapter 3, which attempts to provide a justification of random sampling. The "silly game," despite its appearance of frivolity, aims to answer in simple terms a frequently asked question: Why should any weight be given to a method with desirable properties if applied repeatedly and a large number of times, but which, in reality, is applied only once? The same illustration is also used to introduce the criteria (bias, variability) according to which methods of selection and estimation are compared.

Common to Chapters 4 to 7 (indeed, to the remaining chapters as well) is the "forest-to-tree" approach mentioned earlier. Each chapter begins with a motivation for the method, a summary of its features, and examples illustrating its application. This is followed by an explanation of its derivation and a discussion of special or similar cases. Results are summarized in easy to locate boxes. A final section summarizes the main non-technical points of the chapter.

Chapter 8 is the unifying chapter of random sampling. It is presented last rather than first (as the forest-to-tree approach would imply) because otherwise the target audience has too many new concepts to deal with at the very beginning. Chapter 9 presents the prediction approach to sampling, which is not included in other texts at this level. Chapter 10 examines nonresponse, measurement, and sampling frame errors. Randomized response, telephone surveys, and the measurement of television and radio ratings are also discussed in this chapter.

A case is a description of a real situation that lends itself to the application of methods. A case invites reflection and provides an opportunity for discussion. Unlike a standard problem, but very much like the real world, it may not have a single solution. The cases are based on the author's experience, which is largely

Canadian. For the purposes of this text, actual names, places and data were transformed, for several reasons. In the first place, dates such as 1992 were changed to 19X2 etc. in order to avoid needless obsolescence. Secondly, organization names were changed in order not to subject the original ones to unnecessary criticism (which students tend to dispense ferociously). Thirdly, occasionally the data were changed (for example, by applying uniformly a linear transformation) in order to preserve the confidentiality of original sources. Lastly, there was some concern that frequent references to Canadian places and names would be distracting in a book intended to be read in the United States and other countries. For example, the Maritimes may now be described as "East." Despite these changes, however, the essential integrity of the data, the problem, and the setting has been meticulously preserved.

Appendix B provides a comprehensive glossary and technical summary. In addition to demonstrating the unity of the text, it may also be used as a stand-alone reference by practitioners.

A diskette accompanying the text contains the data files referred to in the text and the cases, a simulation program used in one of these cases, and a "no-frills" computer program designed to assist calculations involving complicated formulas. Appendix C describes the data and programs included in this diskette. Finally, Appendix D provides solutions to selected problems.

An *Instructor's Manual*, available from the publisher on request, includes solutions to all the problems in the text, and teaching notes with sample computer programs describing the author's treatment of the cases, possible extensions, and alternatives.

The author is indebted to IMS of Canada Ltd., NCH Promotional Services, and Nielsen Marketing Research for data and other information. None of the cases or examples in this text, however, should be interpreted as describing the actual practices and policies of these firms.

The author is also grateful to Professor K. H. Chan and Mr. Alan Middleton for suggesting and helping design two of the cases, and to the following people for positive and negative comments on earlier drafts that resulted in significant improvements: Professor Bruce L. Bowerman, Miami University; Professor Sharon Hunter Donnelly, University of Tennessee, Knoxville; Professor John L. Eltinge, Texas A&M University; Professor William A. Erickson, The University of Michigan; Professor Harry Joe, The University of British Columbia; Professor George A. Marcoulides, California State University at Fullerton; Professor Peter Peskun, York University; Professor Gary Simon, New York University; and Professor Arthur B. Yeh, Rutgers University.

The greatest debt, however, is owed to Barbara J. Tryfos, who read, suffered through, and corrected several versions of the manuscript, patiently restoring clarity on the many occasions when it was not present.

If any errors and faults remain despite the help of so many, they are entirely the author's.

CONTENTS

PART TWO: CASES

APPENDICES

PART ONE

TEXT

CHAPTER 1

Introduction

1.1 SAMPLING: THE ISSUES

A *sample* is a part drawn from a larger whole. Rarely is there interest in the sample per se. Almost always a sample is taken in order to learn something about the aggregate (the *population*) from which it is drawn. For instance:

- In an opinion poll, a relatively small number of persons are interviewed, and their opinions on current issues are solicited in order to discover the attitude of the community as a whole.
- The viewing and listening habits of a relatively small number of persons are regularly monitored by ratings services, and, from these observations, projections are made about the preferences of the entire population for available television and radio programs.
- Large lots of manufactured products are accepted or rejected by purchasing departments in business or government following inspection of a relatively small number of items drawn from these lots.
- At border stations, customs officers enforce the laws by checking the effects of only a small number of travelers crossing the border.

- Auditors often judge the extent to which the proper accounting procedures have been followed by examining a small number of transactions, selected from a larger number taking place within a period of time.
- Countless sample surveys are carried out, regularly or occasionally, by marketing and advertising agencies to determine consumers' expectations, buying intentions, or shopping patterns.
- Some of the best known measurements of the economy rely on samples, not on complete enumerations. The weights used in consumer price indexes, for example, are based on the purchases of a sample of urban families; the prices of the individual items are averages established through national samples of retail outlets. Unemployment statistics are based on monthly national samples of households. Similar samples regularly survey retail trade, personal incomes, inventories, shipments and outstanding orders of firms, exports, and imports.

In every case, a sample is selected because it is impossible, inconvenient, slow, or uneconomical to monitor the entire population.

The principal questions in any study involving sampling are: *How should the sample be selected? How large should it be? How should the population characteristics be estimated? How reliable are these estimates?*

These questions are related to one another. Their examination is the objective of this book. Before we begin, however, it will be useful to consider some specific cases involving sampling, and to reflect on the preceding questions in each case. We shall return to some of these cases frequently throughout this book.

1.2 CASE: PRINT MEDIA RESEARCH

Figure 1.1 shows a portion of a questionnaire used in an annual survey[1] conducted by the Institute for Print Media Research. This organization sells information to advertisers, advertising agencies, and publishers concerning the number of readers of each newspaper or magazine and the characteristics of those readers.

The questionnaire consists of about 300 questions. Those shown here deal with the purchase and use of food and household products. Others inquire about other products and services, the newspapers and magazines purchased and/or read by the individual, and the individual's education, income, occupation, and assets.

About 10,000 individuals of age 12 or greater are interviewed for each annual survey from a population of millions of men and women 12 years old or older.

From the responses to the questions shown, Print Media Research could estimate, for example, the proportion of individuals who read *Newsweek* and who buy, use,

---

[1] The term *survey* generally refers to the collection of data by means of interviews, questionnaires, or direct observation. The entities surveyed could form a whole (population survey) or a part (sample survey).

FOOD PRODUCTS

Do you buy, use or serve the following products? If you do, circle the number 8 to indicate YES, and then circle the appropriate number to indicate how frequently the product is used in your household.

	DO YOU BUY, USE, OR SERVE?		More than once a DAY	Once a DAY	4 to 6 times a WEEK	2 or 3 times a WEEK	Once a WEEK	2 or 3 times a MONTH	Once a MONTH or less
	Yes	No							
331) Cereals, cold, unsweetened	8	9	1	2	3	4	5	6	7
333) Cereals, cold, pre-sweetened	8	9	1	2	3	4	5	6	7
335) Cereals, hot	8	9	1	2	3	4	5	6	7
337) Beef, fresh	8	9	1	2	3	4	5	6	7
339) Lamb, fresh	8	9	1	2	3	4	5	6	7
341) Pork, fresh	8	9	1	2	3	4	5	6	7
343) Veal, fresh	8	9	1	2	3	4	5	6	7
345) Ham	8	9	1	2	3	4	5	6	7
347) Bacon	8	9	1	2	3	4	5	6	7
349) Frozen meat	8	9	1	2	3	4	5	6	7

HOUSEHOLD PRODUCTS

	DO YOU BUY OR USE?		In about how many washloads each WEEK, on average, do you use them?					
	Yes	No	Six or more	Five	Four	Three	Two	One or less
351) Soaps and detergents for regular laundry	8	9	1	2	3	4	5	6
353) Soaps and detergents for fine fabrics	8	9	1	2	3	4	5	6
355) Laundry pre-soaks and pre-cleaners	8	9	1	2	3	4	5	6
357) Bleach	8	9	1	2	3	4	5	6
359) Fabric softeners – regular	8	9	1	2	3	4	5	6
361) Fabric softeners – aerosol	8	9	1	2	3	4	5	6
363) Other laundry additives	8	9	1	2	3	4	5	6
365) Automatic dishwashing detergent	8	9	1	2	3	4	5	6
367) Rinse agents for automatic dishwashers	8	9	1	2	3	4	5	6
369) Dishwashing liquid	8	9	1	2	3	4	5	6

Figure 1.1 Extract from survey questionnaire, Print Media Research

or serve ham. Or, the number of *Vogue* readers who use dishwashing liquid once a week.

The information compiled by this survey is useful, for instance, to a manufacturer of cosmetics for young people and the advertising agency handling the account, who wish to determine the magazines in which an advertisement will have the greatest effect. Or, the publisher of a newsmagazine may wish to know the kind

of reader the magazine attracts; it is possible, for instance, to obtain the profile of a reader of a given periodical, to examine whether there is any relationship between income and expenditure on a given product or service, or to determine whether there are substantial differences in the characteristics of the readers of two publications.

How should the sample of 10,000 be selected? How should the characteristics of the population be estimated? How reliable are they?

1.3 CASE: TELEMEDIA INC.

Whether or not a television program or show will continue to be broadcast, and the price it may charge for advertising time, depends on the rating of the program. This *rating* is simply an estimate of the percentage of individuals watching the program at a given time. Table 1.1 is an excerpt from a ratings book available to subscribers only from Telemedia, a company specializing in providing television and radio ratings. The table shows the ratings of programs broadcast in a northeastern city between 7:30 and 8:00 P.M. on a Thursday in September.

We read, for example, that 36% of all persons two years old or older were watching television during this period. 7% of all persons were watching *Jeopardy*, and

TABLE 1.1   Telemedia Ratings, Excerpt

		All 2+		Ratings				
Station	Program	Rating	Share	Adults 18+	Women 18+	Men 18+	Teens 12–17	Children 2–11
	All stations	36	100	37	39	35	32	24
CBLF	(Movie)	-	-	-	-	-	-	1
CBLT	*Cheers*	1	3	1	1	1	-	-
CFMT	*Jeopardy*	7	19	7	8	6	7	4
CFTO	*Campbells*	2	6	3	3	2	1	0
CHCH	*Mannix*	1	3	1	1	1	1	1
CICO	*Nature*	1	3	1	1	1	1	1
CITY	*M.A.S.H.*	6	17	6	6	7	5	1
WGRZ	*Dating Game*	3	8	3	3	4	5	3
WIVB	*Pyramid*	4	11	4	4	3	3	3
WKBW	*Simon*	1	3	1	1	1	1	-
WUTV	*Soap*	1	3	1	1	1	3	1
WBEN	*Jackpot*	4	11	4	4	3	1	4
	Others	5	14	5	6	5	4	5

4% were watching *Jackpot*. The last two figures represent 19% and 11% respectively of people aged two years old or older who were watching television during this period (the *shares* of the programs). Ratings are also presented separately for each age/sex class.

These ratings are based on information provided by a sample of about 1,800, selected from a population of about 2.7 million individuals two years old or older in the city, as shown in Table 1.2.

TABLE 1.2 Population and Sample		
Age/Sex Class	**Sample**	**Population**
All persons 2+	1,778	2,690,830
Adults 18+	1,250	1,910,880
Women 18+	650	981,800
Men 18+	600	929,080
Teenagers 12–17	215	286,250
Children 2–11	313	493,700

Obviously, it would be enormously expensive to monitor regularly the viewing habits of the entire population. This fact is never disputed, but Telemedia finds itself frequently under attack by subscribers and others. Whether motivated by a noble cause ("Why is there so little opera on television?"), or prompted by disagreeable results ("Why is it that the ratings of my station fluctuate so much?"), critics often question the method of sampling used (which consists of selecting approximately 0.07% from each age/sex class), the sample size ("You mean your sample is just seven-hundredths of one percent of the population?"), and the lack of any data by which to judge the accuracy of the ratings.

How should Telemedia's sample be selected? How should the characteristics of the population be estimated? How reliable are these estimates?

1.4 CASE: NORPOWER INC.

NorPower is the exclusive supplier of electricity to 2,819,514 residences (houses and apartments). Over 80% of these residences have individual meters; the remaining 20% are in apartment buildings and share common meters. A list of all individually- and bulk-metered residences is maintained by the company. Each year, the utility's market research department collects information on ownership and use of appliances and of heating and other equipment by means of the questionnaire shown in Figure 1.2.

Mark **X** next to those appliances you own or use in your dwelling.

Your kitchen:
❏ Electric continuous-clean range
❏ Electric self-cleaning range
❏ Other electric range
 (no built-in cleaning)
❏ Gas stove or range, any type
❏ Other cooking facilities
 (describe) _____
❏ Without cooking stove or range
❏ 2- or 3-door combination *Is it frost-free?*
 refrigerator-freezer ❏ Yes ❏ No
❏ Other electric refrigerator ❏ Yes ❏ No
❏ Without electric refrigerator *Number*
❏ Automatic electric dishwasher *in use*
❏ Home food "deep" freezer . . . . ❏
❏ Microwave oven . . . . ❏
❏ Electric kettle . . . . ❏

Your *own* laundry equipment:
(Do not include coin-operated or any other laundry equipment used outside your home.)
❏ Automatic washer
❏ Other electric washer
 (wringer, spin-dry, etc.)
❏ Combination washer-dryer (one unit)
❏ Electric clothes dryer
❏ Gas clothes dryer

Other electric appliances: *Number in use*
❏ Black-and-white television . . . . ❏
❏ Color television . . . . ❏
❏ Swimming pool filter
❏ Electric sauna

Water heating:
Fuel used in your water heater:
❏ Electricity ❏ Gas ❏ Oil
❏ Other or none ❏ Don't know
If swimming pool heated:
❏ Electric ❏ Oil or gas ❏ Solar

Principal heating equipment:
(Check main heating only.)
❏ Forced hot air or hot-water furnace
❏ Heating stove (including space heater)
❏ Electric baseboard or cable
❏ Electric heat pump
❏ Other (describe) _____
❏ Don't know

Fuel for principal heating equipment:
(Check main heating only.)
❏ Oil or other liquid fuel
❏ Piped or bottled gas
❏ Electricity
❏ Other (describe) _____
❏ Don't know
Have you added insulation to the attic in the past two years?
❏ Yes ❏ No
Have you added solar heating equipment in the past two years?
❏ Yes ❏ No
If yes, describe _____

Supplementary equipment:
❏ Central air-conditioning
❏ Electric power humidifier on furnace

Number in use
❏ Plug-in electric space heater . . . . ❏
❏ Built-in electric space heater . . . . ❏
❏ Wood stove . . . . ❏
❏ Electric blanket . . . . ❏
❏ Window air-conditioner . . . . ❏
❏ Humidifier (plug-in) . . . . ❏
❏ Dehumidifier (plug-in) . . . . ❏

Your home:
Piped running water:
❏ From municipal
❏ Your own pressure system
❏ None

Do you:
❏ Own your home? ❏ Rent your home?

Do you live in a:
❏ Single-family house (detached or attached)?
❏ Apartment or flat (including duplex or triplex, etc.)?
❏ (All others; describe) _____

When was your dwelling unit built?
Before 1956   1956–60   1961-65
 ❏   ❏   ❏

1966–70   1971–75   After 1975
 ❏   ❏   ❏

Total number of rooms in your dwelling:
(Do not include bathrooms, halls and unfinished rooms.)
 _____

Number of persons living in your household:
 _____

Figure 1.2 NorPower questionnaire

The results of the survey are published in a 50-page report containing numerous tables and figures. Two of these figures, showing the proportions of households in rural and urban areas owning color and black-and-white television (the so-called "saturation rates" for these appliances), are reproduced in Figure 1.3.

This report is used for planning purposes by the utility, but also by appliance manufacturers, equipment contractors, municipalities, and others. In addition to the published material, users of this survey may request special tabulations of the raw data stored in a computer file.

It is, of course, possible to send the questionnaire to all 2.8-odd million dwellings, but the costs of printing, mailing, editing, and processing the completed questionnaires would be too high (assuming, conservatively, a cost of $1 per questionnaire, the total cost would be $2.8 million). The obvious alternative is to send the questionnaire to a sample of households.

How should that sample be selected? How should the estimates be formed?

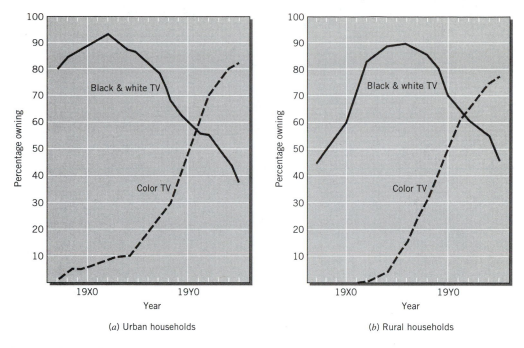

(a) Urban households   (b) Rural households

Figure 1.3 Appliance saturation rates

1.5 CASE: PHARMACOM RESEARCH

Subscribers to Pharmacom's *Pharmaceutical Product Sales (PPS) Report* receive each month a 600-page publication listing the estimated sales of all pharmaceutical products distributed in the country. There are over 100 manufacturers and over 3,200 pharmaceutical products. The *PPS Report* provides estimates of the sales of each product during the month, the year to date, and the most recent 12-month period. These sales are classified by manufacturer, by therapeutic class, and by the amount of sales. An example of these listings is shown in Figure 1.4.

Of particular interest to subscribers is the classification according to therapeutic class or "market." Over 70 principal markets are distinguished, ranging from amoebacides and analgesics (classes 01 and 02) to thyroid preparations and vitamins (classes 72 and 76).

Obviously, manufacturers know accurately and with certainty the sales of *their own* products. They do not know the sales of their competitors' products, the total sales in each market, or the market share of their own or competitors' products. It is for this reason that they are willing to pay the substantial cost of subscribing to the *PPS Report*. No other source for such information is available.

Leading Pharmaceutical Products

	Product	Manufacturer	Class	Sales ($000)	%Chng
1	Ventrolin	Granox	28	130987	−5.55
2	Kapoten	Streb	31	128951	9.89
3	Zantax	Granox	23	99025	3.10
4	Cardizon	Southern	31	89254	1.30
5	Vosotec	First	31	75509	8.03
6	Twophasil	Windham	33	72206	5.41
7	Edilat O	Morris	31	71558	5.59
8	Soltaren SR	Greenock	9	56338	−2.24
9	Seldine	Mersey	14	53677	−0.39
10	Enfilac	Reed	54	53518	−1.33
11	Mevakor	AC&E	32	53262	−3.05
12	Kardizem SR	Southern	31	52744	6.71
13	Soltaren	Greenock	9	49649	4.71
14	Ordec	Ordex	33	47899	4.35
15	Omnipast	Waters	40	47398	6.29
16	Ceclar	Larsen	15	45039	−6.62
17	Standimune	Zondac	30	40074	−9.03
18	Sulrate	Southern	23	39238	−0.28
19	Metoxin	AC&E	15	38541	−2.97
20	Tanormin	IFI	31	37348	5.01
21	Mintovral	Windham	33	36752	−7.41
22	Peptid	AC&E	23	36382	7.53
23	Modaret	AC&E	41	34291	2.82
24	Lopic	Daniels	32	32597	−3.88
25	Theordur	Strong	28	32184	−7.13
26	Isopton	Lerner	31	31410	2.27
27	Natrong SR	Rhinelab	31	31378	8.62
28	Diabita	Charet	39	30551	9.72
29	Minacin	Tanner	15	29323	2.01
30	Zevirax	Burwell	81	28987	−8.33
31	Amaxil	Armstrong	15	28148	8.17
32	Hismanol	Petersen	14	27807	−5.20
33	Atrovint	Mannheim	28	25649	−1.51
34	Premaron	Armstrong	52	25196	−9.02
35	Surgon	Rousseau	9	23959	−4.54
36	Bellovent	Granox	28	23894	8.67
37	Insure	Starr	60	23585	0.68
38	Cinemet	AC&E	20	23373	−8.50
39	Analatin	Morris	31	23015	−0.69
40	Motilion	Petersen	23	22770	1.65

Figure 1.4 *PPS Report*, sample page

As can be seen from Figure 1.5 which illustrates the flow of domestic sales from manufacturers to retailers, nearly all pharmaceutical products are sold through drugstores and hospitals.

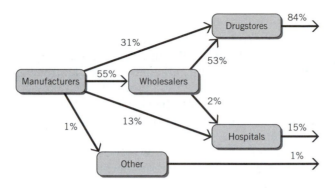

Figure 1.5   Flow of domestic sales, pharmaceutical industry

Pharmacom derives its estimates in the following manner. Once each month, field workers visit a number of selected drugstores and hospitals across the country, and photograph all invoices by drug manufacturers or wholesalers paid since the previous visit. Invoices show the manufacturer, the name of the product, the number of units sold, the unit price, and the dollar amount of the sale. Different sizes or strengths of the same product are treated as different products. The films are processed, and, after editing for accuracy and consistency, their contents are entered into a computer file. On the basis of this set of raw data from the selected drugstores and hospitals, estimates are made of the total sales of pharmaceutical products for the entire country.

How should these surveys be carried out? How should the estimates of product sales be made?

1.6 CASE: CANABAG MANUFACTURING COMPANY

Canabag is a manufacturing company producing plastic bags, such as shopping and garbage bags, and is experiencing a shortage of current capital. This is because the collection of accounts receivable has been slow at a time when supplier credit has been reduced. Since Canabag's accounts receivable comprise nearly one third of total assets, an independent audit is highly desirable.

At the president's request, the auditor spelled out the audit objectives as follows:

1. to determine the proportion of accounts under control
2. to determine if the error rate on accounts receivable has changed substantially from that of previous years
3. to determine if accounts receivable are materially overstated or understated.

To determine if a selected account receivable is under control, the auditor examines the account and establishes whether or not the control procedures were properly followed for all recorded transactions. As a result, the account is classified as either "in control" (that is, all control procedures were properly followed), or "not in control" (at least one control procedure was not complied with).

Each selected account is also to be checked for the numerical accuracy of the recorded transactions, and classified as "in error" if the stated balance is incorrect, or as "correct" if otherwise. An account may be in error for a variety of reasons: a clerical error in calculating the sum of invoiced items, a failure to credit a payment from the customer, a credit to the wrong account, a discrepancy between goods actually shipped out and goods invoiced, and so on. In the past few years, about 25% of Canabag's accounts receivable were in error.

The auditor also estimates the total absolute dollar error in accounts receivable, and indicates whether accounts receivable are "materially" overstated or understated. Based on the size of the accounts receivable balance in relation to total assets, expected income, and outstanding debit, the auditor feels that $20,000 is an appropriate material amount. In other words, if the total absolute dollar error exceeds $20,000, the difference will be judged substantial enough to warrant further investigation and action.

There are currently 988 outstanding accounts receivable. Since it would be impossible to examine all accounts receivable with the limited resources and staff available, the auditor plans to rely on a sample of accounts in order to carry out the audit objectives.

In the past, the auditor, guided by her professional experience, would select accounts that appeared suspicious, audit each for proper authorization and accuracy, and make adjustments whenever an error was found. The auditor would continue selecting accounts for audit until available time had run out. Estimates of the proportion of accounts not under control or in error, and of the total absolute dollar error were based on this sample.

Do you agree with the auditor's past practice? If not, how should the audit be carried out?

1.7 SOME REFLECTIONS

In all these cases, information is desired about a population. For various reasons, it is not possible to examine all the elements of the population. The required information will be supplied by a sample.

We imagine a population as an existing collection of real elements. In principle, and with appropriate resources, each element *could* be identified and located, and inspected, interviewed, questioned, or observed. Put simply, we imagine a population as being "out there," with its elements waiting to be questioned or observed.

For example, a shipment of light bulbs packaged in boxes, cartons, and containers is located somewhere in the warehouse awaiting inspection. On December 31, there are so many outstanding accounts receivable, each showing such and such a book balance, and "carrying" an error amount to be revealed through the audit of transactions. At a given point in time, there are so many people living in a certain region, each of whom either watched or did not watch television between 7:30 and 8:00 P.M. on March 10 of this year; we do not know who did and who did not, but, in principle, and with appropriate resources, we could find out.

Each population element can be imagined to "carry" its features with it. For example, each of the millions of individuals 12 years old and older who form the population of interest to Print Media Research (PMR) can be imagined as having on his or her person a little book (when one imagines, one can take many liberties . . .) on which is written everything that PMR wishes to know: the person's sex, age, marital status, income, . . . , how often the person eats lamb at home, . . . , whether or not the person subscribes to *Vogue*, . . . , the number of washloads per week using liquid bleach, and so on. If only PMR could get hold of all these books, it could compile an accurate description of the entire population (if only, indeed . . .).

Rarely are we interested in a single feature of the population elements. In the case of PMR, for example, the excerpt from the questionnaire suggests that there are hundreds of features PMR is interested in.

It may or may not be possible to observe or measure accurately all the selected population elements. For example, PMR hopes that when the selected individuals are questioned, and perhaps for a suitable consideration, they will release their "little books," that is, they will be willing to respond truthfully to the questionnaire.

However, individuals often refuse to be interviewed or to complete question-naires. The sample, in other words, may have been carefully selected but not all se-lected elements can be measured. If it can be assumed that the two subpopulations—those who respond and those who do not—are similar, then there is no problem. But if this not the case, treating those who respond as representative of the entire population may result in misleading estimates.

Furthermore, individuals may respond, but their responses may not be truthful. For example, PMR may wish to believe that individuals reveal or report their income accurately, but often reported income is at variance with true income, even when people are assured that their responses are treated confidentially.

Some populations change with time. In the PMR case, for example, it is clear that individuals' tastes, attitudes, and preferences change over time. If PMR aims to capture these at a given point in time, then the population must remain unchanged (must "stand still," if you like) while the sample is being taken. To exaggerate the point, there would be hardly any demand for the information supplied by PMR if the survey of 10,000 individuals were carried out by one field worker over a period of years. In the Pharmacom case, product sales change from month to month, requiring a different sample each month.

It is unnecessary to point out that the information furnished by a sample is used by decision makers. But it should be kept in mind that many decision makers are

usually involved. Each may need a different subset of the information. Consider the NorPower survey as an illustration. The information is gathered by the market research department of the utility, but it is used by manufacturers of appliances, equipment contractors, and municipalities, in addition to other departments of the utility itself. Electrical appliance manufacturers, for instance, may have no interest in the part of the questionnaire dealing with home heating equipment.

1.8 A VERY BRIEF HISTORY OF SAMPLING

Sampling must have been practiced much longer than it has been formally studied. We can well imagine a Roman merchant using a sample to evaluate the quality of a boat's cargo of grain, or the dockyard superintendent responsible for provisioning Royal Navy vessels in Nelson's time sampling a shipment of dried meat.

Sampling as a formal field of study did not begin until early in the twentieth century. The standard bearers of the new discipline were government statisticians charged with collecting economic and demographic information. Until then, the prevailing view was that only a complete enumeration of the population (that is, a *census*) could provide reliable information.

During the first quarter of the twentieth century, support gradually grew for samples selected in a purposive fashion, so as to make them "representative" of the population with regard to certain known variables and attributes. Thus, for example, a national sample of farms for the purpose of estimating current crop acreage might be formed by selecting farms from every geographical region in the country in proportion to the region's acreage of farms as ascertained by the most recent census.

From about 1925 on, theory and practice gradually swung in favor of "randomization" as a method for selecting a sample. Random sampling may now be described as the dominant method for estimating the unknown characteristics of a population, not only for government purposes but also in science, the humanities, and business. This swing was helped by, and, in turn, contributed to the growth of a large mathematical literature dealing with the properties of various random sampling methods and estimates based on these methods.

The 1980s have witnessed a reexamination and reevaluation of the foundations of random sampling. New support was found for the long-in-disfavor purposive approach to sampling. Instead of selecting a sample randomly, it now appears desirable under certain circumstances to select it according to certain criteria. It is too early to tell the outcome of the continuing debate, but the recent research has opened up new alternatives to the practitioner.

The structure of this book follows this historical development of sampling theory since 1925. After some preliminary observations in Chapter 2 applying to any populations and to samples of any kind, we explain in Chapters 3 to 8 the logic and principal results of random sampling. In Chapter 9 we consider the purposive

(or prediction, as it is now called) method as an alternative method of sample selection. The final Chapter 10 deals with nonsampling considerations important to both approaches.

1.9 A DISCLAIMER

In the author's experience, there is a widespread belief that sampling (in particular, so-called "scientific" sampling) can reveal the true characteristics of a population. In other words, that there is some way—perhaps not intuitively obvious or simple—of selecting a part of a population so that the resulting estimates are *guaranteed* to be correct. That is not the case, and the reader who expects to find such magic in this book should be advised against reading further. There are no known methods of sample selection and estimation which ensure *with certainty* that the sample estimates will be equal to the unknown population characteristics. The only exceptions apply to situations of no practical interest, such as when the sample is the population, or the population happens to have such extreme features as those described in the next chapter.

It is equally true that there is no known method by means of which the sample estimates can be guaranteed to "bracket" the true population characteristics. Consider, for example, the following newspaper excerpt:

The End of the Recession?

The latest Sverdlup poll shows 45 percent of the sampled companies planning to hire next month. Sverdlup says that this estimate is accurate within 4.5 percentage points. . . .

The impression given by this excerpt is that, although it cannot be guaranteed that 45 percent of all companies intend to hire, it can be guaranteed that the percentage of all companies intending to hire is between 40.5 and 49.5. In other words, that the true percentage is not less than 40.5 or greater than 49.5. This impression is misleading, for there is no known method justifying this statement.

Unfortunately, such statements appear quite frequently in the media. Note that the excerpt does not read: "Sverdlup says that this estimate is accurate within 4.5 percentage points 19 times out of 20," or similar numbers. The latter statement suggests that there is some uncertainty about the estimate, although the meaning of "19 times out of 20" and its justification may not be clear at this stage.

The reader must understand that relying on a sample nearly *always* involves a risk of reaching incorrect conclusions. Sampling theory can assist in reducing that risk, but a certain risk is always present in every sampling study of practical interest.

CHAPTER 2

Preliminaries

2.1 INTRODUCTION AND SUMMARY

In this chapter, we try to establish more clearly the purpose of sampling and the kinds of samples that would be appropriate in some extreme but instructive situations.

We noted earlier that the reason for taking a sample is to obtain estimates of unknown characteristics of a population. The cases examined may have left the impression that there are infinitely many population characteristics that could be of conceivable interest. That is true, but a more careful examination will show that the characteristics of principal interest in applied studies are of four types only, and that these four can in fact be reduced to two types: the average value of a variable and the proportion in a category. This is fortunate and convenient, for it simplifies considerably the task of sampling theory.

Sampling in practice is nearly always without replacement. Under this method, selected population elements are not eligible to be selected again, and a population element can appear at most once in the sample. The opposite method, sampling with replacement, appears to most people wasteful and inefficient. We will show that this impression is justified, that is, that a sample without replacement is better than a sample with replacement of the same size. We shall also confirm an equally appealing impression, that a larger sample without replacement is better than a smaller one, whether or not the sample is randomly, purposively, or arbitrarily selected.

Available information about a population may include not only the list of the population elements, but also known attributes and variables. Such "auxiliary"

information makes it possible to divide the population into groups, or to relate an unknown variable or attribute to known ones, thereby opening up several alternative methods for selecting the sample and estimating the population characteristics.

When the population is homogeneous, that is, when all the elements are identical, it is obvious that a sample of just one element is sufficient to provide correct estimates. A little more interesting is the case of a population consisting of a number of homogeneous groups. In such a case, we will show that it is possible to obtain correct estimates by selecting just one element from each group.

Another extreme but informative case is one in which each group is a "miniature" of the entire population. Here again we intend to show that it is possible to estimate correctly the population characteristics by selecting the smallest group and all its elements.

The last extreme case examined in this chapter involves an exact relationship between an unknown variable and a known auxiliary variable. We shall show that correct estimates are possible, requiring a sample of only as few elements as the number of unknown parameters of the relationship.

In all these cases it does not matter whether or not the sample is selected randomly, purposively, or arbitrarily.

All these are, of course, extreme and unrealistic situations, rarely encountered in practice. They are, however, highly instructive, suggesting how the sample could be selected in situations where the conditions are approximately met.

2.2 POPULATIONS AND THEIR CHARACTERISTICS

A *population* is the aggregate of interest. A population consists of *elements*. Exactly what constitutes a population and its elements will, of course, vary from one case to another.

In the Canabag case of Section 1.6, for example, the population of interest is the collection of 988 accounts receivable. A list of these accounts is available; each entry in this list identifies one such account. In the Print Media Research case of Section 1.2, the population consists of all men and women 12 years old and older in the country; the exact number is perhaps not known, but a reasonably good approximation can be obtained from the most recent population statistics. In the case of NorPower (Section 1.4), the population consists of the 2,819,514 housing units (houses and apartments) served by the utility, while in the case of Pharmacom (Section 1.5) the population consists of the 6,878 drug outlets (5,720 drugstores and 1,158 hospitals).

With each population element, there are associated numerical values of one or more variables, and categories of one or more attributes.

In general, a *variable* is a feature or aspect of an element that lends itself naturally to a numerical description. Height, weight, distance, and income are examples of

variables. On the other hand, only categories of an *attribute* can be distinguished. For example, a person's gender (male, female) is an attribute; so are a person's marital status (single, married, divorced, other) and state of residence (Alabama, Alaska, ..., Wyoming). Categories, of course, can also be formed from variables; for example, persons with incomes under $20,000. A *characteristic* is a summary measure based on a variable or attribute (for example, the average value or total of a variable, the proportion of elements in a category).

The primary objective of sampling is to obtain estimates of the characteristics of the variables and attributes of interest in the particular case.

Let us refer to the Canabag case for an explanation of this terminology. We can *imagine* the population of accounts receivable and the features of this population in which the auditor is interested arranged in the form of Table 2.1.

The control quality of an account is an attribute. With respect to this attribute, an account falls into one of two categories: it is either under control (Yes) or it is not (No).

The error quality of an account is likewise an attribute. An account either has an error (Yes) or does not (No).

The absolute error amount is defined as the absolute difference between the true balance and the book balance of the account. In our terminology, the absolute error amount is a variable, and with each account there is associated a numerical value of this variable. For example, accounts no. 1 and 988 are shown to have zero error; the book balance of account no. 2 is shown to differ from the true balance by $10, while that of account no. 3 differs from the true balance by $25. (We note in passing that the error quality is a redundant attribute, as it can be derived from the error

TABLE 2.1 Canabag Manufacturing, Imaginary Population Structure

Account No.	Under Control?	In Error?	Absolute Error Amount ($)
1	Yes	No	0
2	Yes	Yes	10
3	No	Yes	25
⋮	⋮	⋮	⋮
988	Yes	No	0
Number of accounts under control	920		
Proportion of accounts under control	0.931		
Number of accounts in error		185	
Proportion of accounts in error		0.187	
Total absolute error			12,670
Average absolute error per account			12.824

amount. For if the error amount is $0, the account is not in error; if it is not $0, it is in error.)

If the entries in all 988 rows of the above table were known, it would be a simple matter to derive the characteristics of the variables and attributes shown in the bottom of the table. For example, a count of the number of "Yes" entries in the second column would give the number of accounts under control in the population (920 in the above illustration); dividing this figure by 988, the number of accounts receivable, would yield the proportion of accounts under control (0.931). Likewise, adding up the entries in the last column would yield the total absolute error amount ($12,670), and dividing the last figure by 988 would give the average error per account ($12.824 in the illustration).

We emphasize that the entries in this table are *not* known. If they were known, there would obviously be no point in taking a sample. A realistic depiction of the population of accounts receivable appears in Table 2.2.

The Canabag audit sample is to provide *estimates* of the characteristics shown in italics in Table 2.2.

For another illustration, we consider the Pharmacom case. There are, it will be recalled, 5,720 drugstores and 1,158 hospitals, or 6,878 outlets in total. Also, there are 3,200 different pharmaceutical products, the sales of which are to be estimated each month. For a given month—say, January 1993—the population and characteristics of interest can be imagined as shown in Table 2.3.

A question mark in the body of the table stands for the unknown dollar sales of a given product at a given outlet. The population characteristics of interest are the totals of the 3,200 variables (last row).

TABLE 2.2 Canabag Manufacturing, Population Structure

Account No.	Under Control?	In Error?	Absolute Error Amount ($)
1	?	?	?
2	?	?	?
3	?	?	?
⋮	⋮	⋮	⋮
988	?	?	?
Number of accounts under control	?		
Proportion of accounts under control	?		
Number of accounts in error		?	
Proportion of accounts in error		?	
Total absolute error			?
Average absolute error per account			?

TABLE 2.3 Sales of Pharmaceutical Products, January 1993

Outlet No.	Product 1	Product 2	· · · · · ·	Product 3200
1	?	?	· · ·	?
2	?	?	· · ·	?
⋮	⋮	⋮	⋮	⋮
6,878	?	?	· · ·	?
Total sales	*?*	*?*	*· · ·*	*?*

In this case, the main table entries and the totals vary, of course, from month to month, as do perhaps the numbers of outlets and products. In other words, the population and its characteristics are different every month, requiring that a different sample be taken and estimates made each month.

For a last illustration, we consider the NorPower case. The population elements are the 2,819,514 housing units (houses and apartments) served by the utility. A list of these units is maintained by NorPower. The questionnaire is addressed to the person (owner or tenant) who pays the electricity bill, and is presumably the head of the household. Of course, some households do not have electricity, some have more than one meter, and some share the housing unit. Therefore, the correspondence between electricity-using housing units and households is not exactly one to one. Nevertheless, in light of what we know about the use of electricity in homes, it is reasonable to suppose—as indeed NorPower implicitly does (see Figure 1.3)—that the two populations are, for all practical purposes, one. Later, we shall make a distinction between the population of interest and the one from which a sample is actually selected (the so-called *sampling frame*), but for present purposes the population and characteristics of interest in this case can be outlined as in Table 2.4.

In this case, the number of variables and attributes of interest is large, as can be inferred from the questionnaire. Only a few of the population characteristics of possible interest are shown in Table 2.4. The "number of color TVs in use" is, of course, a variable. "Fuel used in water heater" is an attribute. With respect to this attribute, a household belongs to one of the following categories: "electricity-using," "oil-using," "gas-using," "uses other fuel," "uses no fuel," and "does not know." Incidentally, note the good practice of making the categories of an attribute listed in the questionnaire mutually exclusive (that is, nonoverlapping) and collectively exhaustive (covering all possible responses).

As observed earlier, the purpose of taking a sample is to obtain estimates of certain population characteristics. A careful study of the information sought in the above cases should confirm that *the population characteristics of interest are not of*

TABLE 2.4 NorPower, Population Structure

Unit/Household No.		Number of Color TVs in Use		Fuel Used in Water Heater	
1	⋯	?	⋯	?	⋯
2	⋯	?	⋯	?	⋯
⋮	⋮	⋮	⋮	⋮	⋮
2,819,514	⋯	?	⋯	?	⋯
Total number of TVs in use		?			
Number of households using electricity in water heaters				?	
Number of households using gas in water heaters				?	
Number of households using oil in water heaters				?	

infinite variety but belong instead to one of four general types. We list these below together with the symbols that will identify them in the sequel.[1]

τ: the total value of a given variable in the population (for example, in the Pharmacom case, the total dollar sales of Product 1 by all retail outlets in the country)

μ: the average value of a variable in the population (for example, in the Canabag case, the average absolute dollar error amount per account receivable)

π: the proportion of elements in the population falling into a given category (for example, the proportion of households served by NorPower and using gas for fuel in their water heaters)

τ': the number of elements in the population falling into a given category (for example, the number of households served by NorPower that use gas for fuel in their water heaters).

To this list of symbols, we add for future reference:

N: the population size, i.e., the number of elements in the population

n: the sample size, i.e., the number of elements in the sample.

---

[1]The notation of this text is explained in Appendix B.

One of the main questions in any sampling study, it will be remembered, is how to estimate the unknown population characteristics. In applied studies, the task is substantially simplified because all that need be shown is how to form estimates of *only four* types of population characteristics. In any given study, of course, many proportions, totals, and averages may need to be estimated.

Evidently, the four are not the only types of population characteristics. For example, there may be interest in estimating the variance, median, or modal value of a variable, or the correlation coefficient of two variables. It is, however, fortunate and convenient that the four types listed cover the very large majority of population characteristics of principal interest in practice.

What is more, the list can be further reduced: two of the four types can be calculated from the remaining two. Note the relationships between average or proportion on the one hand, and the corresponding totals on the other:

$$\mu = \frac{\tau}{N}, \qquad \tau = N\mu$$

$$\pi = \frac{\tau'}{N}, \qquad \tau' = N\pi.$$

In words, the population average of a variable is equal to the total value of the variable divided by the population size, which implies that the total equals the average times the population size; and so on.

From now on in this book, *it will be assumed that the number of elements in the population, N (the population size), is known.* This condition is likely to be met in the majority of applications, at least to a good approximation. It requires knowledge, for example, of the number of persons two years old or older in the Telemedia case, the number of residences served by NorPower, the number of drugstores and hospitals in the Pharmacom case, or the number of Canabag's accounts receivable.

Assuming that the population size is known, and given an estimate of a total, the estimate of the corresponding average or proportion can be calculated by dividing the total by N. For example, if the estimate of the total sales of Product 1 by all 6,878 drug outlets is $453,338, the estimate of the average sales per outlet is simply 453,338/6,878 or $65.91. Given an estimate of the average of a variable or the proportion in a category, the total of that variable or the number of elements in that category can be estimated by multiplying the average or proportion by N. For example, if the estimate of the proportion of households using gas among the 2,819,514 households served by NorPower is 0.047, then the estimate of the number of households using gas is (2,819,514)(0.047), or 132,517.

Therefore, if we know how to estimate a population average or proportion, we know how to estimate the corresponding population total as well. The reverse is also true: if we know how to estimate the total of a variable or the number in a category, we also know how to estimate the average or proportion. The choice of emphasis is arbitrary.

In what follows, we shall concentrate on methods for estimating averages and proportions, but shall always bear in mind that these methods also tell us how to estimate totals as well.

2.3 PROPORTIONS AND INDICATOR VARIABLES

There is another way of looking at a proportion, which will be found useful later on.

Suppose a population has five elements, and that each of these elements either does (Yes) or does not (No) belong to a given category (C) of an attribute, as indicated in the first two columns below:

Element No.	In C?	Indicator Variable
1	Yes	1
2	No	0
3	No	0
4	Yes	1
5	Yes	1
		3

The number of elements in C is 3, and the proportion in C is 3/5 or 0.6.

Now consider creating a variable ("Indicator," in the third column) that has the value 1 if the element is in C, or the value 0 if it is not. Observe that the total of the variable is equal to 3, and its average 0.6.

In general, *the number of elements in a given category can be regarded as the total of a variable taking the value 1 if the element is in, or the value 0 if it is not in, the category; likewise, the proportion of elements in the category can be regarded as the average value of this same variable.*[2] Such variables are often called *indicator* or *dummy variables*.

In other words, if we so wish, we can interpret a proportion or the number in a category as the mean or total of an especially created indicator variable. Thus, for the purpose of showing how to form estimates, the list of population characteristics *could* be reduced to a single type, namely, the average of a variable. Despite the conceptual advantages of this last reduction, it will be easier for later reference to continue to treat the mean of a variable and the proportion in a category as two separate types of population characteristics. We shall, however, refer to this section on a few occasions later in this book where it will help explain the derivation of certain results.

---

[2]These observations apply to any list of elements, whether or not it constitutes a population.

2.4 FIRST ASSUMPTIONS

Until further notice, *we shall assume that the population of interest is the one from which the sample is actually selected, that the selected population elements can be measured, and that measurement can be made without error.* By "measuring," we understand determining the true category or value of a variable associated with a population element.

These assumptions are frequently violated in practice. Let us illustrate briefly.

Suppose that a survey requires the selection of a sample of households. As is often the case, there is no list of households from which to select the sample. The telephone book provides a tempting and convenient list. Clearly, though, the telephone-book population and the household population are not identical (there are unlisted numbers, households without telephone or with several telephones, nonresidential telephone numbers, etc.).

Individuals often refuse to be interviewed or to complete questionnaires. The sample may have been carefully selected, but not all selected elements can be measured. If it can be assumed that the two subpopulations—those who would respond and those who would not—have identical characteristics, there is no problem. But if this is not the case, treating those that respond as representative of the entire population may result in misleading estimates.

Measurement error is usually not serious when objects are being measured (although measuring instruments are sometimes inaccurate), but it could be so when dealing with people. For example, we may wish to believe that individuals reveal or report their income accurately, but, often, reported income is at variance with true income, even when participants are assured that their responses are confidential.

Problems created by ill-targeted populations, nonresponse, or measurement error are present in almost every sampling study in practice. There are no simple solutions. Until further notice, we shall ignore these problems in order to concentrate on other important aspects of sampling. We examine nonresponse, measurement error, and ill-targeted populations later in this book.

2.5 FIRST ESTIMATES

You have taken a sample of 50 items from a lot of 1,000 manufactured items. Inspection shows that 2 of the 50 items are defective. What is your estimate of the proportion of defectives in the lot?

Few people would hesitate before answering: 2/50 or 0.04. The problem was to estimate the proportion of elements in the population in the "defective" category. A sample was taken, and the sampled elements inspected. The unknown population proportion was estimated by the observed sample proportion.

Another situation. A sample of accounts receivable was selected from the list of all outstanding accounts of a company. The age (that is, the number of days the current balance is outstanding) of each selected account was determined, and the average age of the selected accounts was found to be 16 days. Question: what is your estimate of the average age of all outstanding accounts receivable? Again, few people would hesitate before replying: 16. The problem was to estimate the average value of a variable in a population. Available is the average value of the same variable in the sample. As in the question involving a proportion, the unknown population characteristic was estimated by the corresponding sample characteristic.

These are indeed reasonable estimates, and could be used unless shown to be faulty in a particular case or until better ones are provided. That we shall do in the following chapters, but not before we explain carefully the sense in which "better" is used (not an easy task, as will be seen). For the time being, the following prescription will serve us well:

To estimate the:	*Use the:*
Population average of a variable	Sample average of the variable
Population proportion in a category	Sample proportion in the category

For lack of a better word, we shall call these the *simple* estimates.

2.6 SAMPLING WITH AND WITHOUT REPLACEMENT

A sample, we noted earlier, is simply a part of the population. *How* the sample is selected is a key issue in sampling, one that we shall address in the following chapters. The sample may be selected arbitrarily and subjectively, perhaps to suit the convenience of the investigator; it may be selected with a certain purpose (for example, to make it representative of the population in a certain sense or to satisfy certain statistical criteria), or it may be selected at random, very much like a hand of cards is dealt in bridge or the winning tickets are drawn in certain lotteries.

At this point, we would like to note the distinction between *sampling with* and *sampling without replacement*—a distinction that applies regardless of the sampling method used.

The distinction is best understood if it is imagined that the elements that will form the sample are selected one at a time. If previously selected elements are not elegible for selection at any one stage, we say that sampling is without replacement. If, on the other hand, all the population elements are eligible for selection at any stage of the selection process, we say that sampling is with replacement.

Think, for example, of a box containing ten small manufactured items. This is the population of our illustration. Imagine selecting a sample of three items in the following manner. The first item is somehow picked up from the box and put aside;

then the second item is selected from the remaining nine items in the box and put aside; finally, the third item is selected from the eight remaining items and put aside. The sample of three items is without replacement. The selected items may then be inspected in order to infer the quality of all the items in the box.

By contrast, imagine that, after the first item is selected, it is inspected and replaced in the box. A second item is selected from the ten items in the box, inspected and replaced. The third item is selected in the same manner. Every selection, therefore, is made from the same population of items. The sample is with replacement. It is possible, in this case, that the sample may consist of the same item that happened to be selected three times in a row.

For sampling to be without replacement it is not necessary that the elements be selected one at a time. Any method of selection qualifies, under which there is no possibility of the same element appearing more than once in the sample. A scoop of beads from a container, for example, is a sample without replacement as it is not possible for the same bead to be a part of the sample more than once.

Intuitively, sampling with replacement seems rather wasteful: once an element is selected and inspected, it appears to yield all the information it could possibly provide. Replacing it in the population, from which it may be selected again, does not seem to serve a useful purpose.

In the next and a following section, we examine if these intuitive notions are correct. In any event, *sampling in practice is nearly always without replacement, and so it will be understood in this book,* unless we explicitly say otherwise.

2.7 IS A LARGER SAMPLE BETTER?

Most would probably answer this question affirmatively. Is this intuition correct?

Imagine a population of five elements with known values of a variable Y of interest, arranged in increasing order of these values as shown below:

Element ID no.:	3	1	5	2	4
Y value:	-10	-1	0	2	5

The population average of Y is $(-10 - 1 + 0 + 2 + 5)/5 = -0.8$.

Imagine selecting some elements from this population, observing their Y values, and estimating the population average using the sample average value of Y.

When the sample is of size one, the sample average is, of course, simply the value of the selected element. The minimum value of the estimate is -10 (the smallest value of Y in the population), and the maximum estimate is 5. The range of the estimate is therefore $5 - (-10)$, or 15—the same as the range of the population values.

Now imagine selecting two elements without replacement, again estimating the population average by the sample average. The minimum estimate is the average of the two smallest Y values, $[(-10) + (-1)]/2$ or -5.5, and the maximum is the

average of the two largest, $(2 + 5)/2$ or 3.5. With a sample of size 2, therefore, the range of the estimate has been narrowed down to 9, from 15, for a sample of size 1.

It is easy to verify that the estimate based on a sample of three elements without replacement will be between -3.67 and 2.33, giving a range of 6.

We see from this numerical illustration that the accuracy of the estimate, as measured by the range of its possible values, improves as the sample size increases. It is not difficult at all to show that, in general, if the sample size is increased from $n - 1$ to n, the range of the sample estimate will be reduced, provided that (a) sampling is without replacement and (b) the n smallest or largest Y values are not all equal.

It can also be verified with this numerical illustration that if sampling is with replacement the minimum and maximum values of the estimate, as well as the range, will be the same regardless of the sample size. If we use the range of the estimate as the measure of accuracy of a sample, therefore, we must conclude that a sample without replacement is better than a sample with replacement of the same size.

These simple results do not depend on the manner in which the sample is selected (arbitrarily, purposively, or randomly). Three observations, however, should be kept in mind.

First, we cannot tell in advance how large the improvement will be, as this depends on the unknown values of Y in the population.

Second, consider estimating a population proportion. Specifically, imagine a population with five elements, each of which either belongs (Y) or does not belong (N) to a certain category (C), as shown below:

Element no.:	1	2	3	4	5
In C?:	Y	N	N	Y	N
Indicator:	1	0	0	1	0

The population proportion in C is $2/5$ or 0.4. We noted in Section 2.3 that a proportion may be viewed as the average value of an indicator variable. Let us arrange the values of this variable in nondecreasing sequence:

$$0 \quad 0 \quad 0 \quad 1 \quad 1$$

The proportion of sample elements in C is equal to the average value of the indicator variable in the sample. The minimum and maximum of the sample estimate, as well as the range, are shown below for each possible size of a sample without replacement:

| | Estimate | | |
Sample Size	Minimum	Maximum	Range
1	0	1	1
2	0	1	1
3	0	2/3	2/3
4	1/4	1/2	1/4
5	2/5	2/5	0

Therefore, the range of the estimate tends to decline, but a decline cannot be guaranteed with every increase in sample size.

Finally, the range is a rather rough measure of accuracy; it depends only on the maximum and minimum, and ignores what happens between these two. In later chapters, we shall examine other measures of sample accuracy.

2.8 SAMPLING FROM A HOMOGENEOUS POPULATION

Perhaps the easiest sampling problem occurs when the population consists of elements that are identical with respect to the values of the variables or the categories of the attributes of interest. We shall call such a population *homogeneous*.

Consider, for example, determining the chemical composition (e.g., the octane rating) of a quantity of gasoline stored in a tank. This task is usually accomplished by dipping a container in the tank and scooping out a quantity of gasoline for testing.

The gasoline is, of course, an undivided liquid whole. Strictly speaking, there are no elements. We can, however, think of the total volume as the aggregate of distinct parts of equal volume—say, equal to one cubic inch each. A scoop then results in the selection of a number of these parts.

If, as is reasonable to suppose in this case, the chemical composition of one part is not different from that of any other part, it is clear that one part is sufficient to provide the needed information about the entire tank; for instance, the octane rating of the gasoline in the tank is that of the part selected.

A sample of size one, therefore, no matter how it is selected, is sufficient to give correct estimates of the characteristics of a homogeneous population.

Thus, for example, a drop of blood from the finger ought to be the same as a drop from any other part of the body. One thimbleful of whiskey, no matter how selected, should tell a taster all she needs to know about the entire keg. Obviously, however, the same does not apply to another liquid, the constituents of which have a tendency to settle in layers of different composition while in storage.

Homogeneous populations are infrequently encountered in practice, although occasionally it is implicitly assumed that the population is homogeneous. Many testing organizations, for example, test a single unit of a given model of a product.

A consumer magazine, for instance, evaluates cars by implicitly assuming that all, say, 1995 Toyota Tercels are identical and testing just one 1995 Toyota Tercel. Likewise, a computer lab appraises computer hardware and software by measuring the performance of just one item of a given brand and model. It is useful, therefore, to bear in mind that in this extreme case of uniformity or assumed uniformity, a sample of just one element is all that is needed to provide an accurate picture of the entire population.

2.9 AUXILIARY INFORMATION

We have assumed so far that the only known information about the population consists of the population size and the identification numbers of the population elements. In the Canabag case, for example, we noted that the population consisted of 988 accounts receivable and that these accounts could be identified as "account no. 1," "account no. 2," etc.

In many cases, however, there is additional information about the population elements. In some cases, this information may be exploited.

In the Canabag case, for example, it can be safely assumed that the name of the customer, the customer's address (hence also the customer's location), and the book balance of the account are known. In addition, it may be possible to tell from various sources how long the customer has been served by Canabag, how frequently and how much the customer has been buying from the company in this and previous years, and so on.

In the Pharmacom case, some known auxiliary information is shown in Table 2.5. Outlets are classified by type into drugstores (D) and hospitals (H), and by region. The size of a hospital is measured by the number of beds, and that of a drugstore is measured by the store floor area. The total sales of each product remain, of course, the unknown population characteristics of main interest.

TABLE 2.5 Sales of Pharmaceutical Products, January 1993

Outlet No.	Type	Region	Size	Product 1	Product 2	· · · · · ·	Product 3200
1	D	East	2300	?	?	· · ·	?
2	H	South	675	?	?	· · ·	?
⋮	⋮	⋮	⋮	⋮	⋮		
6,878	D	West	250	?	?	· · ·	?
Total sales:				?	?	· · ·	?

One use of auxiliary information is to divide the population into groups. For example, the population of drug retail outlets could be divided according to type, according to region, according to size, or according to combinations of the above criteria. Instead of selecting indiscriminately from the population of all retail outlets, Pharmacom could select separate samples of drugstores and hospitals. Alternatively, Pharmacom could decide to select one region and all the drugstores and hospitals in it. And there are many other possibilities.

Auxiliary information opens the door to a variety of different sample designs. In the following sections we consider three admittedly extreme cases where the availability of auxiliary information helps exploit particular population structures.

2.10 SAMPLING FROM HOMOGENEOUS GROUPS

Imagine a population that is not homogeneous when considered as a whole, but that can be divided according to a certain criterion into homogeneous groups of known size. A group is homogeneous if all the group elements have identical values of the variables, and belong to identical categories of the attributes of interest. We would like to show that in such a case correct estimates of the population characteristics can be obtained by selecting a sample of just one element from each group.

To illustrate, suppose that a population has 12 elements that can be divided into five groups according to some criterion based on available auxiliary information. Imagine further that the elements in each group have unknown but identical values of the single variable of interest Y, as illustrated in Figure 2.1.

The population could consist of households in a town, the grouping criterion could be neighborhood of residence, and the variable of interest annual housing expenses. (It is unlikely, of course, that any real population can be divided into strictly homogeneous groups, but let us see what happens if it can be done.)

In this illustration, group 1 (G1) is known to have two elements with the same (but unknown) Y value; group 2 (G2) likewise is known to have three elements with identical but unknown Y value (the common value of the elements in group 2 may be greater, as shown in Figure 2.1, or less than that of the elements in group 1); and so on. Evidently, the population as a whole is not homogeneous because its elements do not have the same value.

Suppose it is desired to estimate the population average of the variable Y. Selecting one element from each group and inspecting it will reveal the Y value common to all the elements in the group. Suppose these revealed values are as shown in Table 2.6.

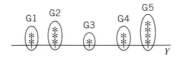

Figure 2.1 Homogeneous groups, first illustration

TABLE 2.6 Calculation of Population Average

Group	Number of Elements	Y value of Selected Element	Group Total Y Value
G1	2	−3.5	−7.0
G2	3	2.0	6.0
G3	1	4.0	4.0
G4	2	10.0	20.0
G5	4	30.0	120.0
	12		Population total = 143.0

It follows that the population average value of Y is 143/12, or 11.92. The calculated group totals and population average are, of course, the true figures. If the assumptions about the population are correct, there is no possibility of making an error in this case. The only uncertainty (the common but unknown Y value of the elements in each group) is removed by inspecting a sample of one element from each group. It does not matter how this element is selected.

The same conclusions apply in the case where it is desired to estimate the proportion of elements in the population that belong to a given category.

For instance, imagine a population with six elements, each of which either belongs or does not belong to a given category (C). The elements could be heads of households, the attribute of interest could be marital status, and the category "married."

Suppose that the population elements can be put into three groups, G1, G2, and G3, with 2, 3, and 1 elements, respectively. The groups are assumed homogeneous, that is, in each group *all* the elements either belong or do not belong to the category C. As illustrated in Figure 2.2, but unknown to us, all the elements of G1 and G3 are in C, all the elements of G2 are not in C. (The horizontal axis in Figure 2.2 has no meaning—it is intended simply to guide the eye.)

One element is selected from each group, its category noted, and the proportion of elements in the population belonging to C is calculated as shown in Table 2.7.

We would conclude that 3 of the 6 population elements belong to the category, and that the proportion of population elements in C is 1/2. This estimate is, of course, correct, as the only uncertainty in this case—whether or not the group

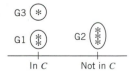

Figure 2.2 Homogeneous groups, second illustration

TABLE 2.7 Calculation of Population Proportion			
Group	Number of Elements	Category of Selected Element	Number in C
G1	2	In C	2
G2	3	Not in C	0
G3	1	In C	1
	6		3

elements belong to C—is removed by inspecting one element from each group. Again, it does not matter how this element is selected.

To sum up then: *If the population can be divided into homogeneous groups, then a sample of size one from each group is sufficient to provide correct estimates of the population characteristics.* Furthermore, if the number of homogeneous groups is small, then only a small sample is needed to provide accurate estimates.

It is not reasonable to suppose there are many situations in the real world where, for some grouping criterion, all the group elements are identical with respect to all the variables and attributes of interest. However, in some cases there may be information suggesting that the situation is approximately that of homogeneous groups. Sometimes the grouping criterion may be chosen from among several available criteria so as to approximate homogeneity in the formed groups. The information could come from an older and similar study, or from a *pilot* sampling study, that is, from a small sample from the current population of interest aimed at guiding the design of the main sample.

The case of homogeneous groups is instructive for it gives intuitive support to a reasonable sampling strategy: *If there is evidence that the population under study approximates the case of homogeneous groups, it seems better to select a sample from each group, as opposed to selecting indiscriminately from the population at large.*

2.11 SAMPLING OF GROUPS

We now consider the case where each group is a "miniature" (or, as some say, a "scaled replica" or "scaled image") of the population as a whole. We would like to show that *in the event the population elements can be divided into groups such that the relative frequency distributions of the variables or attributes of interest are the same for all groups, then, in order to obtain correct estimates of the population characteristics, it is sufficient to select just one group and all its elements.*

Consider Figure 2.3 as an illustration of the population we have in mind.

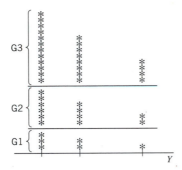

Figure 2.3 Identical group distributions, first illustration

The population consists of 42 elements that, it is assumed, can be divided according to some criterion into three groups (G1, G2, and G3), with 6, 12, and 24 elements respectively. The variable of interest is Y. In this illustration 1/2 of the elements in each group and in the population at large take the first (but unknown) value of Y, 1/3 of the elements the second, and 1/6 the last. Note that the percentage distribution of Y is the same in all groups; we could say that each group is a miniature of the population.

Since the relative frequency distribution of Y is the same in all groups, the average value of Y must also be the same in all groups and must be equal to the population average of Y.

Let us explain. (Appendix A could be consulted if the reader is not familiar with the notation used.) If, in any given group, a percentage $r(y_1)$ of the elements take the value y_1, $r(y_2)$ the value $y_2, \ldots,$ and $r(y_k)$ the value y_k, then, by definition, the average of the Y values in the group is

$$\mu' = y_1 r(y_1) + y_2 r(y_2) + \cdots + y_k r(y_k).$$

Since, by assumption k, the r's and y's are the same in all groups, it follows that the group averages must also be the same.

As for the second part of the above statement, observe that the population average of Y is, again by definition,

$$\mu = \text{(Sum of } Y \text{ values of all } N \text{ population elements)}/N$$
$$= \text{(Sum in group } 1 + \cdots + \text{Sum in group } M)/N,$$

where M is the number of groups. Now, if there are N_i elements in group i and μ' is the common group average value of Y, then

$$\text{Sum in group } i = N_i \mu'.$$

Therefore,

$$\mu = (N_1\mu' + N_2\mu' + \cdots + N_M\mu')/N$$
$$= \mu'(N_1 + N_2 + \cdots + N_M)/N$$
$$= \mu',$$

since $N_1 + N_2 + \cdots + N_M = N$. In other words, the population average of Y, μ, is equal to the common group average, μ'.

Therefore, to estimate the population average of Y all that is needed is to select one group with all its elements, and calculate the group average value of Y. It does not matter how that group is selected. The group average will be the correct estimate as the only uncertainty in this case, the distribution of Y values, is removed by inspecting the elements in the selected group.

Suppose, for example, that group 2 is selected, and all its 12 elements are inspected with the following results:

Y Value (1)	Number of Elements (2)	Proportion of Elements (3)	(4) = (1)×(3)
−2.0	6	0.500	−1.000
1.0	4	0.333	0.333
3.0	2	0.167	0.500
	12	1.000	−0.167

The average value of Y in the selected group is -0.167, and this is the correct estimate of the population average of Y.

As an illustration of a case involving a population proportion, consider Figure 2.4.

We would like to estimate the proportion of elements in the population that belong to a given category, C, of an attribute of interest. The population has 9 elements, which can be divided into two groups, G1 and G2, with 6 and 3 elements, respectively. The proportions of group elements in C and not in C are assumed to be the same for each group. As illustrated in Figure 2.4, 1/3 of the elements in each group belong, and 2/3 do not belong to C.

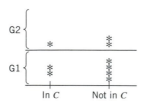

Figure 2.4 Identical group distributions, second illustration

The population proportion in C is equal to the common group proportions in C (1/3 in this illustration).

To show that this holds in all cases where the assumptions of this section are satisfied, let M be the number of groups, and π' the common group proportion in C. Then, by definition, the population proportion in C, π, is

$$\pi = (\text{Number in } C \text{ in population of } N \text{ elements})/N$$
$$= (\text{Number in } C \text{ in group } 1 + \cdots + \text{Number in } C \text{ in group } M)/N$$
$$= (N_1\pi' + N_2\pi' + \cdots + N_M\pi')/N$$
$$= \pi'(N_1 + N_2 + \cdots + N_M)/N$$
$$= \pi'.$$

In order to estimate the population proportion, we need select only one group and all its elements. For example, if G1 is selected in this illustration, inspection will show that of the 3 group elements 1 belongs and 2 do not belong to C. Thus, the (correct) estimate of the population proportion in C is 1/3.

Although there may be some choice among grouping criteria, and therefore some freedom to create a grouping that approximates the situation described earlier, the assumptions of this section are restrictive. In the first place, in order to verify that the assumptions hold, it would be necessary to know the distributions of the variables or attributes of interest, hence also their characteristics. But in such an event, as noted earlier, there would be no point in taking a sample. Rare indeed is a population in practice where the distributions of all the variables or attributes of interest are identical among all groups.

Nevertheless, this extreme case is instructive for it supports a reasonable sampling strategy: *If the population under study approximates the one hypothesized in this section, then it appears better to select one group with all its elements, rather than a number of elements from the population without regard to the group to which they belong, or one element from each group as described in the preceding section.* Any one group will do, but the one with the fewest elements will cost least to inspect. In the event that the number of groups is large, and the number of elements in the smallest group is small, there are likely to be cost advantages to a sample selected in this fashion. If the number of elements in each group is large, a reasonable alternative is to proceed in two stages, first selecting a number of groups, and then a number of elements from each selected group.

2.12 EXPLOITING A RELATIONSHIP

It is often the case that a relationship exists between a variable of interest and another known variable. For example, in the Canabag case, there may be a relationship between the absolute amount of the error and the book balance of the account,

perhaps such that the larger the balance, the greater the error. In the Pharmacom case, there may a relationship between the sales of a pharmaceutical product and the size of the outlet (measured, perhaps, by the floor space of the drugstore or the number of beds in the hospital).

We would like to show that *in the event the relationship is exact and of known form, a sample of only as many elements as there are unknown parameters is sufficient to obtain correct estimates of the population characteristics.*

Let us explain, beginning with a simple relationship. Suppose that a variable of interest, Y, is a certain multiple of another known variable, X:

$$Y = bX,$$

where b is the single unknown parameter of the relationship. For example, the population could be the hospitals in a region, Y the sales of a given pharmaceutical product to a given hospital, and X the number of beds in that hospital.

Imagine a population of three elements, with X and Y values as given in Table 2.8.

In this case, $Y = (0.5)X$, and the relationship between Y and X is as shown in Figure 2.5, that is, a line through the origin.

If indeed the relationship between Y and X is $Y = bX$, but the Y values and b are unknown, selecting and inspecting just one element is sufficient to obtain correct estimates of the population average value of Y. To illustrate, pretend that the entries in the last column of Table 2.8 are not known. Suppose that element no. 2 is selected and inspected, revealing $Y = 5$. It follows that $b = Y/X = 5/10 = 0.5$. Using this figure, the remaining Y values in the last column of Table 2.8 can be calculated, and the population total of Y correctly estimated to be 17.5. It follows that the population average value of Y is 17.5/3, or 5.67.

As a second illustration of a relationship, imagine that a population has three elements with the Y and X values shown in Table 2.9.

The relationship between Y and X is the straight line plotted in Figure 2.6.

It is well-known that a straight line corresponds to the expression

$$Y = a + bX,$$

TABLE 2.8 Simple Linear Relationship

Element No.	Known X	Unknown Y
1	5	2.5
2	10	5.0
3	20	10.0
	35	17.5

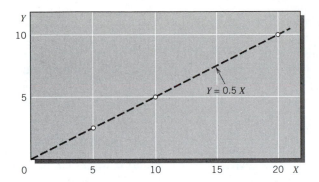

Figure 2.5   Simple linear relationship illustrated

TABLE 2.9	Linear Relationship	
Element No.	Known X	Unknown Y
1	−10	20
2	0	10
3	5	5
	−5	35

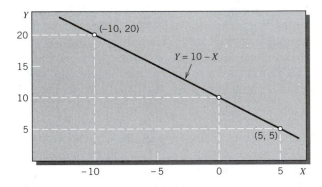

Figure 2.6   Linear relationship illustrated

where a and b are the parameters of the expression. In other words, a straight line is the set of all the pairs of values of X and Y which satisfy the preceding expression for given values of a and b. a is the Y intercept, that is, the value of Y at $X = 0$, and b is the slope of the line.

If the relationship between Y and X is $Y = a + bX$ but the Y values, a and b, are unknown, a sample of two elements is sufficient to provide a correct estimate of the population mean (or total) of Y.

The simplest way to see this is to consider the population of Table 2.9, and to pretend that the entries in the last column are not known. Suppose that the first and last elements are selected and inspected, revealing the two values of Y, 20 and 5. Plot the two (X, Y) pairs: $(-10, 20)$ and $(5, 5)$, and connect them with a straight line (see Figure 2.6). The value of Y corresponding to the known $X = 0$ of the second element can be read from the graph as $Y = 10$. All the entries in the last column of Table 2.9 are now revealed. The population total of Y is estimated to be 35, and the population average of Y 35/3 or 11.67. These estimates are, of course, correct. The uncertainty in this case concerned the unknown values of the parameters a and b, determining the exact location of the line. The plot of the two selected elements removed this uncertainty.

The approach of this section can be easily extended to other types of relationships; for example, $Y = a + bX + cX^2$, $Y = a + b \log X$, etc. Provided that the form of the relationship is known and exact, correct estimates of the population total or average of Y can be obtained with a sample of size equal to the number of unknown parameters of the relationship.

We should not get carried away too far in this direction, however, for rarely in practice are variables related exactly and in a form that is known in advance. But as with the other extreme cases described earlier in this chapter, the analysis provides support for the following reasonable sampling strategy: *If there is evidence of an approximate relationship between a variable of interest Y and a variable X known from available auxiliary information, it may be better to exploit the relationship in order to estimate the population mean or total of Y, rather than ignore it.* We have seen that for simple relationships few elements need be selected, so that the approach of this section may result in lower sampling cost.

To illustrate how one could proceed in the case of an approximate relationship, suppose that a sample of seven elements from a large population was taken. Seven pairs of values of a variable of interest, Y, and a known variable, X, are shown in Figure 2.7.

The relationship between Y and X is not exactly linear, only approximately so. We could draw a straight line through the scatter of points to approximate the relationship. The approximating dotted line shown in Figure 2.7 appears to fit well the observations.

Estimates of the Y values of all uninspected elements of the population may be read from the graph, treating the line as if it represented an exact relationship. For example, an element with known $X = x_0$ would be estimated to have $Y = y_0$, as shown in Figure 2.7. Adding these estimates to the sum of the Y values of the inspected elements would give an estimate of the population total of Y. From this, the population average could be calculated by dividing the total by the number of elements in the population. This estimate utilized the relationship; by contrast,

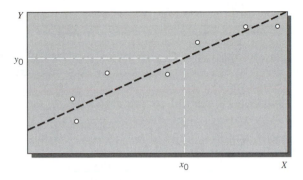

Figure 2.7   Approximate linear relationship

the ordinary sample average of the Y values (the simple estimate) would not have made use of the relationship.

We have not, of course, made precise what constitutes a satisfactory approximation, or whether, how, and why the approach of this section is better than the general. These are issues we shall return to later in this text.

2.13 A NOTE ON TERMINOLOGY

This is perhaps a good time to attach the commonly used labels to the types of samples described in earlier sections.

If the population is divided into nonoverlapping groups according to a certain criterion based on auxiliary information, the sample is called:

- *stratified* if a number of elements are selected from each group separately; the number of selected elements could vary from group to group.
- *two-stage* if, at the first stage, a number of groups are selected; then, in the second stage, a number of elements are selected from each of the groups selected in the first stage. The number of selected elements could vary from group to group. In the event all the elements of each selected group are selected, the sample is called *cluster*.

The samples in Section 2.10, therefore, can be called stratified, and those in Section 2.11 would be called cluster, special cases of two-stage samples.

By contrast, if the population elements are not classified into groups, the sample selected from the undivided population is called *simple*. More complicated samples are, of course, possible.

2.14 TO SUM UP

- A population is the aggregate of interest. A population consists of elements. Exactly what constitutes a population and its elements varies from one case to another.
- With each population element, there are associated numerical values of one or more variables, and categories of one or more attributes of interest. The primary objective of sampling is to obtain estimates of the characteristics of the variables and attributes of interest in the particular case.
- The principal characteristics of interest are the mean and total of a variable, and the proportion and total number of elements in a category. If the population size is known, these totals can be derived from the mean and proportion; it is on these last two types of characteristics that this book concentrates.
- Until further notice, it is assumed that the population of interest is the one from which the sample is actually selected, that all selected population elements can be measured, and that measurement can be made without error.
- Until better estimates are provided, it is reasonable to use the sample average of a variable as an estimate of the population average of the variable, and the sample proportion as an estimate of the population proportion in a category. We call these the simple estimates.
- In practice, sampling is nearly always without replacement. Using the range of the simple estimates as a measure of their accuracy, we concluded that a sample without replacement is better than one with replacement of the same size, and that a larger sample without replacement is better than a smaller one.
- A sample of size one, no matter how selected, is sufficient to give correct estimates of a homogeneous population. If the population can be divided into homogeneous groups, a sample of one element from each group is sufficient to provide correct estimates of the population characteristics. In the event the population elements can be divided into groups such that each group is a miniature of the entire population, then in order to obtain correct estimates of the population characteristics, it is sufficient to select just one group and all its elements. In case the variable of interest is exactly related to a known auxiliary variable, a sample of only as many elements as there are unknown parameters is sufficient to obtain correct estimates of the population mean and total of the variable of interest.
- Such extreme cases are unlikely to be encountered in practice. Nevertheless, they are instructive for they suggest that a simple sample is appropriate when the population is approximately homogeneous, a stratified one when the population can be divided into approximately homogeneous groups, and a cluster or two-stage sample when the population can be divided into like groups. Also, if the variable of interest is approximately related to a known auxiliary variable, it may be better to exploit this relationship as described in Section 2.12 rather than ignore it.

CHAPTER 3

Simple Random Sampling

3.1 INTRODUCTION AND SUMMARY

The sampling problem in general has two components that must be considered together: how to select the elements that will form the sample, and how to form estimates of the population characteristics.

With this chapter we begin examining the dominant method for selecting a sample—randomization. The simplest kind of random sample, the so-called simple random sample, is selected much like a hand of cards is dealt in a card game. The population elements (or such convenient substitutes for these elements as tags, chips, etc.) are mixed thoroughly, and the desired number drawn, together or one at a time. More convenient and efficient procedures for selecting a random sample, involving tables of random numbers or random number generators, are also available.

As for the estimation method, we begin with the simplest possible: we use the simple estimators of Chapter 2, that is, the sample average of a variable and the sample proportion in a category.

We intend to show that these estimators have desirable properties under simple random sampling. First, they are unbiased, that is, they tend neither to underestimate nor to overestimate the population characteristics. Second, their accuracy can be expressed as a function of the sample size, the population size, and population

characteristics. These expressions, in turn, are used to confirm more convincingly that a larger random sample is better than a smaller one, and that random sampling without replacement is preferable to that with replacement. The desirable properties recommend the use of the simple estimators with simple random sampling without replacement.

We also intend to show how to determine the size of a simple random sample needed to meet given accuracy requirements, and how to form interval estimates, that is, estimates intended to "bracket" or "cover" the characteristic of interest with given probability.

The approach of this chapter will serve as the blueprint, and the results as a standard of comparison with other methods of selecting random samples and for estimating unknown population characteristics to be discussed in subsequent chapters.

3.2 A SILLY GAME

The desirable properties we mentioned in the introduction to this chapter are better understood if they can be related to a physical process. The game we describe in this section is artificial and not particularly interesting, but will help us give operational meaning to such abstract concepts as estimation, unbiasedness, and variability.

The objective of this game is to get a hoop as close as possible to a stationary peg located some distance away from the marksman's position. Figure 3.1 illustrates.

The game is played via expert marksmen. You will not be tossing the hoop yourself, but must delegate the toss to an expert marksman. *The designated expert will toss the hoop once.* You and the designated expert will share a prize which is greater the closer the toss is to the peg. There is, therefore, an incentive for the experts to try their best, and for you to choose an expert as wisely as possible.

We shall suppose that the tosses are somehow (magically?) constrained to land along a straight line from the marksman to the peg and beyond, and cannot deviate to the right or left of this line.

Your problem is which expert marksman to choose as your delegate.

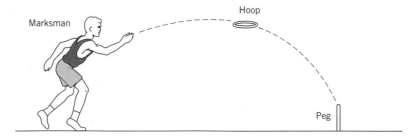

Figure 3.1 Marksman and peg

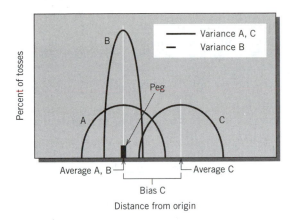

Figure 3.2 Distributions of landings, experts A, B, and C

Four experts are available: A, B, C, and D. For the first three, there is information on their performance in a large number of past tosses under identical conditions. Figure 3.2 shows the distributions of the location of the hoop landings, and the average and variance of these distributions for the three experts.[1]

Before you examine these records of past performance, you should satisfy yourself that they are relevant to the problem of choosing a delegate for the *next* toss. If, for example, B just had an injury affecting her marksmanship, then how well or poorly she performed in the past is irrelevant for forecasting her future performance. You can assume, however, that there is no reason to suspect that the future performance of any marksman will deviate from that in the past as shown in Figure 3.2. In other words, it is reasonable to assume that the distributions of Figure 3.2 also indicate the expected pattern of landings in a large number of *future* tosses.

Let us compare the first three marksmen. Clearly, no marksman is perfect, in the sense of always landing on target. A and C display the same variability in their landings, but A's landings are centered on the target, in the sense that the average location of the landings is that of the peg. C's landings are centered off the target. You could say that, on average, A neither undershoots nor overshoots the target, and C tends to overshoot. For this reason you could even say that C is a *biased* marksman, his *bias* being the difference between the average location of his tosses and the target. A could be described as an *unbiased* marksman, the bias in her case being zero. If a choice had to be made between A and C, you would probably trust your fortunes in this game to A.

---

[1]The principal statistical measures needed for a good understanding of sampling are the mean and variance of a distribution. Appendix A reviews these concepts; it should be read now if the meaning of Figure 3.2 is not clear.

Let us now compare A and B. Both marksmen are unbiased, that is, on average, neither one overshoots or undershoots the target. But clearly, B is the more accurate of the two, in the sense that her tosses tend to be clustered more tightly around the target than A's. This is reflected in the variances of their distributions: B's variance is smaller than A's. If a choice had to be made between A and B, you would probably prefer B.

B, therefore, emerges from this analysis as the favored choice: she is preferred to A, who, in turn, is preferred to C.

What happened to D, you may well ask? For D there is no information at all about his past or future performance. D could be excellent or terrible—there simply is no way of judging how good or bad he is.

Once again: which marksman will you choose—A, B, C, or D?

If your answer is D, you are saying in effect that acting without any relevant information is preferable to acting with some such help. On the other hand, if you prefer to utilize the available information, you will probably rule out D and confine your choice among the other three marksmen, in which case B will probably be the preferred delegate.[2]

Do not lose sight of an important feature of this silly game: you will play it once only. The experts will probably lend their talents on other occasions, but for you it is once only. You realize that the next toss *could* be one of A's or C's best, and one of B's worst. In other words, there can be no guarantee that the marksman with the best track record overall will in fact come closest to the target in the next toss—probably, yes, but certainly, no. After the next and only toss, you may wish in hindsight that your choice had been different, or again you may not. All these worries, of course, are irrelevant, because a decision must be made now.

To sum up, then. It appears reasonable to prefer tried marksmen to untried ones, and to choose an expert marksman on the basis of the expected performance in a large number of future tosses even though the game will be decided by the next single toss. It also appears reasonable to focus in particular on two aspects of expected performance: bias (the difference between the average location of the landings and the target) and accuracy (inversely related to the variance). Among unbiased marksmen the most accurate (that is, the one with the smallest variance) should be preferred.

What, you are probably wondering, does this silly game have to do with sampling? A great deal. In the following sections, we explain that estimating a population characteristic on the basis of a random sample is very much like playing the silly game, and choosing among available sampling methods is like choosing among available expert marksmen. Just as the choice among marksmen was based

---

[2] You may wonder why we have not considered how to choose between two marksmen, one of whom is unbiased but relatively inaccurate, and the other relatively accurate but biased. We do not need this case immediately in our analogy with sampling, but, if we did, it would be reasonable to choose on the basis of the average squared deviation from the target, the average absolute deviation from the target, or some other similar criterion.

on their expected performance in the long run, so will be the choice of sampling method. Just as the silly game is played once, so in reality estimates and decisions must be made on the basis of one sample. And just as it seemed reasonable to rely on absence of bias and accuracy (low variability) for choosing among marksmen, so it will be for choosing among methods for estimating population characteristics.

But there is an important difference. In the silly game, a marksman's expected bias and accuracy could be determined only through observation—by collecting data on past performance. In random sampling, on the other hand, the bias and accuracy of a sampling method can be *inferred* from first principles—without experimentation or observation of past performance.

3.3 SIMPLE RANDOM SAMPLES

If one were to seek professional advice as to how best to select a sample, there is little doubt that the advice would be to make the sample *random*. If one were to ask why random sampling is preferable, one would probably be told that *random samples have desirable properties*. Just what is random sampling? what are the properties? and why are these properties desirable?—these are questions that must be explained with some care. We begin with a working description of a random sample.

A box contains a number of small manufactured items. They form the population of this illustration. To select a random sample without replacement of, say, three items from the box, proceed as follows. First, mix thoroughly the items in the box (shuffling by hand, shaking and turning over the box, or by other similar means). Reach into the box and pick up one item. Put it aside. Mix thoroughly the remaining items; pick up the second item and put it aside. Repeat the procedure to select the third item. The three items put aside constitute a random sample without replacement.

Obviously, this mixing and shuffling cannot be used to select a sample of physically large items or of people. Instead of physical randomization, however, we could use randomization of substitutes for the elements. To do so, all that is needed is a list of population elements, a number of tags, and a box to put them in.

To illustrate, suppose it is desired to select and interview a random sample of three individuals from a population of 73. A list of the latter is available, and looks like this:

Identification No.	Name
1	John Able
2	Andrea Brown
⋮	⋮
73	Maria Crawford

Find 73 small, identical tags, and label them 1, 2, ... , 73. Put the tags in a box. Thoroughly mix the tags. Pick up one tag and observe its label. If it is, say, 49, then the individual with identification number 49 will be one of those interviewed. Put the tag aside. Mix thoroughly the remaining 72 tags. Pick up a tag and observe its label. If it is, say, 2, then Andrea Brown will be one of those interviewed. Repeat the procedure one more time to determine who will be the third individual in the sample.

A sample selected in this fashion, either through randomization of the population elements themselves or of substitutes for these elements, is a random sample. More precisely, it is a *simple random sample without replacement*. "Simple" to distinguish it from stratified, two-stage, cluster or other kinds of random samples. "Without replacement" because previously selected elements cannot be selected again. (For a random sample with replacement, the procedures are identical, except that previously selected population elements or tags are replaced after inspection. As noted earlier, unless otherwise stated, all sampling is understood to be without replacement.)

There are, in fact, more efficient procedures for selecting a random sample that do not require mixing, boxes, or tags. We shall describe them in a later section, but for the time being the two procedures just presented will be adequate.

Let us agree to use the simple estimators of the population characteristics, that is, the sample average (call it $\bar{Y}$) of a variable Y as the estimator of the population mean of the same variable (μ), and the sample proportion (call it P) in a category as the estimator of the proportion of elements in the population that belong to that category (π).

Clearly, the values of $\bar{Y}$ and P cannot be predicted in advance as they depend on the sample elements, and these, in turn, are selected at random. But *imagine* an experiment whereby a large number of simple random samples are selected, each of the same size n, and each selected without replacement from the original population. For each such sample, the values of $\bar{Y}$ and P, the simple estimates, are calculated. The experiment will thus yield a large number of $\bar{Y}$'s and P's. The distributions of $\bar{Y}$ and P can be imagined as illustrated in Figure 3.3.

In the next section we demonstrate with the help of a simple example that the average of these distributions can be expected to equal their target—the population characteristic they attempt to estimate. That is, the long run average value of $\bar{Y}$ is μ, and that of P is π—whatever these characteristics happen to be. In the words of the silly game, $\bar{Y}$ and P can be called unbiased estimators of μ and π, respectively.

In addition, we intend to demonstrate that the variances of $\bar{Y}$ and P can be expressed as simple functions of the population size, the sample size, and the target population characteristic. (Readers who like to know in advance exactly what we intend to show may wish to look ahead at Summary 3.1.) We shall then discuss the implications of these results.

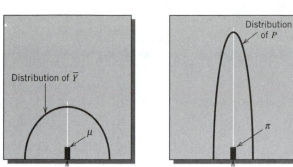

Figure 3.3 Estimators and targets

The key word in the preceding paragraph is "demonstrate," not "prove": the mathematical derivations of these properties can be found in the advanced references listed in the Bibliography.

3.4 THE ACKROYD POPULATION

Imagine that a survey is planned to provide information about employment in the town of Ackroyd. There are, let us suppose, only six firms in the town. These firms make up the population of this example. We are interested in two characteristics of the population: the average number of employees per firm and the proportion of firms which intend to hire. Once we have estimates of these characteristics, we can also estimate the total number of employees in the town as well as the total number of firms intending to hire. Since the number of firms is small, the obvious approach is to visit all the firms, and ask two simple questions:

- How many people do you employ?
- Do you plan to hire more people?

Let us suppose these interviews were made, the firms cooperated, and their responses are as listed in Table 3.1.

It will be noted that the total number of employees in the town is 31, the average number of employees per firm 31/6 or 5.167, the number of firms intending to hire is 2, and the proportion of firms intending to hire is 2/6 or 1/3. For future reference, we calculate also the population variance of the number of employees,

$$\sigma^2 = \frac{1}{N} \sum_{i=1}^{N} (Y_i - \mu)^2 = \frac{1}{6}[(9 - 5.167)^2 + \cdots + (5 - 5.167)^2] = 8.472.$$

TABLE 3.1	Ackroyd Population and Its Characteristics	
Firm	**Number of Employees**	**Intend to Hire?**
A	9	Yes
B	8	Yes
C	6	No
D	2	No
E	1	No
F	5	No

Number of firms, $N = 6$
Average number of employees, $\mu = 31/6 = 5.167$
Proportion of firms intending to hire, $\pi = 2/6 = 1/3 = 0.333$
(Variance of number of employees, $\sigma^2 = 8.472$)

If, as we assume, the information in Table 3.1 were indeed available, no useful purpose would be served by taking a sample. Our objective, however, is instructive rather than realistic: we intend to examine what would happen if we were to select a random sample of firms and attempt to estimate μ and π on the basis of the observations in the sample.

To keep the calculations as simple as possible, we consider selecting only two firms at random (without replacement, of course). The selected firms will be asked the two questions listed above, their responses will be noted, and, utilizing only these responses, estimates will be made of the average number of employees per firm in the town, and of the proportion of firms in the town that intend to hire. At a slight sacrifice in realism, the visits and interviews can be skipped, and the firms' responses read directly from Table 3.1.

Imagine placing into a bowl six identical chips, each marked with the firm's name (letters A through F), the number of its employees, and its hiring intentions ('Yes' for 'intends to hire,' 'No' otherwise). Mix these chips thoroughly and select one chip at random. Note the information on the chip, and put it aside. Mix the remaining chips and select a second one at random, again noting the information on it.

The tree diagram in Figure 3.4 shows that in any such sample, the first firm selected could be A, B, C, D, E, or F. If A is the first selected firm, the second selected firm could be B, C, D, E, or F. If the first is B, the second could be A, C, D, E, or F. And so on, down to the last branch: if the first firm is F, the second could be A, B, C, D, or E.

We see that there are 6×5 or 30 possible sample outcomes: A followed by B (A,B in Figure 3.4), A followed by C (A,C), and so on, down to (F,E).

Now let us imagine selecting in this fashion a large number of random samples of two firms from the Ackroyd population. (After the two chips representing the selected firms are inspected, they are replaced into the bowl, so that every sample

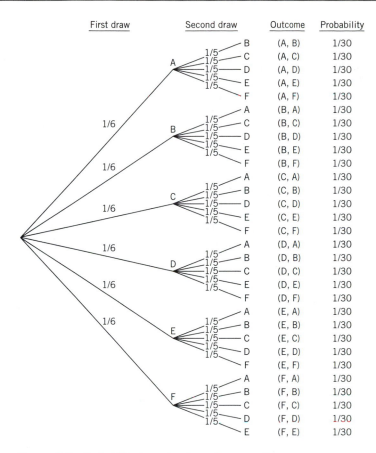

Figure 3.4 Probability tree, simple random sampling

of two firms is selected from the same population.) In view of the thorough mixing employed prior to each draw, most people would agree that each of the six firms would be selected first in 1/6 of a large number of samples. These expected relative frequencies are shown along the first branches in Figure 3.4.

Again because of the thorough mixing prior to each draw on which we have insisted, most people would also agree that, among all the samples in which A was the first selected firm, B would be the second selected in 1/5, C in another 1/5, ..., F in the last 1/5. In general, each of the five firms eligible for selection in any second draw can be expected to appear in 1/5 of the samples, as shown along the second set of branches in Figure 3.4.

We see then that the outcome (A,B), that is, "first A, then B," is expected to occur in 1/30 (i.e., in 1/5 of 1/6) of a large number of samples. For the same reasons, each of the other 29 sample outcomes shown in Figure 3.4 can be expected to appear with the same relative frequency (1/30) in the long run.

The expected long-run relative frequency of the outcome is, of course, the *probability* of that outcome. We could also say, therefore, that the probability that the sample outcome (A,B) (or any other outcome shown in Figure 3.4) will occur is 1/30.[3]

Observe that firm A appears in 10 of the sample outcomes: (A,B) and (B,A), (A,C) and (C,A), . . . , (A,F) and (F,A). Thus, in the long run, firm A can be expected to appear in $10 \times (1/30)$, or 1/3 of the samples. It can be easily verified that the same applies to any other firm: every one of the firms B, C, D, E, and F can be expected to be selected in 1/3 of the samples in the long run. In other words, each of the six firms making up the population of this illustration will appear in the sample with the same probability, 1/3.

This is not an accident. In general, it can be shown that the probability of selecting a specified population element in a simple random sample of size n without replacement is equal to n/N, where N is the population size. In this illustration, $n/N = 2/6$, or 1/3, as observed.

Observe also that the probability of the sample consisting of any two given firms—say, A and B—is 1/15. This, too, is not accidental. It can be shown that the probability that any two distinct elements will appear in a simple random sample of size n from a population of size N is equal to $n(n-1)/N(N-1)$. In the Ackroyd case, $(2)(2-1)/(6)(6-1)$ or 1/15, as observed.

Let us now turn to the problem of estimating the two population characteristics of this illustration: the average number of employees per firm in the town and the proportion of firms in the town that intend to hire. We do, of course, know these in this case, but continue to pretend that we do not.

Let us use the simple estimators: the sample average (in this case the average number of employees per firm in the sample, $\bar{Y}$) and the sample proportion (in this case the proportion of firms in the sample that intend to hire, P). From now on, by *estimator* we shall understand a sample characteristic used to estimate the population characteristic, and by *estimate* the numerical value of an estimator.

The possible sample outcomes, the associated probabilities, and the corresponding estimates are shown in Table 3.2.

Table 3.2 shows 15, rather than 30, sample outcomes. This is because, for the purpose of estimating the population characteristics, one half of the outcomes in Figure 3.4 can be combined with the other half. Consider, for example, two of the 30 possible outcomes: (A,B) and (B,A). The probability of each is 1/30. The two outcomes differ only in the order in which the two firms are selected. Since, in each case, the estimates in which we are interested (the sample average number of employees and the proportion of firms intending to hire) are the same

---

[3] See Appendix A for a review of probability and probability distributions. In general, at each node of a *probability tree*, such as that shown in Figure 3.4, we draw as many branches as there are possible events given that the node is reached. The probability along the branch is the conditional probability of the event given that the node is reached. The probability of each final outcome is the product of the probabilities along the branches leading to the outcome. The reason should be evident if we think of probabilities as long-run relative frequencies.

TABLE 3.2   Simple Random Sample of Size $n = 2$ Without Replacement

Outcome	Number of Employees	Hiring? (Y: Yes, N: No)	Proba- bility	Sample Average[a], $\bar{Y}$	Sample Proportion[b], P
A,B	9,8	Y,Y	1/15	8.5	1.0
A,C	9,6	Y,N	1/15	7.5	0.5
A,D	9,2	Y,N	1/15	5.5	0.5
A,E	9,1	Y,N	1/15	5.0	0.5
A,F	9,5	Y,N	1/15	7.0	0.5
B,C	8,6	Y,N	1/15	7.0	0.5
B,D	8,2	Y,N	1/15	5.0	0.5
B,E	8,1	Y,N	1/15	4.5	0.5
B,F	8,5	Y,N	1/15	6.5	0.5
C,D	6,2	N,N	1/15	4.0	0.0
C,E	6,1	N,N	1/15	3.5	0.0
C,F	6,5	N,N	1/15	5.5	0.0
D,E	2,1	N,N	1/15	1.5	0.0
D,F	2,5	N,N	1/15	3.5	0.0
E,F	1,5	N,N	1/15	3.0	0.0
			15/15		

[a]Average number of employees.   [b]Proportion of firms intending to hire.

($\bar{Y} = 8.5$, $P = 1.0$), the order in which the firms appear in the sample does not matter. The probability (the expected relative frequency in the long run) that the sample will consist of firms A and B, regardless of the order in which A and B are selected, is $(1/30) + (1/30) = 1/15$. The same applies to all other possible outcomes: thus, (A,C) is combined with (C,A), (A,D) with (D,A), etc.

We see from the first line of Table 3.2 that one possible outcome is for the sample to consist of firms A and B. Since A has 9 and B has 8 employees, the average number of employees of the two firms in the sample is $(9 + 8)/2$, or 8.5. Because both A and B plan to hire, the proportion of firms in the sample intending to hire is 1. The estimates corresponding to all other sample outcomes are calculated in the same way.

The probability distribution (the expected relative frequency distribution in the long run) of each estimator can be constructed from Table 3.2 by listing the possible values of the estimator and adding up the associated probabilities.

Let us begin with P. A glance down the last column of Table 3.2 will show that the possible values of P are 0, 0.5, and 1. The value 1 occurs only when the sample outcome is (A,B), and the probability of that outcome is 1/15; therefore, the probability that $P = 1$ is 1/15. In other words, we expect our estimate of π to be equal to 1 in 1/15 of the samples in the long run. Continuing in the same manner,

we observe that $P = 0.5$ in 8 sample outcomes: (A,C) through (B,F). Therefore, the probability that $P = 0.5$ is $8 \times (1/15)$ or 8/15. Finally, $P = 0$ with probability 6/15.

The first two columns of Table 3.3 show the probability distribution of P. The remaining columns show intermediate steps for calculating the mean and variance of this distribution.

TABLE 3.3 Distribution of Sample Proportion, P

P	$p(P)$	$Pp(P)$	P^2	$P^2p(P)$
0	6/15	0	0	0
0.5	8/15	4/15	1/4	2/15
1	1/15	1/15	1	1/15
	1	5/15		3/15

The mean of the probability distribution of P, that is, the average value of P in the long run, is

$$E(P) = \sum Pp(P) = (0)(\frac{6}{15}) + (0.5)(\frac{8}{15}) + (1)(\frac{1}{15}) = \frac{5}{15} = \frac{1}{3} = 0.333.$$

The variance of P, that is, the average squared deviation from the mean in the long run, is

$$Var(P) = \sum P^2 p(P) - [E(P)]^2$$
$$= (0)^2(\frac{6}{15}) + (0.5)^2(\frac{8}{15}) + (1)^2(\frac{1}{15}) - (\frac{1}{3})^2$$
$$= (\frac{3}{15}) - (\frac{1}{3})^2$$
$$= 0.089.$$

Table 3.4 shows the probability distribution of the sample average, $\bar{Y}$. It is easy (although a little tedious) to verify, proceeding as in the case of P, that the mean and variance of $\bar{Y}$, are

$$E(\bar{Y}) = (1.5)(1/15) + (3.0)(1/15) + \cdots + (8.5)(1/15) = 5.167,$$

and

$$Var(\bar{Y}) = (1.5)^2(1/15) + \cdots + (8.5)^2(1/15) - (5.167)^2 = 3.389.$$

Figure 3.5 shows the probability distributions of the two estimators, and the means and variances of these distributions.

TABLE 3.4 Distribution of Sample Average, $\bar{Y}$	
$\bar{Y}$	$p(\bar{Y})$
1.5	1/15
3.0	1/15
3.5	2/15
4.0	1/15
4.5	1/15
5.0	2/15
5.5	2/15
6.5	1/15
7.0	2/15
7.5	1/15
8.5	1/15
	15/15

The sample size ($n = 2$) is quite small, so the jagged appearance of these distributions is not surprising. But we begin to see (at last, you may say) the connection with the silly game. Figure 3.5 shows that if a large number of simple random samples of size 2 are taken and estimates of the average number of employees in

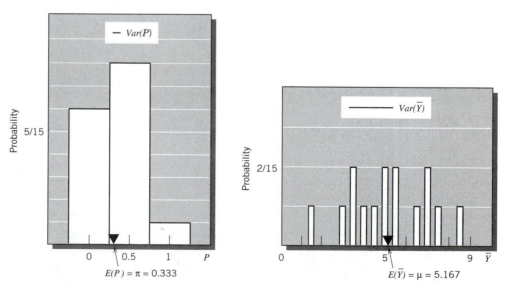

Figure 3.5 Distributions of estimators, Ackroyd case

the town are made after each sample is taken, the average of these estimates can be expected to equal the true population average. Likewise, the average of a large number of estimates of the proportion of firms in the town planning to hire can be expected to be equal to the true proportion. In the long run, therefore, the simple estimators tend neither to underestimate nor to overestimate; by analogy with the silly game, they may be called *unbiased* estimators. The noteworthy feature of this conclusion is that it was *inferred* from first principles—it was not based on experimentation and observation.

In general, we say that *a sample characteristic is an unbiased estimator of a given population characteristic if the mean of its probability distribution equals that population characteristic.*

From Summary 3.1 we see that the sample mean and proportion are unbiased estimators of the population mean and proportion respectively for any sample size and any population, not just those of this illustration.

According to Eq. 3.5, the mean of the probability distribution of P is always equal to π. In the present illustration,

$$E(P) = \pi = 0.333,$$

and this agrees with the direct calculation earlier in this section. According to Eq. 3.6, the variance of P should be equal to

$$Var(P) = \frac{\pi(1-\pi)}{n} \frac{N-n}{N-1} = \frac{(\frac{1}{3})(1-\frac{1}{3})}{2} \frac{6-2}{6-1} = 0.089,$$

a result also in agreement with earlier direct calculations.

According to Eq. 3.2,

$$E(\bar{Y}) = \mu = 5.167,$$

and, according to Eq. 3.3,

$$Var(\bar{Y}) = \frac{\sigma^2}{n} \frac{N-n}{N-1} = \frac{8.472}{2} \frac{6-2}{6-1} = 3.389.$$

The direct numerical calculations of this case, therefore, confirm the theoretical results of Summary 3.1.

3.5 IMPLICATIONS

Estimates of population characteristics, to be made after a simple random sample is taken, cannot be predicted in advance with certainty. The estimates may be close to, or they may be far from the unknown population characteristics. In a realistic situation, we do not know how close or far the estimates are, because we do not know the population characteristics.

SUMMARY 3.1 Simple random sampling, unbiased estimators

- *Method of sample selection:* Simple random sampling.
- An *unbiased estimator of μ, the population mean of a variable Y,* is the ordinary average of the n Y values in the sample:

$$\bar{Y} = \frac{1}{n}(Y_1 + Y_2 + \cdots + Y_n), \tag{3.1}$$

that is,

$$E(\bar{Y}) = \mu. \tag{3.2}$$

The variance of $\bar{Y}$ is given by

$$Var(\bar{Y}) = \frac{\sigma^2}{n} \frac{N-n}{N-1}. \tag{3.3}$$

σ^2 is the variance of variable Y in the population.

- An *unbiased estimator of π, the proportion of elements in the population that belong to a given category,* is

$$P = \text{Proportion of sample elements in the category}, \tag{3.4}$$

that is,

$$E(P) = \pi. \tag{3.5}$$

The variance of P is given by

$$Var(P) = \frac{\pi(1-\pi)}{n} \frac{N-n}{N-1}. \tag{3.6}$$

In the above expressions, N is the population size, and n the sample size.

However, the properties just confirmed, that $E(P) = \pi$ and $E(\bar{Y}) = \mu$ *always,* indicate that the average of estimates based on a large number of simple random samples of the same size can be expected to equal the population characteristic being estimated no matter what the value of this characteristic is. Put differently, the distribution of a large number of estimates can be expected to be centered on the target—in this context, the population characteristic. The simple estimators tend not to underestimate or to overestimate—they are not biased.

Consider now the variances of $\bar{Y}$ and P given by Eqs. 3.3 and 3.6. Let us start with the variance of $\bar{Y}$, which can be written as

$$Var(\bar{Y}) = \frac{\sigma^2}{n} \frac{N-n}{N-1} = \frac{\sigma^2}{N-1} \frac{N-n}{n} = (\frac{\sigma^2}{N-1})(\frac{N}{n} - 1).$$

The first term does not depend on the sample size; the second term becomes smaller and smaller as n approaches N (and equals 0 when $n = N$). In other words, the distribution of $\bar{Y}$ tends to become more and more concentrated around μ as the sample size increases. Figure 3.6 illustrates this.

The same conclusion holds true for $Var(P)$; it, too, decreases as n approaches N. We conclude that *a larger simple random sample tends to be more accurate—and therefore better—than a smaller sample*. This conclusion, it will be noted, confirms a similar one based on the range, as explained in Chapter 2.

The next and final implication of Summary 3.1 is also another confirmation of a preliminary conclusion in Chapter 2: we would like to show that *sampling without replacement is better than sampling with replacement*.

Note first that when the population size is large, N is approximately equal to $N - 1$ (in shorthand: $N \approx N - 1$), and the last term of the variances of $\bar{Y}$ and P given in Eqs. 3.3 and 3.6 can be written as

$$\frac{N - n}{N - 1} \approx \frac{N - n}{N} = 1 - \frac{n}{N}.$$

Therefore, for large populations (large N) and small sample to population size ratios (small n/N), the term $(N - n)/(N - 1)$ is approximately equal to 1 and can be ignored. In such cases, the variances of $\bar{Y}$ and P do not depend appreciably on N.

Sampling with replacement differs from sampling without replacement essentially in that the population remains unchanged throughout the selection of the sample. Sampling without replacement from a very large population when the ratio of sample size to population size is very small has approximately the same feature.

To illustrate, suppose that the number of firms in the Ackroyd case was 100,000 instead of 6. The probability of selecting any two different firms in a given order—say, A then B—in a random sample of size 2 with replacement is

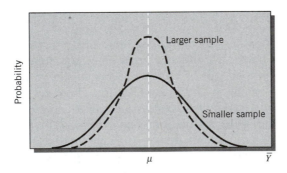

Figure 3.6 Accuracy of $\bar{Y}$ tends to increase as n approaches N

(1/100,000)(1/100,000). The probability of this outcome in a random sample of the same size without replacement is (1/100,000)(1/99,999), which is obviously approximately the same. The same approximate equality holds for the probabilities of all other outcomes and consequently for the probability distributions of all estimators.

Therefore, the variances of $\bar{Y}$ and P for samples with replacement are given by Eqs. 3.3 and 3.6 with $(N - n)/(N - 1)$ replaced by 1. That is, *the variance of the probability distribution of $\bar{Y}$ in samples with replacement is σ^2/n, while that of P is $\pi(1 - \pi)/n$.*

The means of these distributions, of course, are not affected: they are given by Eqs. 3.2 and 3.5 whether sampling is with or without replacement. *The simple estimators*, therefore, *are unbiased also when the sample is with replacement.*

The variances of $\bar{Y}$ and P based on samples without replacement differ from those with replacement by the term $(N - n)/(N - 1)$. Provided that N is large,

$$\frac{N - n}{N - 1} \approx 1 - \frac{n}{N},$$

as noted previously, and this is less than 1 for any n such that $1 < n < N$. Therefore,

$$\underbrace{Var(\bar{Y})}_{\text{without repl.}} = \frac{\sigma^2}{n} \frac{N - n}{N - 1} < \frac{\sigma^2}{n} = \underbrace{Var(\bar{Y})}_{\text{with repl.}}$$

The same conclusion applies with respect to $Var(P)$, as can be easily demonstrated by following the same steps.

In other words, for the same sample size, the simple estimators tend to vary less around the population characteristic under sampling without than under with replacement. In this sense, then, sampling without replacement should be preferred over sampling with replacement.

It should be noted again that the properties in Summary 3.1 are long-run properties; they describe the expected behavior of the simple estimators in a large number of simple random samples of the same size. The two main conclusions of this section (that sampling without is better than sampling with replacement, and that a larger sample is better than a smaller one) were based on these long-run properties, much like choices among expert marksmen relied on their track records. (Those readers who find the analogy with expert marksmen helpful may have observed that, for example, sampling with replacement and sampling without replacement were treated as two marksmen, both unbiased, but one being more accurate than the other in the long run.)

These long-run properties and their implications assist the *design* of the sample, that is, the choice of methods for sample selection and estimation. This will become more clear in the next chapter, where we begin considering alternatives to simple random sampling. But *if, for whatever reason, it has been decided to use a simple random sample in a given case, the above properties and implications together*

imply that the sample should be without replacement, as large as the budget permits, and that the ordinary sample averages and proportions should be used to estimate the population characteristics of interest—at least until better estimators become available. All that remains is to take the sample and perform the calculations.

3.6 SELECTING A RANDOM SAMPLE IN PRACTICE

In general, simple random sampling requires that, in every draw, each eligible population element be given equal probability of selection. In sampling without replacement, eligible are the population elements that were not selected in earlier draws; in sampling with replacement, of course, all population elements are eligible in every draw. The thorough physical randomization of the population elements or of their substitutes is thought to satisfy the equal probability requirement.

Physical randomization will be very awkward if the population size is very large (think of sampling from a population of millions of individuals in a large city). However, if a list of the population elements is available, the selection of a random sample may be easily accomplished with the use of a random device or of random numbers.

To illustrate, suppose that a population consists of 345 elements. These elements are listed in sequence and identified by the numbers 001, 002, ... , 345. Now suppose that we mark ten identical chips with the numbers 0 to 9, and put them in a hat. To select an element from the population, we draw three chips at random from the hat, *replacing them after each draw*. The first chip drawn will establish the first digit, the second and third draws will establish the second and third digits of the identification number. For example, if the chips marked 2, 1, and 8 are drawn, population element no. 218 is selected. If the three-digit identification number formed is 000 or greater than 345, it is ignored. If the sample is without replacement, previously formed identification numbers are also ignored.

To prove that this procedure does indeed yield a simple random sample is not very easy. We can, however, confirm its validity by means of a simple example. The following digression may be skipped by readers who do not care to question the validity of the procedure.

Suppose that the six firms in the Ackroyd population are listed in alphabetical order, so that A is no. 1 in the list, B no. 2, ... , F is listed as no. 6. Suppose the above procedure will be used to select a sample of size *one*. We noted earlier that in a random sample of size n the probability that a given firm—say, B—will be included in the sample is n/N, in this case, 1/6. We would like to confirm that the procedure produces the same result. Figure 3.7 illustrates.

If the number on the first chip is 2, B is selected and we stop; the probability of this occurring is 1/10. If the number is 1, 3, 4, 5, or 6, firm A or C or D or E or F is selected and

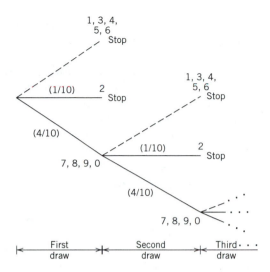

Figure 3.7 Illustration of selection procedure

we stop; the probability of each of these outcomes is 1/10 (to avoid cluttering Figure 3.7, these outcomes are represented by a single dotted line). If the number is a 7, 8, 9, or 0 (and the probability of this "other" outcome is 4/10), the number is ignored and another chip is drawn. The outcomes of the second draw are exactly like those of the first. And so are those of the third, if a third draw is needed. And so on.

We see that it may be necessary to select a second, third, fourth, ... chip before we can stop. Theoretically, there is no end to the tree in Figure 3.7.

What, then, is the probability that the sample will consist of firm B? This will happen if the chip numbered 2 is selected on the first draw, *or* if a chip with an "other" number (7 to 0) is selected on the first *and* one marked 2 is selected on the second, *or* if a chip with an "other" number is selected on the first *and* second draws *and* the chip marked 2 is selected on the third, *or*.... This probability is calculated by multiplying the probabilities along the appropriate branches of the tree and adding the results, as follows:

$$Pr(B) = (\frac{1}{10}) + (\frac{4}{10})(\frac{1}{10}) + (\frac{4}{10})(\frac{4}{10})(\frac{1}{10}) + \cdots$$
$$= (\frac{1}{10}) + (\frac{4}{10})(\frac{1}{10}) + (\frac{4}{10})^2(\frac{1}{10}) + \cdots$$
$$= (\frac{1}{10})[1 + (\frac{4}{10}) + (\frac{4}{10})^2 + \cdots].$$

The term in square brackets is the sum of an infinite series of the form $1 + a + a^2 + a^3 + \ldots$; when $0 < a < 1$, the sum is known to equal $1/(1 - a)$. We conclude that

$$Pr(B) = \frac{1}{10} \frac{1}{1 - (\frac{4}{10})} = \frac{1}{10} \frac{10}{6} = \frac{1}{6}.$$

Under this procedure, therefore, firm B will appear in the sample with probability 1/6, the same as in a simple random sample of size 1 without replacement. This ends our digression.

Random numbers achieve the same objective but eliminate the need to mark and mix chips. The numbers shown in Table 3.5 can be thought of as if generated by repeated draws of chips marked 0, 1, 2, ... , 9 from a hat. They can be read in any orientation (horizontally, vertically, diagonally, etc.), in any direction (forward, backward, etc.), beginning with any number and proceeding one by one, or using every second, third, etc. number encountered. (There is no need, of course, to be overly elaborate. An ordinary reading is easiest and perfectly adequate.) The random numbers can be combined to form identification numbers of any number of digits.

For example, to select a random sample of five three-digit numbers between 001 and 345 without replacement, we may begin with the second row of Table 3.5 and form the following consecutive three-digit numbers: 420, 065, 289, 435, 537, 817, 382, 284, 214, 455, 121. Ignoring numbers greater than 345, the sample consists of the elements numbered 065, 289, 284, 214, 121.

Needless to say, the numbers in Table 3.5 were not generated by drawing chips from a hat. Most computer languages and many programs provide a *random number generator*, a routine for generating any number of random numbers. These routines are usually carefully tested to ensure that the digits 0, 1, ... , 9 they generate behave as we expect the draws of chips with replacement from a hat to behave: in the long run, they appear with equal relative frequency (1/10) and are unrelated to (independent of) one another. One such routine was used for the random numbers shown in Table 3.5.

3.7 HOW LARGE SHOULD A SAMPLE BE?

Twice earlier we concluded that a larger sample is preferable to a smaller one. Therefore, the practical solution to the sample size problem is often to simply select as many elements as the budget and other resources permit. Thus, for example, a budget of $2,000 and a cost per sampled element of $4 suggest a sample of size 500.

Although this approach is followed in the majority of cases in practice, there are situations where it is desirable to quantify the expected improvement in accuracy resulting from increasing the sample size.

The simplest method considers the variance of an unbiased estimator as the measure of accuracy. The sample size, n, affects the variance of the estimators $\bar{Y}$ and P. In the case of $\bar{Y}$,

$$Var(\bar{Y}) = \frac{\sigma^2}{n} \frac{N-n}{N-1}. \tag{3.7}$$

TABLE 3.5 Some Random Numbers

```
0 4 0 4 6 8 4 5 2 0 3 0 2 1 9 1 5 9 6 7 8 0 2 4 3 5 7 7 0 6
4 2 0 0 6 5 2 8 9 4 3 5 5 3 7 8 1 7 3 8 2 2 8 4 2 1 4 4 5 5
1 2 1 3 7 7 0 1 6 5 6 7 9 2 2 4 0 1 3 2 3 6 9 0 3 6 9 7 6 4
4 9 4 2 6 9 1 1 1 5 7 4 2 2 5 2 6 2 2 2 6 4 2 1 7 1 4 1 8 5
7 8 0 6 4 8 7 5 6 6 4 3 9 5 9 5 1 2 0 2 5 8 6 4 3 4 1 3 2 7
7 2 1 4 9 6 8 4 4 5 1 0 4 4 7 3 5 3 6 4 3 8 6 7 0 6 2 6 2 1
6 5 4 9 1 1 9 6 2 6 0 0 0 8 1 2 5 1 5 2 0 8 1 5 6 0 1 3 6 5
7 1 0 0 8 1 6 7 4 6 5 4 4 7 9 8 8 2 6 4 8 1 6 7 6 2 6 0 5 3
6 0 4 0 6 3 3 4 9 4 6 3 7 7 8 2 5 0 9 4 7 6 1 9 4 4 8 0 2 3
9 6 6 7 9 7 3 7 7 1 0 6 1 9 5 8 3 1 0 0 5 9 8 5 9 4 4 7 2 6
4 8 9 3 5 8 5 0 8 4 9 0 8 4 9 2 3 9 1 5 4 5 6 7 8 7 6 2 4 8
3 9 6 0 4 2 5 1 5 3 6 4 8 7 6 0 9 5 1 1 9 9 8 9 9 4 9 3 1 5
3 6 3 3 8 7 2 1 3 8 6 0 0 6 9 1 9 5 4 0 5 5 8 1 7 7 0 8 2 5 0
8 2 5 4 6 8 8 1 3 6 4 8 8 7 4 2 1 7 9 7 8 2 9 6 6 1 2 0 8 8
2 4 3 4 3 7 9 3 6 9 3 9 3 9 2 9 9 9 4 0 7 2 0 0 9 2 0 2 9 6
1 6 7 3 9 7 2 7 0 9 4 4 7 8 2 6 0 4 3 3 1 7 5 2 0 8 7 9 2 6
6 4 1 1 0 2 7 1 8 7 3 7 3 2 1 4 5 7 1 9 4 9 3 7 7 7 9 1 9 5 4
2 5 5 1 5 1 5 7 5 1 0 2 1 0 2 5 0 1 0 3 4 0 7 9 1 0 9 9 1 0
3 3 3 6 7 1 0 6 3 1 2 6 9 1 7 7 6 3 1 1 1 0 5 0 1 4 5 3 4 5
4 3 8 4 5 8 2 2 3 5 6 6 1 0 3 2 8 1 7 7 3 4 8 9 6 2 1 2 2 5
1 3 0 3 8 9 1 3 2 3 9 5 4 8 9 2 9 2 6 7 6 4 8 4 0 8 8 5 7 3
8 3 6 5 2 5 7 5 0 6 4 0 7 5 7 6 7 2 0 9 2 4 5 3 1 5 9 5 2 2
7 7 5 4 1 0 8 7 0 9 1 4 3 0 0 4 1 9 5 4 9 4 9 8 6 1 1 2 0 8
5 1 6 0 8 0 6 5 0 7 3 3 3 8 8 7 1 5 9 1 3 0 1 0 5 8 2 5 9 1
9 9 1 7 7 7 9 4 8 0 9 5 2 5 2 8 0 4 6 3 6 2 9 5 9 4 7 9 0 3
3 4 5 4 2 1 4 5 8 6 0 9 3 3 7 9 8 1 5 6 9 4 2 3 8 7 9 4 0 4
5 2 3 4 1 5 2 3 2 5 5 4 1 5 1 3 5 9 3 8 6 3 8 1 6 1 5 7 4 2
3 5 7 6 0 5 9 3 2 5 1 8 3 2 0 5 1 0 2 6 2 7 4 6 9 1 6 7 7 6
8 4 6 5 2 0 7 8 9 2 2 4 4 0 1 6 9 2 0 1 4 2 2 4 7 1 8 2 8 4
5 9 3 5 5 6 5 1 1 3 1 5 8 5 8 7 7 3 3 2 8 2 7 0 5 0 0 8 9 2
8 6 2 0 6 5 6 6 5 0 7 9 9 9 1 8 7 3 4 1 2 5 3 0 4 5 5 6 8 7
8 9 5 7 4 1 7 9 7 9 0 5 4 5 8 8 0 3 2 6 9 0 9 2 5 1 7 5 6 2
1 3 4 1 5 2 4 4 0 3 5 4 2 7 1 6 6 5 6 4 1 5 0 9 9 3 4 6 6 6
4 7 7 1 1 8 2 3 4 4 6 2 4 8 2 0 3 1 3 2 3 4 0 1 2 1 7 0 0 4
3 5 3 0 7 2 2 7 2 5 2 5 5 1 5 5 1 5 0 7 6 6 5 0 6 4 9 2 5 1
8 0 1 3 2 4 6 6 7 3 5 1 6 1 6 7 1 9 6 3 4 2 7 3 5 6 4 1 4 7
8 1 1 2 2 1 5 0 6 5 9 0 4 7 3 0 7 5 3 8 9 3 1 8 1 0 3 4 5 7
1 1 0 1 1 2 9 4 6 6 7 2 7 9 1 0 0 6 0 1 8 5 6 2 7 4 5 5 6 0
3 5 7 1 3 8 2 2 1 7 7 9 0 3 9 3 6 8 7 0 0 8 0 5 8 5 3 9 3 2
```

If the sample size is increased by a factor k, that is, from n to kn, the variance of $\bar{Y}$ becomes

$$\frac{\sigma^2}{kn} \frac{N - kn}{N - 1}. \tag{3.8}$$

The difference between Eqs. 3.7 and 3.8, the improvement in accuracy, can be written as

$$\frac{\sigma^2}{kn(N - 1)}[k(N - n) - (N - kn)] = \frac{N\sigma^2(k - 1)}{kn(N - 1)}. \tag{3.9}$$

Equation 3.9 can be divided by Eq. 3.7 to give the *relative improvement in accuracy*, that is, the relative reduction in variance,

$$(1 - \frac{1}{k})(\frac{N}{N - n}), \tag{3.10}$$

as a result of increasing the sample size from n to kn (we assume, of course, that $n < kn < N$).

The relative improvement in accuracy depends on the population size N, the original sample size n, and the factor k. For example, if $N = 500$ and the sample size is doubled from 100 to 200 ($k = 2$), the relative improvement in accuracy is

$$(1 - \frac{1}{2})(\frac{500}{500 - 100}) = \frac{1\,500}{2\,400} = 0.625,$$

that is, accuracy is improved by 62.5%.

It is easy to show (see Problem 3.11) that Eq. 3.10 also gives the relative improvent in accuracy in the case of $Var(P)$. We may conclude, therefore, that the relative improvement in accuracy in general resulting from increasing the sample size is given by Eq. 3.10. To calculate this improvement, it will be noted, it is not necessary to know either the σ^2's (the population variances) of the variables of interest, or the π's, the proportions of population elements in the categories of interest.

To determine the size of the sample needed to estimate a population characteristic with given accuracy—for example, so that $Var(\bar{Y}) = C$, where C is a given number—we may solve $Var(\bar{Y})$ or $Var(R)$ for n. If it is required that $Var(\bar{Y}) = C$, then the sample size can be easily shown to be

$$n = \frac{N\sigma^2}{(N - 1)C + \sigma^2}.$$

The sample size required to make $Var(R) = C$ can likewise be shown to be

$$n = \frac{N\pi(1 - \pi)}{(N - 1)C + \pi(1 - \pi)}.$$

There are two difficulties with this approach. The obvious one is that the required sample size depends on σ^2 or π, and these population characteristics are unknown. This problem is unavoidable when we seek to satisfy given (as opposed to relative) accuracy requirements, but, as we discuss below, can be sidestepped if an earlier

similar sample or a pilot sample suggests reasonable estimates of σ^2 or π. The second problem with this approach is that it is usually difficult to defend a particular choice of C (recall that the variance of a variable depends on the units in which a variable is measured).

For the latter reason, another approach is often used to give an idea of the sample size needed to meet given accuracy requirements.

To begin with, assume that the purpose of the sample is to estimate the proportion (π) of elements in the population belonging to a given category. Naturally, one wants the estimator P to be close to π with high probability. Specifically, suppose it is desired that P be in the interval from $\pi - c$ to $\pi + c$ with probability $1 - \alpha$, whatever the value of π happens to be. c is a given number defining "closeness," and $2c$ is the width of the interval. Together, c and $1 - \alpha$ specify the accuracy requirements.

Provided that the accuracy requirements are suitably demanding (this will be explained below), it can be shown that *the size of a sample without replacement needed for the estimate of π to be in the interval from $\pi - c$ to $\pi - c$ with probability $(1 - \alpha)$ is approximately*:

$$n = \frac{N\pi(1 - \pi)}{(N - 1)D^2 + \pi(1 - \pi)}, \tag{3.11}$$

where $D = (c/Z_{\alpha/2})$ and $Z_{\alpha/2}$ is a known factor depending on $1 - \alpha$. The factors $Z_{\alpha/2}$ corresponding to commonly used $1 - \alpha$ are listed in Table 3.6.

TABLE 3.6 $Z_{\alpha/2}$ for Selected $1 - \alpha$			
$1 - \alpha$	$Z_{\alpha/2}$	$1 - \alpha$	$Z_{\alpha/2}$
0.99	2.576	0.80	1.282
0.95	1.960	0.60	0.842
0.90	1.645	0.50	0.674

The calculations will be illustrated in a moment, but first we must explain how Eq. 3.11 may be used since it depends on π, and π is, of course, unknown.

As demonstrated in the following table, the term $\pi(1 - \pi)$ increases as π approaches 0.5 from either direction and reaches its maximum 0.25 when $\pi = 0.5$.

π	$\pi(1 - \pi)$
0.1 or 0.9	0.09
0.2 or 0.8	0.16
0.3 or 0.7	0.21
0.4 or 0.6	0.24
0.5	0.25

Other things being equal, n in Eq. 3.11 is largest when $\pi(1 - \pi)$ is largest (divide both numerator and denominator by $\pi(1 - \pi)$ to see this more clearly); therefore, *the required sample size is largest when $\pi = 0.5$.*

If there is no information at all concerning π, it could be assumed conservatively that $\pi = 0.5$ and the required sample size calculated under this assumption. That large a sample will be greater than needed no matter what the value of π happens to be.

If available information suggests that π is no closer—in either direction—to 0.5 than π_0, then π_0 could be used in place of 0.5 in the calculations. For example, if it is thought that the population proportion cannot possibly exceed 0.20, $\pi = 0.20$ should be used; if it is thought that this proportion cannot possibly be less than 0.60, $\pi = 0.60$ should be used.

■ EXAMPLE 3.1

How large a random sample without replacement should be taken of a district's $N = 50,000$ households so that the estimate of the proportion of households buying a given product is in the interval from $\pi - 0.01$ to $\pi + 0.01$ with probability 95%?

For this problem we have $c = 0.01$ and $1 - \alpha = 0.95$. From Table 3.6 we find $Z_{\alpha/2} = 1.96$.

A survey taken some two years ago indicated the product was used by 40% of the households at that time. Using 0.40 as an estimate of π, the required size of the sample is

$$n = \frac{(50{,}000)(0.4)(1 - 0.4)}{(50{,}000 - 1)(0.01/1.96)^2 + (0.4)(1 - 0.4)} = 7{,}784,$$

or, in round figures, about 7,800 households. ■

From now on, it will be convenient to use the phrase "within $\pm c$ of π" for the longer "in the interval from $\pi - c$ to $\pi + c$." (In general, "within $\pm c$ of A" should be understood as "in the interval from $A - c$ to $A + c$.")

If, as happens very frequently in practice, the sample is planned so as to estimate several π's, and if the sample size is calculated using $\pi = 0.5$ (under the assumption that some π is likely to be 0.5 or close to that number), then the probability is *at least* $1 - \alpha$ that *any* P will be within $\pm c$ of the corresponding π.

The above formula also yields the sample size needed to estimate the total number in a category, $\tau' = N\pi$. The estimator of τ' is, of course, $T' = NP$. The inequality

$$\tau' - c' \le T' \le \tau' + c'$$

can be written as

$$N\pi - Nc \le NP \le N\pi + Nc,$$

where $c = c'/N$. Dividing through by N, the same inequality becomes

$$\pi - c \leq P \leq \pi + c.$$

We see that for T' to be in the interval from $\tau' - c'$ to $\tau' + c'$, P must be in the interval from $\pi - c$ to $\pi + c$. Therefore, *the sample size required to estimate the total in a category within* $\pm c'$ *with probability* $1 - \alpha$ *is given by Eq. 3.11 with* $c = c'/N$.

If the primary objective of the sample is to estimate the mean or total of a variable, then, provided again that the accuracy requirements are suitably demanding, it may be shown that *the sample size required for the estimate of the population mean* (μ) *of a variable* Y *to be in the interval from* $\mu - c$ *to* $\mu + c$ *with probability* $(1 - \alpha)$ *is approximately*

$$n = \frac{N\sigma^2}{(N-1)D^2 + \sigma^2}, \tag{3.12}$$

where, as before, $D = (c/Z_{\alpha/2})$. *Equation 3.12 with* $c = c'/N$ *also gives the sample size needed for the estimate of the population total of a variable,* $\tau = N\mu$, *to be in the interval from* $\tau - c'$ *to* $\tau + c'$ *with probability* $(1 - \alpha)$.

Just as an estimate of π is necessary to calculate the required sample size for proportions, so an estimate of the population variance (σ^2) of the variable is needed for the application of Eq. 3.12. In practice, such an estimate may be suggested by a pilot study, an earlier similar sample, or it could be just a guess so as to get a rough idea of the size of the sample. There is no "conservative" solution for this problem, as when estimating a proportion, because the required sample size increases with σ^2.

EXAMPLE 3.2 A telephone company plans to ascertain the condition of telephone poles in the region it services and the cost of their repair.[4] There are altogether 10,000 poles, a list of which is maintained by the company. From this list, a simple random sample of 100 poles was selected without replacement. Crews were sent out to examine the condition of the poles selected, and to calculate the cost of needed repairs. The results of this inspection were as follows: sample average repair cost, \$83; sample variance of repair costs, 121 (the latter, of course, is the variance of the 100 sample observations, the average of which is 83).

How many additional poles must be sampled if the estimate of the total cost of repairing all telephone poles (formed by pooling the observations in the pilot and the planned sample) is to be within ±\$5,000 of the true total cost with probability 90%?

Here, $N = 10{,}000$, $1 - \alpha = 0.90$, $Z_{\alpha/2} = 1.645$. For the estimate of the total cost to be within $\pm c = 5,000$ of the true value, the estimate of the average cost

[4] Suggested by a case study in W. E. Deming, *Some Theory of Sampling*, Wiley, 1950.

must be within $\pm c = (5,000/10,000)$ or \$0.50 of the population average. Using the variance of the pilot sample as an estimate of the population variance and $c = 0.50$, the required sample size is

$$n = \frac{(10,000)(121)}{(10,000 - 1)(0.5/1.645)^2 + 121} = 1,158.$$

Therefore, $1,158 - 100 = 1,058$, or, in round figures, 1,060 additional poles must be inspected. ◾

We can combine Eqs. 3.11 and 3.12 into a single formula and state the sample size prescriptions as shown in Summary 3.2. We shall use this terse description in subsequent chapters as well, where it will help save space.

When applying these formulas, one should always bear in mind that they are approximate. Their derivation, explained in mathematical statistics texts, relies on a variant of the Central Limit Theorem, according to which, as n and $N - n$ become large, the probability distributions of the P and $\bar{Y}$ approach the normal distribution (indeed, the factor $Z_{\alpha/2}$ is the $100(\alpha/2)$ upper percentile of the standard normal distribution). This result does not say how large n must be for a good approximation, and, of course, in determining the sample size, n is the unknown rather than the given of the problem. What should be said is that *the accuracy requirements (the values of c and $1 - \alpha$) must be demanding enough (that is, c must be small enough and/or $1 - \alpha$ must be large enough) so that the application of these formulas yields large n and $N - n$, which, in turn, justify the use of the formulas in the first place.* As to how large is "large enough" there can be no precise answer without knowledge of the population characteristics and the empirical requirements. As a rule of thumb, it is suggested that n and $N - n$ be greater than 50—a rule of thumb that is easily satisfied in many sampling applications.

3.8 INTERVAL ESTIMATES

We have explained why it is reasonable to use the sample mean ($\bar{Y}$) as an estimator of the population mean (μ) of a variable Y, and the sample proportion (P) as an estimator of the population proportion (π) in a category. After the sample is taken, the numerical values of $\bar{Y}$ and P are the estimates of μ and π respectively.

We realize, of course, that in general it is unlikely these estimates will equal μ or π exactly. Although we may say, for example, "we estimate μ and π to be 10.3 and 0.12," we certainly do not consider these to be necessarily the exact values of μ and π.

Instead of saying "μ is estimated to be $\bar{Y}$," we may say "μ is estimated to be *in the interval* from $\bar{Y} - c$ to $\bar{Y} + c$." This is an interval around $\bar{Y}$, usually abbreviated as $\bar{Y} \pm c$, with c to be specified. $\bar{Y} \pm c$ is an *interval estimator* of μ, and we may

---

SUMMARY 3.2 Simple random sampling, sample size

Provided that the accuracy requirements are demanding enough, the size (n) of a simple random sample needed to estimate μ or π within $\pm c$ with probability $(1 - \alpha)$ is

$$n = \frac{NA}{(N-1)D^2 + A},\qquad (3.13)$$

where $D = (c/Z_{\alpha/2})$, and $A = \pi(1 - \pi)$ when estimating π, or $A = \sigma^2$ when estimating μ. For estimating $\tau' = N\pi$ or $\tau = N\mu$ within $\pm c'$ with probability $(1 - \alpha)$, apply Eq. 3.13 with $c = c'/N$. Selected values of $Z_{\alpha/2}$ can be found in Table 3.6.

---

on occasion prefer it to the ordinary estimator $\bar{Y}$ (a *point estimator*, as an ordinary estimator is called when contrasted to an interval estimator).

For given values of $\bar{Y}$ and c, the statement "μ is in the interval $\bar{Y} \pm c$" is either correct (that is, the interval contains μ), or incorrect. Forming an arbitrary interval is not at all difficult, but forming an interval having a given probability of containing μ is not easy.

An interval estimator of a population characteristic which contains the population characteristic with given probability is called a *confidence interval*, and the given probability the *confidence level*. Approximate confidence intervals for the population mean of a variable and for the population proportion of a category, applicable when the sample size (n) and population remainder ($N - n$) are large, are described in Summary 3.3.

Note that the intervals (3.14) and (3.15) are of the form $\bar{Y} \pm c$ and $P \pm c$, and can be calculated once the sample observations are available. The probability $1 - \alpha$ is specified in advance. It can be large (e.g., 99%, 90%) or small (e.g., 50%, etc.), as desired.

Before commenting on the features of these intervals, let us illustrate the calculations.

☑ EXAMPLE 3.3   A simple random sample of $n = 800$ households in a town revealed that the average weekly household expenditure on food was $\bar{Y} = \$95$ with a variance $S^2 = 1{,}156$. (Just in case the meaning of S^2 is not clear, 1,156 is the variance of the 800 responses to the question, "How much is the weekly food expenditure of your household?" The average of these 800 responses is, of course, 95.) There are $N = 10{,}000$ households in the town.

SUMMARY 3.3   Simple random sampling, confidence intervals

- When n and $N - n$ are large, an approximate $100(1 - \alpha)\%$ confidence interval for μ, the population mean of a variable Y, is

$$\bar{Y} \pm Z_{\alpha/2}\sqrt{\widehat{Var(\bar{Y})}}, \tag{3.14}$$

while one for π, the population proportion in a given category, is

$$P \pm Z_{\alpha/2}\sqrt{\widehat{Var(P)}}. \tag{3.15}$$

In the above expressions, $\widehat{Var(\bar{Y})}$ is an unbiased estimator of $Var(\bar{Y})$ given by

$$\widehat{Var(\bar{Y})} = \frac{S^2}{n-1}\frac{N-n}{N}, \tag{3.16}$$

while $\widehat{Var(P)}$ is an unbiased estimator of $Var(P)$ given by

$$\widehat{Var(P)} = \frac{P(1-P)}{n-1}\frac{N-n}{N}. \tag{3.17}$$

In Eq. 3.16 S^2 is the sample variance of the variable Y:

$$S^2 = \frac{1}{n}\sum_{i=1}^{n}(Y_i - \bar{Y})^2 = \frac{1}{n}\sum_{i=1}^{n}Y_i^2 - \bar{Y}^2. \tag{3.18}$$

- For large n and $N - n$, approximate $100(1 - \alpha)\%$ confidence intervals for the total of a variable ($\tau = N\mu$) and the total in a category ($\tau' = N\pi$) are obtained by multiplying by N the limits of the above intervals, to get

$$N\left[\bar{Y} \pm Z_{\alpha/2}\sqrt{\widehat{Var(\bar{Y})}}\right] \tag{3.19}$$

and

$$N\left[P \pm Z_{\alpha/2}\sqrt{\widehat{Var(P)}}\right], \tag{3.20}$$

respectively.

- $Z_{\alpha/2}$ for selected $(1 - \alpha)$ are given in Table 3.6.

Suppose that a 95% confidence interval is desired for the average weekly food expenditure by *all* households in the town. Thus, $1 - \alpha = 0.95$, $Z_{\alpha/2} = 1.96$, and the desired interval is

$$(95) \pm (1.96)\sqrt{\frac{1{,}156}{800 - 1}\frac{10{,}000 - 800}{10{,}000}}.$$

This is the interval from $(95 - 2.26)$ to $(95 + 2.26)$, or from \$92.74 to \$97.26.

A 95% confidence interval for the total household expenditure on food in the town is $(10,000)(95 \pm 2.26)$, or from about \$927,400 to \$972,600.   ☑

☑ EXAMPLE 3.4   A simple random sample of $n = 500$ from a population of $N = 100,000$ households was taken; 42% of the sampled households said they would buy an experimental product. We calculate

$$\sqrt{\widehat{Var(P)}} = \sqrt{\frac{(0.42)(1 - 0.42)}{500 - 1} \frac{100,000 - 500}{100,000}} = 0.022.$$

Since $P = 0.42$, a, say, 90% confidence interval for the proportion of all households who intend to buy is

$$(0.42) \pm (1.645)(0.022),$$

that is, from $(0.420 - 0.036)$ to $(0.420 + 0.036)$, or from 0.384 to 0.456.

A 90% confidence interval for the number of households who intend to buy is $(100,000)(0.420 \pm 0.036)$, or from about 38,400 to 45,600.   ☑

One should not interpret a *calculated* confidence interval as containing the population characteristic with the stated probability. For instance, in Example 3.4, it is incorrect to say that the probability is 90% that the population proportion is between 0.384 and 0.456. The population proportion is a given number, and either lies in that interval or does not—we do not know which is true. It is the *procedure* by which the intervals are calculated that is correct with probability 90%.

To understand this more clearly, imagine it is possible to select a large number of samples from the given population, all of the same size n. After each sample is taken, imagine calculating a, say, 90% confidence interval using Eq. 3.14 and stating that μ lies in the calculated interval. Because $\bar{Y}$ and $\widehat{Var(\bar{Y})}$ will vary from sample to sample, the location and width of the intervals will vary, as illustrated in Figure 3.8.

The claim is that, in the long run, 90% of these intervals will contain (will "bracket," "cover") μ, and the statement that μ lies in the calculated interval will be correct in 90% of the samples in the long run. (In Figure 3.8, all but one of the five confidence intervals shown—the exception being that for sample no. 4—contain μ.)

In the mathematical derivation of these results (which can be found in advanced mathematical statistics texts), it is shown that the probability that the interval estimator will contain the population characteristic (the coverage probability) approaches the nominal probability $(1 - \alpha)$ as n and $N - n$ become infinitely large. From the practical standpoint, this result does not provide any guidance as to how large n and $N - n$ must be for the coverage probability to be approximately equal to the nominal (the confidence level). Studies show that the approximation can

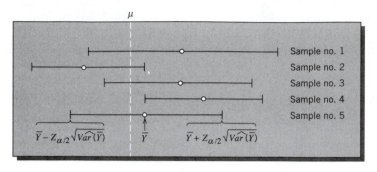

Figure 3.8 Interval estimates

be very good with very small samples from some populations, and very poor with even very large samples from others. The consensus appears to be that n and $N - n$ should be at least 50, but that should be treated purely as a rule of thumb. Undue reliance on this rule should be avoided.

These remarks should be kept in mind when using confidence intervals under more complicated sampling methods, where the quality of the approximation for finite n and $N - n$ is even more difficult to judge. A confidence interval is not a *necessary* appendage to the sample estimate. In most samples designed to obtain estimates of a large number of population characteristics, a routine reporting of all possible confidence intervals would most certainly confuse rather than enlighten.[5]

3.9 TO SUM UP

- The silly game suggests it is reasonable to prefer tried marksmen to untried ones, and to choose a marksman on the basis of his or her expected performance in a large number of future tosses, even though the game is decided by the next single toss. It also suggests that it is reasonable to focus on two aspects of expected performance: bias (the difference between the average location of the landings and the target), and accuracy (as measured by the variance of the landings), and to prefer among unbiased marksmen the most accurate (that is, the one with the

---

[5]We note in passing that the sample variance (S^2) of a variable Y as defined by Eq. 3.18 is not an unbiased estimator of the population variance σ^2. An unbiased estimator of σ^2 is

$$\hat{S}^2 = \frac{n}{n-1} \frac{N-1}{N} S^2.$$

This is why $\widehat{Var}(\bar{Y})$ and $\widehat{Var}(P)$ are not $Var(\bar{Y})$ and $Var(\bar{Y})$ with S^2 and P replacing σ^2 and π, respectively, as intuition may suggest. For large n and N, however, the difference between S^2 and $\hat{S}^2$, as well as the bias of the former may be overlooked. See also Section A.8 of Appendix A for comments regarding a different definition of variance used in some other texts.

smallest variance). These conclusions carry over to choices involving sampling methods and estimators.

- An estimator is said to be unbiased if the mean of its probability distribution equals the population characteristic it is targeting.
- Under simple random sampling, the sample mean of a variable and the sample proportion in a category are unbiased estimators of the population mean of the variable and of the population proportion in the category.
- The accuracy of an unbiased estimator is measured by its variance. Statistical theory provides expressions for the variances of the simple estimators. A study of these variances shows that a larger simple random sample tends to be more accurate—and therefore better—than a smaller sample, and that sampling without replacement is better than sampling with replacement.
- The above suggest that if, for whatever reason, it has been decided to use simple random sampling, the sample should be without replacement and as large as the budget permits, and that the ordinary sample averages and proportions should be used to estimate the population characteristics of interest—at least until better estimators become available.
- A simple random sample may be selected by physical randomization of the population elements or their substitutes, or, if a list of the population elements is available, with the use of a random device or of random numbers.
- In this chapter there may also be found approximate formulas for determining the size of a simple random sample needed to meet given accuracy requirements, and approximate confidence intervals (that is, interval estimates having a given probability of bracketing a population characteristic).
- Equally importantly, the approach of this chapter serves as a model for, and the results as a basis of comparison with, other sampling methods. Each of the following chapters will deal with a different method for selecting a random sample or for estimating population characteristics. For each method, in addition to formulas for sample size and confidence intervals, we intend to present unbiased estimators and expressions for their variances. The latter, in turn, will be used for comparing different methods of selection or estimation, with a view toward establishing the best method of selection and estimation in a given situation.

PROBLEMS

3.1 *Computing exercise:* Using the program SCALC, verify the numerical results presented in this chapter. Note any deviations, and determine if these are the result of rounding at intermediate stages.

3.2 A random sample of 6 light bulbs was selected from a lot of 100 for the purpose of estimating its quality. The life of each selected light bulb was measured by letting the bulb burn until it burned out. The test results were as follows:

Bulb No.	Life Duration (hours)
1	950
2	1,210
3	1,070
4	840
5	1,420
6	980

(a) Estimate the average and the total life of the bulbs in the lot.

(b) Estimate the proportion and the number of bulbs in the lot with life exceeding 1,000 hours.

(c) Briefly defend your choice of these estimates.

3.3 A simple random sample of 5 students was selected without replacement from 50 names appearing in a class list. The questionnaire had only two questions:

Q1: Do you smoke? Yes___ No___
Q2: What was your total income last year? ___ ($000)

The responses were as follows:

Student No.	Q1	Q2
1	No	15
2	Yes	7
3	No	10
4	No	2
5	Yes	9

(a) Estimate the average income and the total income of the 50 students in the class.

(b) Estimate the proportion and the number of smokers in the class.

(c) Briefly defend your choice of these estimates.

3.4 A population consists of 5 individuals. It is known that 1 of these has an annual income (Y, in $000) of 30, while each of the other 4 has an income of 20. You *plan* to take a random sample without replacement of 3 individuals from this population.

(a) Calculate the population mean (μ) and variance (σ^2) of Y.

(b) Determine the probability distribution of the average income ($\bar{Y}$) of the 3 individuals in the sample. (Express $\bar{Y}$ in $000. List results as fractions—do not convert to decimals.)

(c) Calculate the mean, $E(\bar{Y})$, and the variance, $Var(\bar{Y})$, of this distribution.

(d) Show that $E(\bar{Y}) = \mu$ and $Var(\bar{Y}) = \sigma^2(N - n)/n(N - 1)$, where μ and σ^2 are the population mean and variance of Y.

(e) Determine the probability distribution of the proportion of individuals in the sample (P) having income greater than 25 ($000).

(f) Calculate the mean, $E(P)$, and the variance, $Var(P)$, of this distribution.

(g) Show that $E(P) = \pi$ and $Var(P) = \pi(1 - \pi)(N - n)/n(N - 1)$, where π is the proportion of individuals in the population with income greater than 25 ($000).

3.5 A lot consists of 10 ball bearings. Specifications call for these bearings to be one-quarter of an inch in diameter. The inspection supervisor had all ball bearings measured, and the results are shown below.

Diameter	Number of Ball Bearings
−1	3
0	5
+1	2
Total	10

The diameter is measured as the difference from the specification (0.250 inches), in thousandths of an inch.

(a) Construct the probability distribution of the average diameter ($\bar{Y}$) of two ball bearings drawn from the lot at random *without* replacement. Calculate *by two different methods* the mean and variance of this probability distribution.

(b) Same as (a) except that the sample is *with* replacement.

(c) Construct the probability distribution of the proportion of ball bearings in a random sample of size $n = 2$ *without* replacement which do *not* meet the specification (i.e., whose diameter is not equal to 0). Calculate *by two different methods* the mean and variance of this distribution.

(d) Same as (c), except that the sample is *with* replacement.

3.6 A population has 8,943 elements, numbered serially from 0001 to 8943. Select a random sample of 10 elements without replacement. Describe clearly your method.

3.7 A population has 200 elements, numbered serially from 001 to 150 and 251 to 300. Select a random sample of 10 elements without replacement. Describe clearly your method.

3.8 Many firms are able to maintain fairly accurate lists of customers by utilizing warranty registration cards. Manufacturers of watches, calculators, washers, dryers, toasters, refrigerators, blenders, and many other appliances usually guarantee the product against defects in manufacture for a certain number of years from the date of purchase. In order to register the warranty, buyers must fill out and return a card showing their name and address, the model purchased, the type of store from which it was purchased, the method of payment, etc. These cards provide useful information to the company when they are first received (for example, in

determining the distribution of sales by type of store or location, or of the time elapsing between the date of sale and the date of shipment), and a list of customers which may be used in later surveys.

Yoshita is a manufacturer of television and other electronic equipment. On the basis of warranty registration cards, it maintains a file of 21,528 customers who purchased a TV set. From this file, a random sample of 500 customers was selected (without replacement, of course) for the purpose of estimating the proportion owning a VCR and the average age of these VCRs. Of the 500 sampled customers, 361 stated they owned a VCR. The average age of these VCRs was 14.3 months.

(a) Using a table of random numbers, show how the first 5 customers should be selected.

(b) Estimate the number of customers owning a VCR. Is yours an unbiased estimator? Why?

(c) Estimate the average age of all VCRs owned by customers. Is yours an unbiased estimator? Why? Can you write the variance of this estimator?

3.9 Among the questions in the NorPower questionnaire (Figure 1.2 and Section 1.4) were the following (slightly paraphrased):

> Do you own a color television set? Yes___ No___ If 'Yes,' indicate the number of sets in use: ___
>
> Do you own a 2- or 3-door combination refrigerator–freezer? Yes___ No___ If 'Yes,' is it frost-free? Yes___ No___

Suppose that a simple random sample of 500 from the population of 2.8 million electricity-using residences was selected. Suppose further that the responses to these questions were as follows:

> Own a color television set: 400; do not own: 100. Total number of color television sets in use: 600.
>
> Own refrigerator–freezer: 450; do not own: 50. Frost-free: 200; not frost-free: 250.

(a) Estimate the average number of color television sets in use per residence owning a color television set.

(b) Estimate the total number of color television sets in use in residences having a color television set.

(c) Estimate the proportion and number of residences having a frost-free refrigerator-freezer among residences having a 2- or 3-door combination refrigerator-freezer.

(d) Are your estimators unbiased? Explain why.

3.10 A prominent cable news network regularly reports opinion poll results in the following format. "The latest poll has 58% approving the President's actions. . . . *The margin of error for this poll is ±5 points.*"

How do you understand the sentence in italics? Is something missing? How should the sentence read if the poll constitutes a simple random sample of eligible voters?

3.11 Show that Eq. 3.10 also gives the relative improvement in $Var(P)$ resulting from increasing the sample size from n to kn ($n < kn < N$).

3.12 **(a)** A population consists of $N = 5,000$ elements. How large a sample without replacement should be drawn from this population if it is desired to estimate all population relative frequencies within ± 0.075 with probability 90%?

(b) A population consists of $N = 7,500$ elements. How large a sample without replacement should be drawn from this population if it is desired to estimate the population mean within ± 20 with probability 99%? A pilot survey suggests that the population variance is likely to be in the range 80 to 100.

3.13 A random sample of size $n = 200$ was drawn without replacement from a population of size $N = 1,000$. The sample mean of a variable Y is $\bar{Y} = 150$, and the sample variance $S^2 = 280$. Calculate 95% confidence intervals for the population mean and total of Y. Briefly interpret these intervals.

3.14 A random sample of size $n = 120$ was drawn without replacement from a population of size $N = 1,800$. The proportion of elements in the sample falling into a certain category is $P = 0.43$. Calculate 90% confidence intervals for the proportion and number of elements in the population falling into the category. Briefly interpret these intervals.

3.15 The following is an excerpt from a recent newspaper article:

One in Three Believes Education Has Slipped

Nationally today, 49 percent think children are better educated than they were, while 33 percent believe their education is worse. Another 13 percent feel education is much the same today as it was in their time, while 5 percent are unsure. . . .

The recent findings are based on personal in-home interviews with 1,017 adults, 18 years and over during the first week of February. The sample size is accurate within four percentage points, 19 in 20 times.

The question was: "Do you think children today are being better educated or worse than you were?"

Comment. Assuming that the poll was based on a random sample of size 1,017, *interpret and check* the statement: "The sample size is accurate within four percentage points, 19 in 20 times."

3.16 Refer to the Ackroyd population described in Table 3.1. Note that the total number of employees in Ackroyd is $\tau = N\mu = 31$, and the number of firms intending to hire is $\tau' = N\pi = 2$.

(a) Use Table 3.3 to show that $T' = NP$ is an unbiased estimator of τ', and that its variance is given by $Var(T') = N^2 Var(P)$.

(b) Use Table 3.4 to show that $T = N\bar{Y}$ is an unbiased estimator of τ, and that its variance is given by $Var(T) = N^2 Var(\bar{Y})$.

3.17 The question can be stated pithily as follows: Is an SRS of an SRS an SRS? In less cryptic terms: Is a simple random sample from a simple random sample itself a simple random sample?

The question has practical relevance because, among other reasons, it is a common practice of surveys monitoring radio and television audiences to select a "master" sample of individuals once every year, and to draw from that pool subsamples of individuals for their weekly, monthly, or quarterly ratings.

(a) Specifically, consider the Ackroyd population described in Table 3.1. Suppose that a sample of one firm will be selected first by drawing a simple random sample of two firms from the Ackroyd population, and then a simple random sample of one firm from the first sample. With the help of Table 3.2, show that the probability that any given firm will appear in the final sample is the same as in an ordinary simple random sample of size 1. In view of this, is an SRS of an SRS indeed an SRS?

(b) Same situation as in (a), except that a sample of two firms will be selected by drawing first a simple random sample of three firms from the Ackroyd population, and then a simple random sample of two firms from the first sample. Is indeed an SRS of an SRS an SRS in this situation?

3.18 GreenTurf is a company making more than 100 garden and agricultural fertilizers, consisting of different combinations of three basic ingredients. One of these ingredients is nitrogen. The company estimates the total quantity of nitrogen used on the basis of a random sample of production orders. Each order shows the quantity of nitrogen and of the other ingredients used for a particular job. There were 4,000 production orders issued during the year, numbered from 0001 to 4000.

(a) Describe briefly but precisely how you would select a simple random sample without replacement of 100 production orders.

(b) The auditor prefers to draw a sample in a different way. He would select at random one production order among those numbered 0001 to 0040, and every 40th order in sequence thereafter. For example, if the first order selected is numbered 0005, the remaining selected orders would have numbers 0045, 0085, ..., 3965. This method (known as *systematic random sampling*) will indeed produce a sample

of 100. Is a systematic sample a random one? Under what conditions can such a sample be treated as random for all practical purposes?

(c) A random sample without replacement of 100 production orders was taken. The average quantity of nitrogen per order was 150 lb, and the standard deviation of the quantities of nitrogen in the sample was 40 lb. Estimate the total quantity of nitrogen used.

(d) How large should *next* year's sample be so that the estimated total quantity of nitrogen used will be within ±10,000 lb. of the true total with probability 99%? Assume this year's number of production orders (4,000).

(e) Calculate 95% confidence intervals for the average and the total quantity of fertilizer used.

3.19 The accounting firm of McDermott and McDermott (M&M) recently undertook the audit of Smith's, a large department store with branches in several northeastern cities. In part, the audit involves an examination of Smith's accounts receivable. Nearly all these accounts represent outstanding balances on credit card charge accounts. On December 31, 19X9, there were altogether 25,423 accounts with reported balances varying from $10.45 to over $2,000, and a total book value of $1,525,608.

In order to audit an account, M&M must examine all debits and credits made in 19X9, and make sure that these are correctly reflected in Smith's sales and cash accounts. Clearly, it is very expensive to audit all 25,000-odd accounts. The audit must be based on a sample.

(a) There is some difference of opinion between two members of the audit team as to exactly how a sample of accounts should be taken. Mr. Audrey is in favor of selecting a random sample of accounts, on the grounds that such a sample would be (in his words) "unbiased." Mr. Jones, on the other hand, argues for auditing the largest accounts. "If there are any errors," he says, "especially errors intended to deceive the stockholders, these are more likely to be found among accounts with large balances. Auditing small accounts is a waste of time." Briefly explain how a sample should be selected, and why.

(b) The accounts are arranged alphabetically and numbered serially from 1 to 25,423. All the information for an audit can be pulled out of the appropriate files once the account is specified. Using a table of random numbers, select a random sample of 10 accounts without replacement.

(c) The auditors would like to know how large a random sample without replacement should be drawn if the estimate of the proportion of accounts with error is to be within ±0.01 of the true proportion with probability 95%, assuming that the true proportion of accounts with error is unlikely to exceed 10%.

(d) A random sample of 5 accounts was selected without replacement, and the sampled accounts were audited with the following results:

Account No.	Book Value ($)	Audit Value ($)	Error ($)
9866	45.23	42.23	+3.00
15217	28.95	28.95	0
2982	76.32	77.08	−0.76
17413	37.86	32.31	+5.55
20094	82.47	82.47	0

Error is the difference between Book and Audit ("true") value. An account is said to have an error if the book value is not equal to the audit value.

(i) Calculate the mean and variance of the sample book and audit values. (ii) Calculate the proportion of accounts in the sample that have an error.

(e) Suppose that a random sample of 300 accounts without replacement was selected and audited, with the following results:

Number of accounts sampled	300
Proportion of accounts with error	0.08
Sample mean of book values	60.67
Sample mean of audit values	59.45
Sample variance of book values	2,725.12
Sample variance of audit values	2,632.75

(i) Estimate the total audit ("true") value of Smith's accounts receivable. (ii) Estimate the total error in accounts receivable. *Note*: Total error = Total book − Total audit.

(f) Calculate 90% confidence intervals for the population characteristics listed in (e).

3.20 Every year, Eastern Hydro, an electric utility, sends a questionnaire to a random sample of its private customers and requests information on various aspects of appliance ownership and use.

The private customers (as distinct from commercial and industrial customers) are owners or tenants of detached houses, semi-detached houses, townhouses, or apartments. In Hydro's records, they are numbered sequentially with 6-digit identification numbers from 000001 to 845600.

Three of the questions in the questionnaire are as follows:

Q1. Do you own an electric washing machine? Yes__ No__
Q2. How many years have you lived at the same address? __years
Q3. How many persons live in the dwelling? __ persons

(a) Describe precisely how Hydro should select a random sample of 5 customers from its records. List the identification numbers of the selected customers, quoting the exact source and explaining in detail the procedure used.

(b) A random sample of 5 customers yielded the following responses to the three preceding questions:

Customer No.	Q1 (Y=Own; N=Do not own)	Q2 (Number of years)	Q3 (Number of persons)
1	Y	1	2
2	N	5	2
3	Y	4	1
4	N	2	3
5	Y	3	4

Estimate (i) the proportion of all customers who own an electric washing machine; (ii) the average number of years at same address by Hydro customers; and (iii) the total number of persons served by Hydro.

(c) Same as (b), except that this year's survey was based on a random sample of 3,000 customers and yielded the following results:

Sample proportion owning electric washing machine	63%
Sample average number of years at same address	1.75 years
Sample average number of persons in dwelling	2.15 persons
Sample variance of number of years at same address	0.09
Sample variance of number of persons in dwelling	1.2

(d) Eastern Hydro is planning *next year's* survey of customers. In answering the following questions, you may use whatever information from Question (c) you consider relevant. (i) How large a random sample without replacement should be taken so that the estimate of the proportion of customers owning an electric washing machine will be within ±0.01 of the true proportion with probability 90%? (ii) How large a random sample without replacement should be taken so that the estimate of the average number of years at the same address is within ±0.5 year of the true average with probability 95%? (iii) How large a random sample without replacement should be taken so that the estimate of the total number of persons served by Hydro is within ±10,000 of the true total with probability 90%? (iv) Are your answers for (i) to (iii) consistent? Should they be?

(e) Calculate 99% confidence intervals for the population characteristics listed in (b).

3.21 The following is an excerpt from a recent newspaper article:

Rules Are Broken in 25% of Hirings
For Civil Service

Government departments broke the law or federal regulations in one of every four appointments they made last year, the annual report of the Public Service Commission says.

The commission chose for intensive scrutiny a random sample of 7,082 of the 96,749 appointments made by departments. Infractions were found in 1,946 or 25.2 per cent of the cases studied. In addition, the auditors found 2,754 errors.

"Infractions" were defined to be violations of the regulations and the law. "Errors" were procedural matters such as not ranking the unsuccessful candidates in descending order of success, failing to give reasons why they were passed over, or not stating in writing why a candidate was chosen.

(a) Describe briefly how you would have selected "a random sample of 7, 082 of the 96, 749 appointments." You may assume that a file is kept for each appointment and that the files are numbered serially.

(b) Briefly discuss the advantages and disadvantages of sampling as a procedure for investigating the hirings and promotions of government departments.

(c) Do you think the sample size was large enough? After all, the sample is only 7% of the total number of appointments.

3.22 *Work sampling* refers to the practice of observing a worker at randomly selected times in order to estimate how his or her time is allocated among various activities.

As an illustration, consider a lathe operator. Most of her time is spent operating the machine but occasionally she is idle because the lathe is being repaired, the raw material is depleted, assistance is required, or for personal reasons. Suppose an inspector observes the operator at 200 randomly selected times (for example, at 8:37, 9:02, ..., 14:57, ...) and notes the activity in which the operator is engaged at the instant of the observation. The results are as follows:

Activity	Number of Observations
Normal work	145
Idle—avoidable delay	32
Idle—unavoidable delay	23
Total	200

Since the operator was engaged in normal work on 145 of the 200 observations, it is estimated that 72.5% of her time is spent at normal activity; similarly, since the operator was idle on 55 of the 200 observations, it is estimated that she is idle

27.5% of the time, of which a substantial part is due to delays that can perhaps be reduced with better organization.

In work sampling, time is viewed as a series of short time intervals (e.g., 1 minute, 1 second, etc.), so that it is reasonable to assume that an operator's activity at the instant of observation describes the activity throughout the interval.

(a) How would you determine the "randomly selected times" for observing the operator in a given working day? Assume an 8-hour (480-minute) work day. The observations will be made at the start of each selected 1-minute interval.

(b) Estimate the total idle time (in minutes) of this worker in one working day.

3.23 The National Automobile Association's (NAA) *Used Car Buyer's Guide* is "a compilation of consumer-oriented information designed to reduce the frustration and uncertainty often associated with buying a car." It is published annually and includes frequency of repair records, average annual mechanical repair cost estimates, and statistics detailing owner satisfaction (including how many would buy the same car again)—all these by make, model, and model year.

A questionnaire is distributed to all NAA members, asking such questions as:

> What type of car do you own?
>
> What distance do you drive annually?
>
> What problems have you experienced?
>
> Who does your maintenance and repairs?
>
> How satisfied are you with your car?

This questionnaire is also distributed to motorists through consumer magazines, random mailings, newspaper ads, and public press releases. A total of 15,446 responses were received and analyzed this year.

(a) You are the spokesperson for a car manufacturer that is not rated well in this guide. A reporter is doing a story on the quality of cars, and will use NAA's survey. Your company's comments on the survey are invited. Prepare a statement clearly and thoughtfully expressing your company's reservations about the survey results.

(b) Same as (a), but your company's cars are rated well in the guide. Prepare a statement clearly and thoughtfully outlining the reasons why the survey results should be considered reliable.

(c) As a person knowledgeable in sampling methods, what advice will you give NAA concerning the design of next year's survey? The advice must be not only theoretically correct but also practical and implementable.

3.24 The Commercial Bank is a financial institution with branches covering a large region. A number of customers recently received a questionnaire and the following covering letter.

Dear Customer:

On a frequent basis, we randomly select customers to take part in surveys to provide us with information on how they feel about certain products and services.

You have been selected for the current survey and we would very much like to hear from you. By taking a few minutes to complete the enclosed questionnaire, you will be letting us know your views about the quality of service at your Branch.

It is only by knowing how you and other customers feel that we can improve our services in the way you prefer. Your comments will receive careful consideration and your participation in the study will be completely anonymous. We would be grateful if you would return your completed questionnaire in the enclosed postage paid envelope by May 9.

Yours sincerely,

(signed) Bernie F. Agostini,

Director, Financial Services Division.

The questionnaire was stamped with the identification and address of the customer's branch and included 80 questions. Following are three of these.

1. How long have you been a customer at this branch of the Commercial Bank? Please circle the appropriate number below.

Less than 1 year	1	10 to 14 years	4
1 to 4 years	2	15 years or more	5
5 to 9 years	3		

20. How would you rate the overall quality of service you receive from this branch? Please use the scale below where "0" means that you think the service is Poor and "10" means you think the service is Excellent.

Poor					Average					Excellent
0	1	2	3	4	5	6	7	8	9	10

80. What is your household's total annual income before taxes? Please circle closest answer.

Up to $30,000	1	$70,000 - $79,999	6
$30,000 - $39,999	2	$80,000 - $89,999	7
$40,000 - $49,999	3	$90,000 - $99,999	8
$50,000 - $59,999	4	$100,000 and over	9
$60,000 - $69,999	5		

A simple random sample of about 10% of customers of each branch was selected. Branch A has 1,580 customers; 160 of these were randomly selected for the May survey. Their responses to the above questions are summarized in Table 3.7.

TABLE 3.7 Responses to Questions, Problem 3.24

Code	Number of Responses		
	Question 1	Question 20	Question 80
0	-	3	-
1	10	5	3
2	20	7	4
3	50	10	5
4	15	13	6
5	5	15	8
6	-	12	7
7	-	9	4
8	-	8	2
9	-	6	2
10	-	4	-
No response	60	68	119

(a) Estimate the average length of time customers have been with Branch A, the average rating of customer satisfaction and the average annual household income before taxes of customers in Branch A. State and justify any assumptions you are forced to make.

(b) Calculate 95% confidence intervals for the population characteristics listed in (a).

CHAPTER 4

Stratified Random Sampling

4.1 INTRODUCTION AND SUMMARY

Auxiliary information may allow us to divide the population elements into groups (or "strata," as they are often called). Stratified random sampling consists of selecting a simple random sample of elements from each group.

Since the stratified sample is selected randomly, albeit in a different fashion from a simple random sample, the first issue to be settled is whether or not stratified is essentially the same as simple random sampling. If it is the same, then there is no need for special estimators, since the simple estimators of Chapter 3 would serve here as well. But if it is not, how should the population characteristics be estimated? Further, if it is possible to take either a simple or a stratified random sample, which method of sample selection is better? Is one method better under all circumstances, or does the choice depend on the population being sampled?

In this chapter we intend to show that stratified is not the same as simple random sampling, and that the simple estimators, which were unbiased under simple random sampling, are no longer unbiased under stratified random sampling. We show how to form unbiased estimators when the sample is stratified; these estimators are weighted averages of the group sample means or proportions.

We shall show further that a stratified random sample is not necessarily better than a simple random sample of the same size. But a special type of stratified

random sample, the proportional stratified sample, is nearly always better, and should be the preferred choice whenever it can be implemented.

We shall also show that under certain circumstances it is possible to do even better than with a proportional stratified sample, by allocating the total sample to the groups in such a way as to minimize variability or cost.

Finally we intend to describe how to determine approximately the size of the total stratified sample to meet specified accuracy requirements, and how to form approximate confidence intervals using the stratified estimators in large samples.

4.2 STRATIFIED SAMPLING—A FIRST VIEW

We have in mind a population that can be divided into nonoverlapping groups according to a certain criterion. We assume that the number of population elements in each group is known.

For example, the population of interest could consist of all the households in a city. A list of the households and their addresses may be available (such lists are often compiled and sold by market research firms). On the basis of the addresses, households can be grouped by postal code.

Stratified random sampling consists of selecting a simple random sample from each group separately.[1]

In the above example, a stratified random sample would be selected by drawing at random and without replacement a number of households from each separate list of households having the same postal code.

A stratified sample is sometimes selected because it is desired to estimate the characteristics of each group separately in addition to those of the population at large. At other times, a stratified sample is believed to be more accurate than a simple one: we observed in Section 2.10 that if the groups are approximately homogeneous, stratified sampling can be expected to be better than simple sampling.

With the following section we begin the study proper of stratified random sampling. Because it will be a little while before we can justify the main practical prescriptions from this study, it may be helpful to start at the end as it were and assume that, for whatever reasons, it has been decided to take a stratified random sample. How are the population characteristics to be estimated?

Imagine a population of N elements, divided into M groups with $N_1, N_2, \ldots, N_M$ elements, respectively $(N_1 + N_2 + \cdots + N_M = N)$. We shall show that an unbiased estimator of the population mean of a variable Y is

$$\bar{Y}_s = w_1 \bar{Y}_1 + w_2 \bar{Y}_2 + \cdots + w_M \bar{Y}_M,$$

---

[1]The term "stratum" is often used as a synonym for "group," and it is to this practice that the term "stratified" owes its origin. Occasionally, the term "cluster" is also used to describe a group of elements, especially if the criterion of grouping is geographical proximity.

and an unbiased estimator of the proportion of population elements in a category is

$$P_s = w_1 P_1 + w_2 P_2 + \cdots + w_M P_M.$$

In these expressions, $w_i = N_i/N$ is the fraction of population elements in the ith group, $\bar{Y}_i$ is the average of the variable Y of interest in the sample from the ith group, and P_i is the proportion of elements that fall into the category of interest in the sample from the ith group.

An unbiased estimator of the population total of a variable is $T_s = N\bar{Y}_s$, and one of the number of elements in a category is $T_s' = N P_s$.

The subscript s in $\bar{Y}_s$, T_s, P_s, and T_s' stands, of course, for "stratified." We shall call these the *stratified estimators*. They are the recommended estimators when sampling is stratified.

Two examples will illustrate the calculation of the stratified estimators.

EXAMPLE 4.1   Books Research specializes in providing estimates of retail sales of books and other publications. These estimates are based on stratified samples and are sold to subscribing publishers, newspapers, and magazines. The latter publish the estimates in the form of "bestseller lists."

The 10,000 bookstores across the country are classified by Books Research according to floor space into three groups—Large, Medium, and Small. Table 4.1 contains the relevant data for estimating the total number of copies sold of a given title.

To understand the entries in Table 4.1, consider the first line. Of the 10,000 bookstores, 1,000 or 10% are classified as Large. A list of these bookstores was available. A simple random sample of 500 Large bookstores was selected. Each selected store reported the number of copies of the given title that were sold during a given period of time. The average of these numbers was 240. Simple random samples of 200 Medium and 100 Small stores yielded the averages shown in the

TABLE 4.1 Estimation of Book Sales, Example 4.1

		Population		Sample	
i	Group	Size, N_i	Percent of Elements, $w_i = N_i/N$	Size, n_i	Mean, $\bar{Y}_i$
1	Large	1,000	0.1	500	240
2	Medium	3,000	0.3	200	120
3	Small	6,000	0.6	100	40
	Total	$N = 10,000$	1.0	$n = 800$	

last column of the table. The stratified estimate of the average number of copies sold per bookstore in the country is

$$\bar{Y}_s = w_1\bar{Y}_1 + w_2\bar{Y}_2 + w_3\bar{Y}_3 = (0.1)(240) + (0.3)(120) + (0.6)(40) = 84.$$

The estimate of the total number of copies of this title sold in the country is $N\bar{Y}_s = (10,000)(84) = 840,000$.

Books Research estimates the sales of other titles in the same fashion, and ranks them as "#1 National Bestseller," "#2 National Bestseller," and so on. ◪

■ **EXAMPLE 4.2**   The Department of Revenue audits only a small fraction of the millions of individual income tax returns submitted every year. The number of audits is dictated by the size of the staff; usually, only about 1/4 of 1% of the returns are audited. The returns to be audited are selected on the basis of a stratified sample.

The Department's computers are programmed to classify individual files into a number of groups according to a set of prescribed criteria. These include the amount of income, the type of deductions claimed (tax shelters, home office, travel and entertainment, etc.), the individual's occupation (doctor, lawyer, artist, etc.), the sources of income (wages, business, dividends, interest, etc.), and so on. For the purposes of this illustration, we assume there are four groups, and consider how to estimate the population proportion of income tax returns "in error." A return is "in error" if the tax payable differs from that calculated by the individual. The data shown in Table 4.2 refer to files handled by one district taxation office.

For example, group 1 consists of 10,000 individual files. Four hundred of these files were randomly selected and the returns audited; 8.5% of these returns were judged to be in error.

The stratified estimate of the proportion of returns in error for the district office is

$$P_s = w_1 P_1 + \cdots + w_4 P_4 = (0.067)(0.085) + \cdots + (0.467)(0.010) = 0.0291.$$

TABLE 4.2 Estimation of Proportion of Returns in Error, Example 4.2

Group, i	N_i	$w_i = N_i/N$	n_i	P_i
1	10,000	0.067	400	0.085
2	30,000	0.200	200	0.051
3	40,000	0.267	100	0.032
4	70,000	0.467	100	0.010
Total	$N = 150,000$	1.000	$n = 800$	

The estimate of the total number of returns in error at the district office is $NP_s = (150,000)(0.0291) = 4,365.$ ◾

The stratified estimators are not arbitrary, but can be justified in intuitively appealing terms. To see this, suppose the objective is to estimate the population total of a variable Y, τ, where

$$\tau = \text{(Sum of } Y \text{ values of all population elements)}$$
$$= \text{(Sum of } Y \text{ values of elements in group 1)}$$
$$+ \text{(Sum of } Y \text{ values of elements in group 2)}$$
$$\cdots + \text{(Sum of } Y \text{ values of elements in group } M).$$

Each group can be considered a separate population, from each of which a separate simple random sample is selected. The simple estimator of the sum in group 1 is $N_1 \bar{Y}_1$, that in group 2 $N_2 \bar{Y}_2$, and so on, that in group M $N_M \bar{Y}_M$. Therefore, a reasonable estimator of the total of the group sums is the sum of the estimators:

$$N_1 \bar{Y}_1 + N_2 \bar{Y}_2 + \cdots + N_M \bar{Y}_M.$$

This is the stratified estimator T_s, since

$$T_s = N\bar{Y}_s$$
$$= N(w_1 \bar{Y}_1 + w_2 \bar{Y}_2 + \cdots + w_M \bar{Y}_M)$$
$$= N(\frac{N_1}{N} \bar{Y}_1 + \frac{N_2}{N} \bar{Y}_2 + \cdots + \frac{N_M}{N} \bar{Y}_M)$$
$$= N_1 \bar{Y}_1 + N_2 \bar{Y}_2 + \cdots + N_M \bar{Y}_M.$$

It follows that a reasonable estimator of the population mean of the variable Y is T_s/N, which is the stratified estimator $\bar{Y}_s$.

The same interpretation can be given to the stratified estimators T_s' and P_s (see Problem 4.2).

We shall now go back to the beginning and establish with some care the properties of the stratified estimators.

4.3 THE ACKROYD CASE CONTINUED

Let us suppose that four of the six firms in the Ackroyd case (A, B, C, and F) are known to be located on the east side of town, and the remaining two (D and E) are located on the west side. In this case, the population of firms can be divided into two groups according to geographic location. The composition and characteristics of the two groups are shown in Table 4.3.

Obviously, neither the population nor its characteristics have changed. The population mean number of employees (μ) is still 5.167, the population variance of the number of employees (σ^2) is still 8.472, and the proportion of firms intending

TABLE 4.3 Ackroyd Population and Its Characteristics, by Groups Formed According to Geographic Location

		Group 1 (East)			Group 2 (West)	
	Firm	**Employees**	**Hiring?**	**Firm**	**Employees**	**Hiring?**
	A	9	Y	D	2	N
	B	8	Y	E	1	N
	C	6	N			
	F	5	N			
Group size		$N_1 = 4$			$N_2 = 2$	
Group mean[a]		$\mu_1 = 7$			$\mu_2 = 1.5$	
Group variance[a]		$\sigma_1^2 = 2.5$			$\sigma_2^2 = 0.25$	
Group proportion[b]		$\pi_1 = 0.5$			$\pi_2 = 0.0$	

[a]Number of employees.
[b]Proportion of firms intending to hire.

to hire (π) is still 0.333. The group mean, variance, and proportion (μ_i, σ_i^2, and π_i, as shown in Table 4.3) are calculated exactly like their population counterparts, except that they are based only on the elements in each group. For example, μ_1 is the average of the numbers 9, 8, 6, and 5; σ_1^2 is the variance of these numbers; and π_1 is the proportion of Y's in the set Y, Y, N, and N. The reader should understand the meaning of these symbols and the manner in which they are calculated because they shall be used often in this chapter and the next.

Let us consider what might happen if we were to draw from this known population a stratified random sample of two firms, one firm to be selected at random from group 1, and the other, also at random, from group 2.

Place into one bowl four identical tags marked A, B, C, and F, and into another bowl two identical tags marked D and E. Select at random one tag from the first bowl, and, also at random, one tag from the second bowl. Figure 4.1 shows that

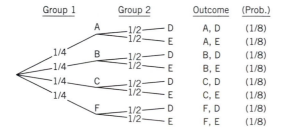

Figure 4.1 Probability tree, stratified sampling

the selected firm from group 1 could be A, B, C, or F; the probability of each outcome is 1/4. No matter what that outcome is, the selected firm from group 2 will be either D or E, with probabilities 1/2 and 1/2, respectively. There are, therefore, eight possible sample outcomes in this case: A from group 1 and D from 2 (A,D), A from 1 and E from 2 (A,E), and so on, down to (F,E). The probability of each outcome is $(1/4)(1/2)$, or 1/8. In other words, each outcome is expected to appear in 1/8th of the samples in the long run.

Observe that the number of sample outcomes is now 8, compared to 15 for a simple random sample of the same size. This observation is sufficient to conclude that, in general, stratified random sampling is not the same as simple random sampling, since the possible sample outcomes and their probabilities differ under the two methods.

Observe also that the probability that a given firm will be included in the sample is not equal to n/N as in simple random sampling. In fact, the probability is different for different firms: it is 1/4 for the firms in group 1, and 1/2 for the firms in group 2. In general, if group i has N_i elements, and a simple random sample of size n_i is drawn from that group, then the probability that a given element of that group will be included in the sample can be shown to be equal to n_i/N_i. In our illustration, $n_1/N_1 = 1/4$, and $n_2/N_2 = 1/2$, as observed.

Suppose that we use the same simple estimators as in simple random sampling: the sample average for estimating the population average number of employees, and the sample proportion for estimating the proportion of firms in the town that intend to hire. Table 4.4 shows the values of the simple estimators for each sample outcome.

		Hiring?			
Outcome	Number of Employees	(Y: Yes, N: No)	Proba-bility	Sample Average[a], $\bar{Y}$	Sample Proportion[b], P
A,D	9,2	Y,N	1/8	5.5	0.5
A,E	9,1	Y,N	1/8	5.0	0.5
B,D	8,2	Y,N	1/8	5.0	0.5
B,E	8,1	Y,N	1/8	4.5	0.5
C,D	6,2	N,N	1/8	4.0	0.0
C,E	6,1	N,N	1/8	3.5	0.0
F,D	5,2	N,N	1/8	3.5	0.0
F,E	5,1	N,N	1/8	3.0	0.0
			8/8		

TABLE 4.4   Stratified Random Sampling, Simple Estimators

[a] Average number of employees.
[b] Proportion of firms intending to hire.

Are the simple estimators unbiased? It is easy to check this by calculating directly the expected average value of the estimators in Table 4.4:

$$E(\bar{Y}) = (5.5)(1/8) + \cdots + (3.0)(1/8) = 4.25,$$

and

$$E(P) = (0.5)(1/8) + \cdots + (0.0)(1/8) = 0.25.$$

Recall (see Table 3.1) that the population mean of the number of employees is 5.167, and the population proportion of firms intending to hire is 0.333. Therefore, *the simple estimators, which were unbiased in simple random sampling, are not unbiased under stratified sampling.*

How, then, are we to form unbiased estimators of the population characteristics when using stratified sampling?

We would like to show that *a weighted average of the sample group means or proportions, in which the weights equal the relative sizes of the groups, is an unbiased estimator of the population mean or proportion.*

When there are two groups, as is the case in this example, this *stratified estimator* of the population mean of a variable Y can be written as

$$\bar{Y}_s = w_1 \bar{Y}_1 + w_2 \bar{Y}_2,$$

where $\bar{Y}_1$ and $\bar{Y}_2$ are the sample means of the variable in groups 1 and 2, and $w_1 = N_1/N$ and $w_2 = N_2/N$ are the fractions of elements in the population belonging to groups 1 and 2 respectively. Similarly, the stratified estimator of the proportion of elements in the population that belong to a given category is

$$P_s = w_1 P_1 + w_2 P_2,$$

where P_1 and P_2 are the sample proportions in groups 1 and 2.

For each sample outcome listed in Table 4.5, there correspond values of the sample group means, $\bar{Y}_1$ and $\bar{Y}_2$ (in this case, the average number of employees in the samples from groups 1 and 2), and of the sample group proportions, P_1 and P_2 (in this case, the proportion of hiring firms in the samples from groups 1 and 2). In this simple example, of course, where only one firm is drawn from each group, the average is equal to the sample observation and the proportion is either 0 or 1. For example, if A is selected from the first group and D from the second group, the sample mean for group 1 is 9, that for group 2 is 2, the sample proportion in group 1 is 1, and that in group 2 is 0.

The first group consists of 4 and the second of 2 firms. In forming the weighted sample mean, $\bar{Y}_s$, we multiply the sample mean of group 1 by 4/6 and that of group 2 by 2/6, and add the results. The weighted sample proportion, P_s, is formed similarly. The values of the weighted sample mean and proportion are shown in Table 4.5.

TABLE 4.5 Stratified Random Sampling Without Replacement Using a Sample of Size 1 from Each Group in Table 4.3

Outcome	Probability	Sample Group Mean $\bar{Y}_1$	$\bar{Y}_2$	Weighted Sample Mean[a], $\bar{Y}_s$	Sample Group Proportion P_1	P_2	Weighted Sample Proportion[a], P_s
A,D	1/8	9	2	6.67	1	0	2/3
A,E	1/8	9	1	6.33	1	0	2/3
B,D	1/8	8	2	6.00	1	0	2/3
B,E	1/8	8	1	5.67	1	0	2/3
C,D	1/8	6	2	4.67	0	0	0
C,E	1/8	6	1	4.33	0	0	0
F,D	1/8	5	2	4.00	0	0	0
F,E	1/8	5	1	3.67	0	0	0
	8/8						

$$E(\bar{Y}_s) = 5.167, \quad Var(\bar{Y}_s) = 1.139$$
$$E(P_s) = 0.333, \quad Var(P_s) = 0.111$$

[a] $\bar{Y}_s = w_1\bar{Y}_1 + w_2\bar{Y}_2$, $P_s = w_1 P_1 + w_2 P_2$ $w_1 = N_1/N = 4/6$, $w_2 = N_2/N = 2/6$.

The probability distributions and the characteristics of the distributions of these estimators can be calculated in the usual manner. The reader can easily verify that the mean and variance of each estimator are as shown in Table 4.5.

We see that $E(\bar{Y}_s) = 5.167 = \mu$ and $E(P_s) = 0.333 = \pi$. We conclude that the stratified estimators are indeed unbiased estimators of the population mean and proportion respectively.

It is interesting to compare the properties of the stratified estimators ($\bar{Y}_s$ and P_s) in the stratified random sample of size 2, with those of the simple estimators ($\bar{Y}$ and P) in the simple random sample of the same size examined in Section 3.4.

All four estimators are unbiased. Between two unbiased estimators of the same population characteristic, we should prefer the one with the smaller variance. The variance of $\bar{Y}_s$ (1.139) is smaller than that of $\bar{Y}$ (3.389), but the variance of P_s (0.111) is greater than that of P (0.089). If that were always true, one would prefer $\bar{Y}_s$ to $\bar{Y}$, but P to P_s. But does this apply to all stratified samples, or is it merely a feature of this particular example?

To investigate this question, let us suppose that the six firms in the Ackroyd case can also be divided into two different groups according to whether or not they are manufacturing firms. (In general, it may be possible to stratify a population according to several criteria.) Suppose that firms A, F, and D are manufacturing firms, and firms B, C, and E are not. Two different groups can thus be formed according to

the nature of the firms' operations; their composition and characteristics are shown in Table 4.6.

The question, once again, is: What will happen if we were to select a stratified sample of size 2 by drawing at random one firm from group 1 and another from group 2? The results, calculated exactly as in the previous case, are shown in Table 4.7.

$\bar{Y}_s$ and P_s are the stratified estimators. As in the previous illustration, the sample group mean is equal to the observation, and the sample group proportion is equal to either 0 or 1. Since the groups have the same size, the weights are equal to 3/6 each. Table 4.7 also shows the mean and variance of the probability distributions of $\bar{Y}_s$ and P_s. As in the previous case, these estimators are unbiased. Unlike the previous case, however, the variance of $\bar{Y}_s$ is greater than that of $\bar{Y}$, the estimator based on a simple random sample of the same size.

These simple examples illustrate three features of stratified random sampling that can be shown to hold in general:

- The simple estimators, which are unbiased under simple random sampling, are not unbiased under stratified sampling.
- The stratified estimators of a population mean or proportion, that is, the weighted averages of the sample group means and proportions, are unbiased; the weights should equal the fraction of elements in the population belonging to the groups.
- The variances of the stratified estimators $\bar{Y}_s$ and P_s are sometimes smaller and sometimes larger than those of the simple estimators for simple random samples of the same size. Therefore, a stratified sample is *not* necessarily preferable to a simple random sample of the same size.

TABLE 4.6   Ackroyd Population and Its Characteristics, by Groups Formed According to Nature of Operations

	Group 1 (Manufacturing)			Group 2 (Nonmanufacturing)		
	Firm	Employees	Hiring?	Firm	Employees	Hiring?
	A	9	Y	B	8	Y
	F	5	N	C	6	N
	D	2	N	E	1	N
Group size	$N_1 = 3$			$N_2 = 3$		
Group mean[a]	$\mu_1 = 5.333$			$\mu_2 = 5.000$		
Group variance[a]	$\sigma_1^2 = 8.222$			$\sigma_2^2 = 8.667$		
Group proportion[b]	$\pi_1 = 0.333$			$\pi_2 = 0.333$		

[a]Number of employees.
[b]Proportion of firms intending to hire.

TABLE 4.7 Stratified Random Sample Without Replacement Using a Sample of Size 1 Selected from Each Group in Table 4.6

Outcome S1	Outcome S2	Probability	Sample Group Mean $\bar{Y}_1$	Sample Group Mean $\bar{Y}_2$	Weighted Sample Mean[a], $\bar{Y}_s$	Sample Group Proportion P_1	Sample Group Proportion P_2	Weighted Sample Proportion[a], P_s
A	B	1/9	9	8	8.5	1	1	1.0
A	C	1/9	9	6	7.5	1	0	0.5
A	E	1/9	9	1	5.0	1	0	0.5
F	B	1/9	5	8	6.5	0	1	0.5
F	C	1/9	5	6	5.5	0	0	0.0
F	E	1/9	5	1	3.0	0	0	0.0
D	B	1/9	2	8	5.0	0	1	0.5
D	C	1/9	2	6	4.0	0	0	0.0
D	E	1/9	2	1	1.5	0	0	0.0

9/9

$$E(\bar{Y}_s) = 5.167, \qquad Var(\bar{Y}_s) = 4.222$$
$$E(P_s) = 0.333, \qquad Var(P_s) = 0.111$$

[a] $\bar{Y}_s = w_1\bar{Y}_1 + w_2\bar{Y}_2$, $P_s = w_1 P_1 + w_2 P_2$, $w_1 = w_2 = 3/6$.

4.4 PROPERTIES OF STRATIFIED ESTIMATORS

Let us generalize. Imagine a population consisting of N elements. As usual, let μ be the population mean and σ^2 the population variance of a given variable. Also, let π be the proportion of elements in the population belonging to a given category. The purpose of the sample is to provide estimates of μ and π. In practical applications there are, of course, numerous μ's and π's to be estimated; whatever we say about one μ or π applies to all.

Suppose that the population can be stratified into M groups according to a given criterion. (There could be, as we have seen, several criteria of stratification.) Let N_i be the number of elements, μ_i and σ_i^2 the mean and variance of a given variable Y, and π_i the proportion of elements in a given category C in the ith group $(i = 1, 2, \ldots, M)$. The notation is outlined in Table 4.8.

A stratified random sample will be selected by drawing from *each* group a simple random sample without replacement. Let n_i, $\bar{Y}_i$, and P_i be respectively the size of the sample drawn from the ith group, the sample mean of the variable, and the proportion of sampled elements in the category of interest. Table 4.9 clarifies this notation.

The principal properties of the stratified estimators are listed in Summary 4.1.

SUMMARY 4.1   Stratified random sampling, unbiased estimators

- *Method of sample selection:* Stratified random sampling.
- An *unbiased estimator of the population mean, μ of a variable Y* is the stratified estimator of μ:

$$\bar{Y}_s = w_1 \bar{Y}_1 + w_2 \bar{Y}_2 + \cdots + w_M \bar{Y}_M = \sum_{i=1}^{M} w_i \bar{Y}_i, \qquad (4.1)$$

and its variance is given by

$$Var(\bar{Y}_s) = \sum_{i=1}^{M} w_i^2 \frac{\sigma_i^2}{n_i} \frac{N_i - n_i}{N_i - 1}. \qquad (4.2)$$

- An *unbiased estimator of the proportion, π, of population elements belonging to a given category C* is the stratified estimator of π:

$$P_s = w_1 P_1 + w_2 P_2 + \cdots + w_M P_M = \sum_{i=1}^{M} w_i P_i, \qquad (4.3)$$

and its variance is given by

$$Var(P_s) = \sum_{i=1}^{M} w_i^2 \frac{\pi_i(1 - \pi_i)}{n_i} \frac{N_i - n_i}{N_i - 1}. \qquad (4.4)$$

The notation is explained in Tables 4.8 and 4.9.

TABLE 4.8   Stratified Sampling, Notation for Population

	Assumed Known		Unknown		
				Variable Y	
Group	Number of Elements	Percent of Elements	Mean	Variance	Proportion of Elements in C
1	N_1	$w_1 = N_1/N$	μ_1	σ_1^2	π_1
2	N_2	$w_2 = N_2/N$	μ_2	σ_2^2	π_2
$\vdots$	$\vdots$	$\vdots$	$\vdots$	$\vdots$	$\vdots$
M	N_M	$w_M = N_M/N$	μ_M	σ_M^2	π_M
Overall	N	1	μ	σ^2	π

TABLE 4.9 Stratified Random Sampling, Notation for Sample			
Group	Sample Size	Variable Y, Sample Mean	Category C, Sample Proportion
1	n_1	$\bar{Y}_1$	P_1
2	n_2	$\bar{Y}_2$	P_2
⋮	⋮	⋮	⋮
M	n_M	$\bar{Y}_M$	P_M
	n		

We have already confirmed for the Ackroyd case (see Tables 4.5 and 4.7) that $E(\bar{Y}_s) = \mu$ and $E(P_s) = \pi$. Let us confirm Eq. 4.2, using the geographic stratification of Table 4.3, and leave the confirmation of Eq. 4.4 for Problem 4.7. According to Eq. 4.2, we should have:

$$
\begin{aligned}
Var(\bar{Y}_s) &= w_1^2 \frac{\sigma_1^2}{n_1} \frac{N_1 - n_1}{N_1 - 1} + w_2^2 \frac{\sigma_2^2}{n_2} \frac{N_2 - n_2}{N_2 - 1} \\
&= (\frac{4}{6})^2 \frac{2.5}{1} \frac{4 - 1}{4 - 1} + (\frac{2}{6})^2 \frac{0.25}{1} \frac{2 - 1}{2 - 1} \\
&= 1.139.
\end{aligned}
$$

This agrees with the direct calculation shown in Table 4.5.

4.5 PROPORTIONAL STRATIFIED SAMPLING

We noted previously that a stratified sample is not necessarily better than a simple random sample of the same size. We would like to show, however, that a special kind of a stratified sample, called *a proportional stratified sample, is nearly always better than a simple random sample of the same size.*

In a *proportional stratified sample* (or *stratified sample with proportional allocation*), the size of the sample from group i, n_i, is in the same proportion to the total sample size, n, as N_i is to N, that is, $n_i/n = N_i/N$. For example, if a group contains 15% of the population elements, the sample from this group will be 15% of the total sample size. In such a case, $n_i = (N_i/N)n = w_i n$, for every group i. We assume, of course, that the $w_i n$'s are integers.

Like all stratified estimators, $\bar{Y}_s$ and P_s of a proportional stratified sample are unbiased. So are $\bar{Y}$ and P, the simple estimators in simple random sampling. In

essence, the proof that follows establishes that the estimators based on a proportional stratified sample have smaller variance than the simple estimators based on a simple random sample of the same size. The proof is instructive, but may be omitted without penalty.

Let us begin by comparing the variance of $\bar{Y}_s$ with that of $\bar{Y}$, assuming the total sample size n is the same. Substituting $n_i = nw_i$ in Eq. 4.2, the variance of $\bar{Y}_s$ under proportional stratified sampling can be written as

$$\underbrace{Var(\bar{Y}_s)}_{prop.} = \sum_{i=1}^{M} w_i^2 \frac{\sigma_i^2}{nw_i} \frac{N_i - nw_i}{N_i - 1} = \frac{1}{n} \sum_{i=1}^{M} w_i \sigma_i^2 \frac{N_i - nw_i}{N_i - 1}. \tag{4.5}$$

If the N_i are large (as is likely in practice) then $N_i - 1 \approx N_i$, in which case

$$\frac{N_i - nw_i}{N_i - 1} \approx \frac{N_i - nw_i}{N_i} = 1 - n\frac{w_i}{N_i} = 1 - n\frac{N_i}{NN_i} = 1 - \frac{n}{N},$$

and Eq. 4.5 becomes

$$\underbrace{Var(\bar{Y}_s)}_{prop.} \approx \frac{1}{n}\Big(1 - \frac{n}{N}\Big) \sum_{i=1}^{M} w_i \sigma_i^2. \tag{4.6}$$

Recall that the variance of the simple estimator $\bar{Y}$ based on a simple random sample of size n is, in the case of large N,

$$Var(\bar{Y}) = \frac{\sigma^2}{n} \frac{N - n}{N - 1} \approx \frac{\sigma^2}{n} \frac{N - n}{N} = \frac{1}{n}\Big(1 - \frac{n}{N}\Big)\sigma^2. \tag{4.7}$$

We see that Eqs. 4.6 and 4.7 differ only with respect to the last term. It so happens that there is a relationship between the population variance on the one hand and the group variances and means on the other. This is

$$\sigma^2 = \sum_{i=1}^{M} w_i \sigma_i^2 + \sum_{i=1}^{M} w_i (\mu_i - \mu)^2. \tag{4.8}$$

There is also a relationship between the population mean, on the one hand, and the group means on the other:

$$\mu = w_1 \mu_1 + w_1 \mu_1 + \cdots + w_1 \mu_1 = \sum_{i=1}^{M} w_i \mu_i.$$

The proof of these relationships is a little difficult and will not be attempted here. However, they can be easily confirmed numerically. Consider Table 4.3 as an example. Applying 4.8,

$$\sigma^2 = w_1\sigma_1^2 + w_2\sigma_2^2 + w_1(\mu_1 - \mu)^2 + w_2(\mu_2 - \mu)^2$$
$$= (\frac{4}{6})(2.5) + (\frac{2}{6})(0.25) + (\frac{4}{6})(7 - 5.167)^2 + (\frac{2}{6})(1.5 - 5.167)^2$$
$$= 8.472.$$

This result agrees with the direct calculation of σ^2, as shown in the explanation of Table 3.1. According to Eq. 4.8, therefore,

$$\sigma^2 = \sum_{i=1}^{M} w_i\sigma_i^2 + \sum_{i=1}^{M} w_i(\mu_i - \mu)^2 = \sum_{i=1}^{M} w_i\sigma_i^2 + c,$$

where $c \geq 0$; $c = 0$ only when all $\mu_i = \mu$, that is, when all group means are equal (a rather rare case in practice). It follows then that

$$Var(\bar{Y}) \approx \frac{1}{n}\left(1 - \frac{n}{N}\right)\left(\sum_{i=1}^{M} w_i\sigma_i^2 + c\right) \geq \frac{1}{n}\left(1 - \frac{n}{N}\right)\left(\sum_{i=1}^{M} w_i\sigma_i^2\right) \approx \underbrace{Var(\bar{Y}_s)}_{prop.},$$

that is,

$$Var(\bar{Y}) \geq \underbrace{Var(\bar{Y}_s)}_{prop.}.$$

By similar calculations, it is possible to show that the variance of the estimator of a population proportion (P) based on a simple random sample is at least as great as that of the stratified estimator (P_s) based on a proportional stratified sample of the same total size. The difference between these variances may be small, but it is always nonnegative for large N_i.

We conclude that, for large N_i, the stratified estimators based on a proportional stratified sample have no greater (and, in practice, smaller) variances than the simple estimators of a simple random sample of the same size. Put simply, proportional stratified sampling is better than simple random sampling.

Three noteworthy features of this conclusion are: (a) it applies to estimators of *all* population means or proportions; (b) it does not require for its implementation any information concerning the population beyond knowledge of the fractions $w_i = N_i/N$; and (c) it holds for *any* criterion of stratification.

For these reasons, *proportional stratified sampling should be preferred to simple random sampling in all cases where it is feasible*, that is, in all cases where the group sizes (N_i) are known.

		Population		Sample
		Size,	Percent of Elements,	Size,
i	Group	N_i	$w_i = N_i/N$	n_i
1	Large	1,000	0.1	80
2	Medium	3,000	0.3	240
3	Small	6,000	0.6	480
	Total	$N = 10,000$	1.0	$n = 800$

TABLE 4.10 Proportional Allocation, Example 4.1

■ EXAMPLE 4.2
(Continued)

If Books Research were to make the next sample of 800 bookstores proportional stratified, the allocation of the total sample among the three groups would be as shown in the last column of Table 4.10.

Evidently, this allocation is quite different from the original one in Table 4.1.   ◪

When the stratified sample is proportional, the stratified estimator ($\bar{Y}_s$) is equal to the ordinary average of all the sample observations, $\bar{Y}$, and P_s is equal to P, the ordinary proportion of all sample observations belonging to the category of interest. Instead of calculating weighted averages of sample averages or proportions, therefore, it may be more convenient to pool all the sample observations and calculate instead just one average or proportion.

To show that $\bar{Y}_s = \bar{Y}$ when the stratified sample is proportional, let Y_{ij} be the jth observation of variable Y in the sample from the ith group. Assuming $n_i = nw_i$ are integers,

$$\bar{Y}_s = \sum_{i=1}^{M} w_i \bar{Y}_i = \sum_{i=1}^{M} w_i \frac{1}{n_i} \sum_{j=1}^{n_i} Y_{ij} =$$

$$= \sum_{i=1}^{M} w_i \frac{1}{nw_i} \sum_{j=1}^{n_i} Y_{ij} = \frac{1}{n} \sum_{i=1}^{M} \sum_{j=1}^{n_i} Y_{ij} = \bar{Y}.$$

In exactly the same manner, we can show that $P_s = P$ when the sample is proportional stratified.

4.6 "OPTIMAL" STRATIFIED SAMPLING

Let us now consider how a given total sample size, n, should be allocated among the groups so that the stratified estimator ($\bar{Y}_s$) of the population mean of a variable

Y will have the smallest possible variance. Formally, the problem is to determine $n_1, n_2, \ldots, n_M$ so as to minimize

$$Var(\bar{Y}_s) = \sum_{i=1}^{M} w_i^2 \frac{\sigma_i^2}{n_i} \frac{N_i - n_i}{N_i - 1}, \qquad (4.9)$$

subject to the constraint that the total sample size equals n, i.e., $n_1 + n_2 + \cdots + n_M = n$.

This is a constrained optimization problem familiar to students of advanced calculus. It can be shown that the solution to this problem is given by

$$n_i^* = n \frac{N_i \sigma_i}{\sum_{j=1}^{M} N_j \sigma_j}. \qquad (4.10)$$

Thus, other things equal, the sample from a group should be relatively larger the larger the size of the group and the greater the variability of the Y values in the group.

If the objective is to minimize the variance of the stratified estimator P_s of the proportion of elements in the population belonging to a given category, subject to the same total sample constraint, the optimal solution can be shown to be

$$n_i^* = n \frac{N_i \pi_i (1 - \pi_i)}{\sum_{j=1}^{M} N_j \pi_j (1 - \pi_j)}. \qquad (4.11)$$

Equations 4.10 and 4.11 do not necessarily give integer values, but, when the n_i^* are large, the nearest integers could be considered optimal for all practical purposes.

In order to apply these formulas, the group sizes (N_i) and variances (σ_i^2) or proportions (π_i) must be known. In situations where a stratified sample is considered, the group sizes (N_i) are assumed to be known, but the σ_i^2 or π_i are invariably unknown. Thus, strictly speaking, an optimal allocation can never be determined.

In the event, however, that an earlier or similar stratified sample is available, it seems reasonable to substitute for σ_i in Eq. 4.10 the observed sample group standard deviations, S_i. Similarly, intuition, common sense, or a pilot study may suggest some reasonable first estimates of the π_i in Eq. 4.11. The results of these calculations, obviously, do not provide an optimal allocation, but rather reasonable guidelines as to how the next sample could be allocated.

◪ EXAMPLE 4.3 The Gateville Plant is one of six large facilities operated by the Post Office exclusively for the handling of parcels. The plant serves a large geographic area in the East. Local post offices in the area forward all parcels to Gateville, where they are sorted and subsequently shipped to other plants elsewhere in the country or abroad, or back to the local post offices. All parcels are shipped in containers of two kinds: "monotainers" (wire cages about $50 \times 42 \times 40$ in.) and bags (canvas bags secured with a string).

The unpacking, sorting, and repacking of containers at Gateville is a continuous and fast-moving operation. A count is kept of the total number of containers entering and leaving the plant, but neither the number of parcels handled nor the destination of containers is accurately recorded. This information is not essential for the day-to-day operation of the plant, but it is useful for planning, reviewing parcel rates and revenues, and understanding the national patterns of parcel traffic. A stratified random sample is regularly used to obtain estimates on a weekly basis. The data shown in Table 4.11 will illustrate the estimation of the total number of parcels leaving Gateville, and the proportion of containers shipped to Western (Gateville's sister plant on the West Coast) during one week of operation.

During the week under study, Gateville shipped out a total of 47,500 containers, of which 2,500 were monotainers and 45,000 bags. One hundred monotainers and 450 bags were randomly selected.[2] The average number of parcels per sampled monotainer was 92, and the standard deviation of the number of parcels per monotainer in the sample (that is, the standard deviation of the 100 sample observations) was 46.26; 17% of the sampled monotainers were destined for the Western plant. The corresponding figures for the sample of 450 bags were 12, 9.28, and 12%.

The stratified estimate of the average number of parcels per container is

$$\bar{Y}_s = (0.053)(92) + (0.947)(12) = 16.24.$$

The estimate of the total number of parcels shipped out of Gateville during the week is (47,500)(16.24), or 771,400.

The estimate of the proportion of containers shipped to Western is

$$P_s = (0.053)(0.17) + (0.947)(0.12) = 0.1226,$$

or 12.26%. The estimate of the total number of parcels shipped to Western is (47,500)(0.1226), or 5,824.

Let us now consider how *next week's* sample should be taken, assuming that the total sample size will be the same ($n = 550$).

If next week's sample is *proportional stratified*, $n_1 = w_1 n = (0.053)(550) \approx 29$, and $n_2 = w_2 n = (0.947)(550) \approx 521$. In other words, about 29 monotainers and

TABLE 4.11 Sampling at Gateville Plant, Example 4.3

Containers	N_i	$w_i = N_i/N$	n_i	$\bar{Y}_i$	S_i	P_i
Monotainers	2,500	0.053	100	92	46.26	0.17
Bags	45,000	0.947	450	12	9.28	0.12
Total	$N = 47,500$	1.000	$n = 550$			

[2]In fact, the samples were systematic, but considered random for all practical purposes for reasons to be explained in Section 5.5.

521 bags should be sampled. Assuming next week's w_i remain the same, such a sample will be better than a simple random sample of size 550, but it will not necessarily be the best stratified sample.

If the primary purpose of the sample is to estimate the average (μ) or total ($N\mu$) number of parcels shipped out of Gateville, Eq. 4.10 may be applied with σ_i replaced by S_i. The calculations are shown in Table 4.12. The optimal solution calls for samples of 119 monotainers and 431 bags.

If the main purpose of the sample is to estimate the proportion (π) or total number ($N\pi$) of containers shipped to Western, Eq. 4.11 may be used with π_i replaced by P_i. Table 4.13 details the calculations.

The "optimal" allocation for this second purpose is quite different from that of Table 4.12. This should not be surprising—different objectives require different solutions. ◪

A proportional stratified sample is optimal (that is, the best possible stratified sample) if all the σ_i or π_i are the same. This is easy to see. If all σ_i are equal—say, equal to σ'—then Eq. 4.10 becomes

$$n_i^* = n\frac{N_i\sigma'}{\sum N_j\sigma'} = n\frac{N_i\sigma'}{\sigma'\sum N_j} = n\frac{N_i}{\sum N_j} = n\frac{N_j}{N} = nw_i,$$

TABLE 4.12   First Optimal Allocation, Example 4.3

Group	N_i	S_i	$N_i S_i$	$\dfrac{N_i S_i}{\sum N_j S_j}$	n_i^*
Monotainers	2,500	46.26	115,650	0.2169	119
Bags	45,000	9.28	417,600	0.7831	431
Total	47,500		533,250	1.0000	550
			$(= \sum N_j S_j)$		

TABLE 4.13   Second Optimal Allocation, Example 4.3

Group	N_i	P_i	$N_i P_i(1 - P_i)$	$\dfrac{N_i P_i(1 - P_i)}{\sum N_j P_j(1 - P_j)}$	n_i^*
Monotainers	2,500	0.17	352.75	0.0691	38
Bags	45,000	0.12	4,752.00	0.9309	512
Total	47,500		5,104.57	1.0000	550
			$[= \sum N_j P_j(1 - P_j)]$		

which is, of course, the proportional allocation. The same result applies to Eq. 4.11 when all $\pi_i = \pi'$.

An allocation that is optimal for one variable or category need not be optimal for another variable or category. The allocation obtained by applying 4.10 will, in general, be different from that dictated by 4.11, and, since the σ_i or π_i of different variables or categories are different, the allocations will also be different. In practice, you will recall, there are usually many variables and categories of interest. The above formulas, therefore, are useful primarily when the main purpose of the sample is to estimate a single population characteristic. (It is possible, of course, to determine an allocation that minimizes the sum—possibly the weighted sum—of the variances of estimators of a number of population means or proportions, but we shall not pursue this approach here.)

In the formulation of the optimization problem at the beginning of this section, the requirement that the group sample size not exceed the group size (that is, $n_i \leq N_i$) was not explicitly taken into account. It could thus happen that the optimal n_i^*, calculated from Eq. 4.10 or 4.11, exceeds N_i for a particular group i, which is, of course, impossible in a sample without replacement. In such a situation, one could experiment with feasible values near the n_i^*, using Eq. 4.9 to evaluate these alternatives. For example, if $n_k^* > N_k$ for one group k, one alternative is to sample all the elements in that group (that is, make $n_k = N_k$), and reduce the total sample size by the difference $n_k^* - n_k$.

Other formulations lead to different "optimal" stratified samples. For example, if it costs $\$c_i$ to sample one element from group i, one may wish to find $n_1, n_2, \ldots, n_M$ so as to minimize $Var(\bar{Y}_s)$, subject to the constraint that the total sampling cost does not exceed a given budget,

$$c_0 + c_1 n_1 + c_2 n_2 + \cdots + c_M n_M = C,$$

where c_0 is the fixed (overhead) cost of taking a sample and C is the given budget. This problem is also a constrained optimization problem, and may be easily solved. The optimal solution can be shown to be

$$n_i^* = n^* \frac{N_i \sigma_i / \sqrt{c_i}}{\sum_{j=1}^{M} N_j \sigma_j / \sqrt{c_j}}, \tag{4.12}$$

where n^* is the total sample size, given by

$$n^* = \frac{(C - c_0) \sum_{j=1}^{M} N_j \sigma_j / \sqrt{c_j}}{\sum_{j=1}^{M} N_j \sigma_j \sqrt{c_j}}. \tag{4.13}$$

Several other "optimal" solutions are possible, depending on the objective and the constraints. We shall not elaborate here. The same cautionary remarks apply to all these solutions as for (4.10) and (4.11).

4.7 HOW LARGE A STRATIFIED SAMPLE?

There is a problem when we attempt to extend the sample size prescriptions of Section 3.7, because the variance of the stratified estimators depends not only on the total sample size but also on its allocation among the groups. One possible approach is to assume that this allocation has been decided upon; specifically, that a given proportion v_i of the total sample size n will be allocated to each group i. The size of the sample from group i is therefore nv_i. We then ask how large n must be to meet given accuracy requirements. The prescription is given in Summary 4.2.

This result is based on a variant of the Central Limit Theorem, according to which for large n_i and $N_i - n_i$ the probability distributions of $\bar{Y}_s$ and P_s are approximately normal. By "demanding enough" we mean that c must be small enough and/or $(1 - \alpha)$ large enough so that all n_i and $N_i - n_i$ calculated by applying (4.14) are large (greater than 50, according to the rule of thumb of Chapter 3). Equation 4.14 involves unknown population characteristics and is thus of no practical use as it stands. However, it can provide a rough indication of the required sample size if estimates of σ_i or $\pi_i(1 - \pi_i)$ are available from a pilot or earlier similar sample.

◢ EXAMPLE 4.3
(Continued)

How large should be next week's sample of containers at Gateville so that the estimate of the proportion (π) of containers shipped to Western is within ± 0.02 with probability 90%?

By "within ± 0.02" we mean, of course, "in the interval from $\pi - 0.02$ to $\pi + 0.02$," whatever the value of π happens to be. Assume that the numbers of monotainers and bags that will be processed at Gateville next week are $N_1 = 2{,}500$ and $N_2 = 45{,}000$, and that 40% of next week's sample of containers will consist of monotainers and 60% of bags (these assumptions can, of course, be altered). We have, therefore, $v_1 = 0.4$, $v_2 = 0.6$, $N = 47{,}500$, $c = 0.02$, $1 - \alpha = 0.90$, $Z_{\alpha/2} = 1.645$.

SUMMARY 4.2   Stratified random sampling, sample size

Provided that the accuracy requirements are demanding enough, the total size (n) of a stratified sample needed to estimate μ or π within $\pm c$ with probability $(1 - \alpha)$ and to be allocated to the groups according to $n_i = nv_i$ can be shown to be

$$n = \frac{\sum_{i=1}^{M}(N_i^2 A_i/v_i)}{N^2 D^2 + \sum_{i=1}^{M} N_i A_i} \qquad (4.14)$$

where $D = (c/Z_{\alpha/2})$, and $A_i = \pi_i(1 - \pi_i)$ when estimating π, or $A_i = \sigma_i^2$ when estimating μ. For estimating $\tau' = N\pi$ or $\tau = N\mu$ within $\pm c'$ with probability $(1 - \alpha)$, apply (4.14) with $c = c'/N$. Selected values of $Z_{\alpha/2}$ can be found in Table 3.6.

Using the most recent sample estimates P_i (Table 4.11) in place of the unknown π_i, we calculate first the numerator of Eq. 4.14:

$$\frac{(2{,}500)^2(0.17)(1-0.17)}{0.4} + \frac{(45{,}000)^2(0.12)(1-0.12)}{0.6} = 358{,}604{,}687.5,$$

and then the denominator:

$$(47{,}500)^2(\frac{0.02}{1.645})^2 + (2{,}500)(0.17)(1-0.17) + (45{,}000)(0.12)(1-0.12) = 338{,}619.78.$$

The required sample size is therefore

$$n = \frac{358{,}604{,}687.5}{338{,}619.78} = 1{,}059,$$

which is about double the size of the most recent sample.

How large should be next week's sample of containers so that the estimate of the average number of parcels per container is within ± 1 parcel of the true average with probability 95%?

Under the same assumptions as in the earlier question, we apply again Eq. 4.14 using the most recent sample estimates S_i in Table 4.11 in place of the unknown σ_i. The numerator of Eq. 4.14 is now

$$\frac{(2{,}500)^2(46.26)^2}{0.4} + \frac{(45{,}000)^2(9.28)^2}{0.6} = 3.240869 \times 10^{11},$$

and the denominator is

$$(47{,}500)^2(\frac{1}{1.96})^2 + (2{,}500)(46.26)^2 + (45{,}000)(9.28)^2 = 596{,}545{,}684.$$

The required sample size is therefore

$$n = \frac{3.240869 \times 10^{11}}{596{,}545{,}684} = 543,$$

or about half that needed to meet the first question's requirements and about the same as that of the most recent sample. There is no reason, of course, for the two problems to have the same solution. ■

4.8 INTERVAL ESTIMATES

Just as with simple random samples, it may be desirable to supplement the point (single-number) stratified estimates of the population characteristics with confidence intervals, that is, interval estimates having a given probability of containing the population characteristic. These confidence intervals are listed in Summary 4.3.

SUMMARY 4.3 Stratified random sampling, confidence intervals

- When all n_i and $N_i - n_i$ are large, an approximate $100(1 - \alpha)\%$ confidence interval for μ, the population mean of a variable Y, is

$$\bar{Y}_s \pm Z_{\alpha/2}\sqrt{\widehat{Var(\bar{Y}_s)}}, \tag{4.15}$$

and that for π, the population proportion in a given category, is

$$P_s \pm Z_{\alpha/2}\sqrt{\widehat{Var(P_s)}}. \tag{4.16}$$

$\widehat{Var(\bar{Y}_s)}$ is an unbiased estimator of $Var(\bar{Y}_s)$ given by

$$\widehat{Var(\bar{Y}_s)} = \sum_{i=1}^{M} w_i^2 \frac{S_i^2}{n_i - 1} \frac{N_i - n_i}{N_i}, \tag{4.17}$$

while $\widehat{Var(P_s)}$ is an unbiased estimator of $Var(P_s)$ given by

$$\widehat{Var(P_s)} = \sum_{i=1}^{M} w_i^2 \frac{P_i(1 - P_i)}{n_i - 1} \frac{N_i - n_i}{N_i}. \tag{4.18}$$

In Eq. 4.17, S_i^2 is the variance of Y in the sample from group i.
- Approximate $100(1 - \alpha)\%$ confidence intervals for the population totals $\tau = N\mu$ and $\tau' = N\pi$ are obtained by multiplying the upper and lower limits of (4.15) and (4.16), respectively, by N.
- Selected values of $Z_{\alpha/2}$ are given in Table 3.6.

◪ EXAMPLE 4.3
(Continued)

We use the data in Table 4.11 to illustrate the calculation of a 90% confidence interval for the average number of parcels per container (μ), and the proportion of containers shipped to Western (π). The point estimates of these two population characteristics, it will be recalled, were $\bar{Y}_s = 16.24$ and $P_s = 0.1226$, respectively.

We calculate first

$$\widehat{Var(\bar{Y}_s)} = w_1^2 \frac{S_1^2}{n_1 - 1} \frac{N_1 - n_1}{N_1} + w_2^2 \frac{S_2^2}{n_2 - 1} \frac{N_2 - n_2}{N_2}$$

$$= (0.053)^2 \frac{(46.26)^2}{100 - 1} \frac{2,500 - 100}{2,500} + (0.947)^2 \frac{(9.28)^2}{450 - 1} \frac{45,000 - 450}{45,000}$$

$$= 0.2286,$$

and

$$Var(\widehat{P_s}) = w_1^2 \frac{P_1(1-P_1)}{n_1-1} \frac{N_1-n_1}{N_1} + w_2^2 \frac{P_2(1-P_2)}{n_2-1} \frac{N_2-n_2}{N_2}$$

$$= (0.053)^2 \frac{(0.17)(1-0.17)}{100-1} \frac{2{,}500-100}{2{,}500}$$

$$+ (0.947)^2 \frac{(0.12)(1-0.12)}{450-1} \frac{45{,}000-450}{45{,}000}$$

$$= 0.000213.$$

An approximate 90% confidence interval for μ is

$$16.24 \pm (1.645)\sqrt{0.2286},$$

or from about 15.45 to 17.03 parcels. An approximate 90% confidence interval for π is

$$0.1226 \pm (1.645)\sqrt{0.000213},$$

or from about 9.86 to 14.66%.

An approximate 90% confidence interval for the total number of parcels is from (47,500)(15.45) to (47,500)(17.03), or from 733,875 to 808,925. A 90% confidence interval for the number of containers shipped to Western is from (47,500)(0.0986) to (47,500)(0.1466), or from about 4,683 to 6,963. ◾

4.9 TO SUM UP

- Stratified random sampling consists of selecting a simple random sample from each group.
- A stratified sample is selected either because it is desired to estimate the characteristics of each group in addition to those of the population at large, or because stratified is believed to be more accurate than simple random sampling.
- Stratified sampling is not the same as simple random sampling, and the simple estimators are not unbiased under stratified random sampling. Unbiased are the stratified estimators, which are weighted averages of the sample group means or proportions with weights equal to the fractions of elements in each group.
- Statistical theory provides expressions for the variances of the stratified estimators. A study of these variances shows that a stratified random sample is not necessarily better than a simple random sample of the same size. However, a proportional stratified sample is nearly always better and should be preferred to a simple random sample of the same size.
- A proportional stratified sample is not necessarily the best stratified sample of given total size. The "optimal" stratified random sample depends on the purpose of the sample and the unknown characteristics of the population. However, one

may use estimates of these characteristics based on a pilot, earlier, or similar sample to approximate the optimal allocation.
- Approximate formulas giving the total sample size for a specified allocation and large-sample confidence intervals were also provided.

PROBLEMS

4.1 *Computing exercise:* Using the program SCALC, verify the numerical results presented in this chapter. Note any deviations, and determine if these are the result of rounding at intermediate stages.

4.2 In the manner of Section 4.2, show that the stratified estimators T'_s and P_s can be interpreted as the total and average, respectively, of simple estimators of the group totals and means.

4.3 A population of 8,000 elements is divided into two groups according to a certain criterion with 5,000 and 3,000 elements respectively. A simple random sample of 100 elements was selected from the first group, and one of 200 from the second. The sample averages of a variable Y and the sample proportions in a category were as follows:

Group, i	Number of Elements, N_i	Sample Size, n_i	Sample Mean of Y, $\bar{Y}_i$	Sample Proportion in Category, P_i
1	5,000	100	400	0.25
2	3,000	200	450	0.36

(a) Calculate the stratified estimators, $\bar{Y}_s$ and P_s, of the population mean of Y and of the population proportion in the category, respectively.

(b) Calculate the stratified estimators, T_s and T'_s, of the population total of Y and of the number of elements in the category, respectively.

4.4 A population of 9,000 elements is divided into three groups according to a certain criterion with 1,000, 2,000, and 6,000 elements respectively. Simple random samples of 100, 150, and 200 elements were selected from these groups. The sample averages of a variable Y and the sample proportions in a certain category were as follows:

Group, i	Number of Elements, N_i	Sample Size, n_i	Sample Mean of Y, $\bar{Y}_i$	Sample Proportion in Category, P_i
1	1,000	100	10.2	0.071
2	2,000	150	−15.7	0.052
3	6,000	200	20.5	0.049

(a) Calculate the stratified estimators, $\bar{Y}_s$ and P_s, of the population mean of Y and of the population proportion in the category, respectively.

(b) Calculate the stratified estimators, T_s and T'_s, of the population total of Y and of the number of elements in the category, respectively.

4.5 Consider the population of firms in Table 4.3 classified into two groups according to geographic location.

(a) In the manner of Table 4.5, show what would happen if you were to select a stratified sample of size 3 by drawing at random and without replacement two firms from group 1 (East) and one firm from group 2 (West). Calculate the mean and variance of the distribution of the stratified estimators $\bar{Y}_s$ and P_s.

(b) Same as (a), except the stratified sample is of size 3, and consists of one firm from group 1 and two firms from group 2.

4.6 Consider the population of six firms stratified according to the nature of operations as shown in Table 4.6.

Suppose that a stratified random sample of size 2 will be drawn by selecting one firm from each group (manufacturing, nonmanufacturing). Suppose also that an estimator $(\bar{Y})$ of the population mean, μ (in this case, the average number of employees per firm in the town), will be formed by averaging all the observations in the sample regardless of the group to which they belong; for example, if firms A (with 9 employees) and B (with 8 employees) are selected, the estimate of μ will be $\bar{Y} = (9 + 8)/2 = 8.5$.

Similarly, an estimator (P) of the population proportion, π (in this case, the proportion of firms in the town which intend to hire), will be formed as $P = (Y_1 + Y_2)/2$, where Y_1 and Y_2 are 1 or 0 depending on whether or not the selected firms intend to hire; for example, if firms A (will hire) and C (will not hire) are selected, the estimate will be $P = (1 + 0)/2 = 0.5$.

(a) Determine the probability distributions of $\bar{Y}$ and P.

(b) Calculate $E(\bar{Y})$, $E(P)$, $Var(\bar{Y})$, and $Var(P)$.

(c) Are $\bar{Y}$ and P unbiased estimators of μ and π respectively? Are they better or worse than the stratified estimators for the same sample size ($n = 2$) shown in Table 4.7?

4.7 (a) In the manner of Section 4.4, confirm Eq. 4.4 using the data of Tables 4.3 and 4.5.

(b) In the manner of Section 4.4, confirm Eqs. 4.2 and 4.4 using the data of Tables 4.6 and 4.7.

4.8 In the manner of Section 4.5, show that in the case of proportional stratified sampling $Var(P) \geq Var(P_s)$. *Hint:* Confirm first that $\pi(1 - \pi) = \sum w_i \pi_i (1 - \pi_i) + \sum w_i (\pi_i - \pi)^2$ using the data of Tables 4.3 or 4.6.

4.9 We noted in Section 4.6 the difficulty of applying the "optimal" allocation formulas when the group standard deviations (σ_i) of a variable Y are unknown. In some cases, however, Y could be related to a known auxiliary variable X, and that relationship could be of some assistance.

Specifically, suppose Y is linearly related to X, that is, $Y = a + bX$, where a and b are given constants.

(a) It can be shown that $\sigma_y^2 = b^2 \sigma_x^2$, where the σ_y^2 and σ_x^2 are the variances of Y and X, respectively.

Confirm this relationship using $a = 2$, $b = -1$, and the following data for X,

$$X : \quad 1 \quad 2 \quad 3.$$

(b) Imagine a population divided into M groups according to some criterion. If the variable of interest Y is linearly related to a known auxiliary variable X, what should be the optimal group sample sizes n_i^* given by Eq. 4.10? Can these be calculated?

4.10 Refer to Example 4.3, and assume that it costs \$9 to sample one monotainer and \$1 to sample one bag (the cost is essentially that of inspecting and counting the parcels in the container). The fixed cost is \$100, and the budget \$2,000.

Apply Eqs. 4.12 and 4.13 to calculate the total sample size and its allocation which minimize $Var(\bar{Y}_s)$ subject to the budget constraint.

4.11 (a) Refer to Example 4.3 in Section 4.7. How large should be next week's sample of containers so that the estimate of the proportion of containers shipped to Western is within ± 0.01 of the true value with probability 95%? Assume $N_1 = 5,000$, $N_2 = 45,000$, 60% of the containers will be monotainers, and 40% bags. State clearly any other assumptions you are forced to make.

(b) Refer to Example 4.3 in Section 4.7. How large should be next week's sample of containers so that the estimate of the average number of parcels per container is within ± 0.5 parcel of the true value with probability 90%? Assume $N_1 = 5,000$, $N_2 = 45,000$, 60% of the containers will be monotainers and 40% bags. State clearly any other assumptions you are forced to make.

4.12 Show that the relative improvement in accuracy expected as a result of increasing the total sample size from n to kn when the planned sample is proportional stratified is the same as in simple random sampling, that is,

$$(1 - \frac{1}{k})(\frac{N}{N - n}).$$

4.13 Refer to Example 4.3. Suppose that the sample results were as follows, rather than as given in Table 4.11:

Containers	N_i	n_i	$\bar{Y}_i$	S_i	P_i
Monotainers	5,000	200	95	50	0.15
Bags	45,000	300	15	10	0.11
Total	$N = 50,000$	500			

(a) Calculate 95% confidence intervals for the average number of parcels per container and the total number of parcels.

(b) Calculate 95% confidence intervals for the proportion and number of containers shipped to Western.

4.14 Suppose it is possible to divide the elements of a population into M *homogeneous* groups (that is, such that all elements in the group have the same value of the variable of interest or all the elements of the group either belong or do not belong to the category of interest).

(a) Show that *any* stratified sample is better than a simple random sample of the same size.

(b) What is the optimal stratified sample in this case?

4.15 "The grouping of population elements is normally based on a known auxiliary variable or attribute, but grouping is possible also when only a list of the population elements is available. For example, the first group could consist of the first 10% of the elements listed, the second group of the next 20%, and so on. This means it is always possible to take a proportional stratified sample. Since such a sample is always better than a simple random sample of the same size, we need never consider taking a simple random sample." Comment.

4.16 An electric utility company regularly surveys its customers to establish the availability of certain appliances, their type, age, etc. A certain city is divided into three service areas. These areas and the total number of customers in each area are shown in columns (1) and (2) of Table 4.14.

In the most recent survey, random samples of sizes 500, 300, and 200 were drawn without replacement from the North, Southeast, and Southwest customer lists respectively, as shown in column (3). Interviewers questioned each selected customer. The questionnaire included the following questions:

- How many years have you been living in this house? ___ years.
- Do you own a washing machine? ___ Yes ___ No

The responses to these questions are summarized in columns (4) to (6). For example, in the North, the average length of stay of the sampled customers was 4.12 years, and the standard deviation of their length of stay was 0.81 years; in addition, 45% of these customers owned a washing machine.

TABLE 4.14	Data for Problem 4.16				
Service Area (1)	Customers, N_i (2)	Sample Size, n_i (3)	Average Length of Stay, $\bar{Y}_i$ (years) (4)	Standard Deviation of Length of Stay, S_i (5)	Proportion of Customers Owning a Washer, P_i (6)
North	125,000	500	4.12	0.81	0.45
Southeast	75,000	300	3.24	1.03	0.62
Southwest	50,000	200	2.53	1.15	0.85
Total	250,000	1,000			

(a) Estimate the average length of stay of customers in the city, and the proportion of customers in the city who own washing machines.

(b) Estimate the total length of stay of customers in the city, and the number of customers in the city who own washing machines.

(c) Calculate 90% confidence intervals for the population characteristics in (a) and (b).

(d) A new survey is planned. The planned total sample size is 1,000—the same as in the previous survey. If it is assumed that the purpose of the new survey is to obtain an estimate of the average length of stay of customers in the city with the minimum possible variance, how many customers should be sampled from each area? Use the estimates of standard deviations from the recent survey.

(e) Suppose that the purpose of the planned survey is to estimate as precisely as possible the proportion of customers in the city that own washing machines. What should be the sample size in each service area? Use the estimates of ownership from the recent survey.

(f) Compare your answers in (b) and (c). Comment.

4.17 Marketing Surveys Inc. is a market research firm commissioned to select a sample of automobile body repair shops operating in cities across the state. The Yellow Pages directories will provide lists of body shops in each city. The selected body shops will be visited and interviews conducted for the purpose of determining the current volume of business, prices, etc.

Two methods for selecting the sample are considered. The first is to consolidate the directories into a master list of body shops in all cities, and then to select a simple random sample of shops from that list. The second alternative is to select a simple random sample of body shops from *each* city directory separately. In either case, sampling will be without replacement.

Suppose there are four cities in the state with the following number of body shops in each:

City	Number of Body Shops
A	400
B	200
C	300
D	100
	1,000

Marketing Surveys Inc. has funds for selecting a sample of no more than 200 body shops. Which of the two alternatives do you recommend? How many shops do you propose to select in each city? Explain clearly your reasoning.

4.18 Each spring, the Asphalt Shingles Manufacturers' Association estimates the year's market for roofing materials in a certain city by means of a survey of houses. In this city, asphalt shingles are used as roofing material in nearly all houses.

On the basis of information from the latest census, the city is divided into low-, average-, and high-income areas. An essentially random sample of houses (detached, semi-detached, and townhouses) is selected from *each* area. An experienced estimator inspects from street level the condition and size of the roof of each selected house and records the inspection results as follows:

Condition of Roof

A	=	New roof needed this year
B	=	New roof needed but can wait until next year
C	=	New roof not needed in next two years
NE	=	Condition cannot be determined

Size of Roof

S	=	Small (less than 300 sq ft)
A	=	Average (300 to 500 sq ft)
L	=	Large (over 500 sq ft)
NE	=	Size cannot be determined

Population and sample data are as follows.

Income Area	Number of Houses	Number Selected
Low	10,000	100
Average	15,000	150
High	5,000	50
Total	30,000	300

The sample results are shown in Tables 4.15 to 4.17.

(a) Estimate the proportion of houses in the city in need of a new roof this year.

(b) Estimate the proportion of houses in the city having an average-size roof.

(c) Estimate the proportion of houses in the city with an average-size roof *and* requiring a new roof this year.

(d) Using 250, 400, and 600 sq ft as representative of all small, average, and large roofs respectively, estimate the total area (in sq ft) in the city in need of new roofing this year.

(e) Assuming that roofing contractors can charge low-, average-, and high-income areas $2, $2.50, and $3.50 respectively per square foot of roof area, estimate this year's potential revenue to roofing contractors in the city. Use the representative sizes in (d) and assume all houses in need of a new roof this year—and only these houses—will be re-roofed at the stated prices.

4.19 The following sampling problem arises in several kinds of audits, but it is best explained with reference to *tax audits*.

A district taxation office has received the income tax returns ("files") of the taxpayers in the district for the latest taxation year. These files are classified into a number of groups according to a set of prescribed criteria. The basis of this stratification is the belief that the proportion of files "in error" varies among the groups. For example, professionals—doctors, lawyers, accountants, etc.—may be considered more likely to report income "incorrectly" than salaried people.

To be specific, let us suppose there are three groups, with the following characteristics:

Group	Number of Files	Estimated Proportion in Error
1	10,000	0.7
2	30,000	0.3
3	60,000	0.1
	100,000	

The last column shows the estimated proportion of files in error for each group (the numbers are, of course, exaggerated) based on the office's experience in previous years.

The office plans to select a random sample of files from each group and to audit these in order to find out which ones are in error. The total sample will consist of 600 files—a constraint dictated by the staff available.

(a) If the sample is to be a *proportional stratified sample*, how many files should be selected from each group?

TABLE 4.15 Sample from Low-Income Area

| | Condition of Roof | | | | |
Size	A	B	C	NE	Total
S	8	10	32		50
A	5	5	20		30
L	2	1	7		10
NE				10	10
Total	15	16	59	10	100

TABLE 4.16 Sample from Average-Income Area

| | Condition of Roof | | | | |
Size	A	B	C	NE	Total
S	5	9	16		30
A	15	30	45		90
L	4	7	14		25
NE				5	5
Total	24	46	75	5	150

TABLE 4.17 Sample from High-Income Area

| | Condition of Roof | | | | |
Size	A	B	C	NE	Total
S	1	3	6		10
A	3	6	11		20
L	2	4	9		15
NE				5	5
Total	6	13	26	5	50

(b) Suppose that 200 files are randomly selected from each group and audited. The sample proportions of files in error for groups 1, 2, and 3 are 0.60, 0.35, and 0.20 respectively. Calculate the stratified estimate of the proportion of all files in error.

(c) Mr. Fixedview has been with this taxation office for more years than anyone can remember. In his opinion, the whole point of the tax audit is misunderstood. "Our business," he scoffs, "is not to get 'an unbiased estimate of the proportion of files in error,' but to maximize the effectiveness of the office in detecting incorrect files. We are paid to find errors. I agree that the sample from each group should be random (since each taxpayer in the group is equally likely to be selected), and I agree that at least one file should be selected from each group (because otherwise it would appear that we are neglecting some groups and persecuting others). But I think we should allocate the total sample of 600 among the groups so as to maximize *the expected number of files in error in the sample*.... You don't understand why? Suppose we audit 600 files and find 200 are in error. Suppose another district office audits 500 files but finds only 50 in error. Our success rate is 1 out of 3. Theirs is 1 out of 10. Which office is more effective?"

(i) How would you implement Mr. Fixedview's wish?

(ii) Does the allocation that satisfies him preclude an unbiased estimator of the proportion of all files in error?

(iii) If your answer in (ii) is negative, is there any reason for preferring a proportional stratified sample to Mr. Fixedview's sample?

4.20 A market research firm conducted a survey in a city for the purpose of estimating the total monthly household expenditures on compact discs (CDs) and the total number of households owning a compact disc player (CDP).

The city was divided into four geographic areas. A random sample of households was selected from each area. The results of the survey are shown below:

Area	Number of Households	Number Sampled	Sample Average Monthly Expenditure ($)	Sample Proportion Owning a CDP
Northeast	20,000	100	10.40	0.150
Southeast	10,000	100	6.10	0.083
Southwest	35,000	100	4.05	0.042
Northwest	15,000	100	8.24	0.075
Total	80,000	400		

(a) Estimate the average monthly household expenditure on CDs in the city, and the proportion of households in the city that own a CDP. *Your estimators must be unbiased.* Briefly explain why they are unbiased.

(b) On the basis of your calculations in (a), estimate the total monthly expenditure on CDs and the total number of households owning a CDP in the city.

(c) How many households should be sampled from each area if the next sample of 400 is to be *proportional stratified*?

(d) How many households should be sampled from each area if the next sample of 400 is to estimate as accurately as possible the proportion of households owning a CDP?

4.21 A box contains 10 items in two compartments, A and B. These items form the population of this exercise. They are inspected for quality and declared to be Good or Defective as follows:

Compartment	Number of Good Items	Number of Defective Items	Total
A	3	1	4
B	1	5	6
Total	4	6	10

Obviously, the proportion of defective items (π) in the box is 0.6.

You plan to draw a random sample of *two* items by selecting at random one item from compartment A and one from compartment B. You plan to form an estimate of π based on this stratified sample as follows:

$$P_s = (\frac{4}{10})P_1 + (\frac{6}{10})P_2,$$

where P_1 is the proportion of defective items in the sample from compartment A, and P_2 that from B (clearly, both P_1 and P_2 will be either 0 or 1).

(a) With the help of a probability tree, determine the probability distribution of P_s. Express probabilities as fractions, not in decimal form.

(b) Show that P_s is an unbiased estimator of π. (If you cannot show this, you have made a mistake. Fix it before proceeding further.)

(c) Determine *by two different methods* the variance of P_s.

(d) For this particular population, is this stratified sample better than a simple random sample of the same size? Explain clearly.

4.22 The regular audit of a bank consists in part of selecting a random sample of accounts of a certain type and requesting the account owner to confirm or correct the account's balance as of a certain date. The form letter sent to a daily interest savings account owner by the bank in a recent audit is shown in Figure 4.2.

Any reported discrepancies (usually few in number) are carefully investigated and the "true" account balance determined.

For the purpose of this exercise, suppose that a certain bank has 80,000 daily interest savings accounts. The recorded total balance of these accounts (the "total book balance") is $362,560,000. The average book balance is thus $4,532. The standard deviation of these book balances has been found to be $24.50.

The Commercial Bank

Your account has been selected at random for confirmation in connection with our regular audit. The information shown relates only to the specific account and branch designated below.

SHOULD YOU NOTE ANY DISCREPANCY AS OF DATE OF AUDIT, PLEASE ADVISE US AT THE ABOVE ADDRESS.

CB values our continuing business association and the opportunity to be of service to you.

MR. JOHN SMITH
40 MAIN STREET, WILLOWDALE

TYPE OF ACCOUNT	ACCOUNT NO.
DAILY INTEREST	153670

DATE OF AUDIT	BALANCE
SEP 16, 19X5	$20,777.28

Figure 4.2 Bank form, Problem 4.22

The purpose of the audit is to obtain an estimate of the true total balance for this type of account.

(a) The accounts have a six-digit identification number ranging from about 100,000 to 300,000. The accounts are not numbered in sequence because some have been closed. Using a table of random numbers, illustrate the random selection of five accounts and describe in detail the rules for the selection of the remaining ones.

(b) Assume that a simple random sample of 1,000 accounts was selected. Twenty discrepancies were reported and found to be valid; the remaining 980 accounts may be assumed to have no discrepancy. Discrepancy is the difference between true and book balance. The average discrepancy of these 20 accounts was $103. Estimate the total discrepancy of all 80,000 accounts. Estimate the number of accounts with a discrepancy.

(c) Suppose that the auditor selected instead a *stratified* random sample of accounts based on book balance, as detailed in Table 4.18.

Estimate the total number of accounts with a discrepancy. Is this estimator unbiased? Carry your calculations to at least four decimal places.

TABLE 4.18   Stratified Sample Results, Problem 4.22

Account Balance	Number of Accounts	Number of Selected Accounts	Number of Accounts with Discrepancy	Average Discrepancy of Accounts with Discrepancy ($)
Under $2,000	40,000	500	10	− 46
$2,000 to 5,000	20,000	250	6	102
$5,000 to 10,000	15,000	187	4	−160
Over $10,000	5,000	63	1	255
Total	80,000	1,000	21	

(d) Estimate the total discrepancy. Clearly explain your procedure and why your estimator is unbiased. Carry your calculations to at least four decimal places.

(e) Estimate the true total balance of the bank's daily interest savings account.

(f) "This type of audit has one serious logical—not statistical—flaw." Comment.

CHAPTER 5

Two-Stage Random Sampling

5.1 INTRODUCTION AND SUMMARY

As with stratified sampling, in order to implement two-stage sampling it is necessary that the population elements can be divided into nonoverlapping groups according to a certain criterion.

A two-stage sample, as the term implies, is implemented in two stages. In the first stage, a simple random sample of groups is selected; in the second one, a simple random sample of elements is drawn from each of those groups.

We shall show that, despite similarities, two-stage sampling is not the same as simple or stratified sampling. However, unbiased estimators of the population mean of a variable or the proportion in a category do exist, as will be shown in this chapter, and their variances can be determined. Thus, two-stage estimators can be compared with those for simple and stratified sampling in order to determine whether or not a particular sampling method has any advantages over another.

Two-stage sampling has an important practical advantage over simple or stratified sampling. In order to implement the latter methods, it is necessary to have a list of *all* population elements. For two-stage sampling, by contrast, it is only necessary to have a list of the groups and of the elements in the *selected* groups. This is frequently a critical advantage favoring two-stage sampling even in situations where two-stage is suspected to be less accurate than another sampling method.

There is no simple and universally applicable rule regarding when to apply two-stage sampling similar to that concerning proportional stratified sampling. There are guidelines, however, but these require considerable information about the population.

Two special cases of two-stage sampling will be examined in this chapter. The first is *cluster sampling*, when *all* the elements of each group drawn at the first stage are selected in the second stage. The second case is *systematic sampling*, which is effectively cluster sampling but is often treated in practice as a convenient substitute for simple random sampling under certain conditions.

The chapter concludes with approximate large-sample confidence intervals for the population characteristics based on the two-stage estimators, and some observations on sampling plans that involve a combination of stratified and two-stage sampling or sampling in more than two stages.

5.2 SOME ADVANTAGES OF TWO-STAGE SAMPLING

Imagine that a survey is planned of households in a certain city. The city is divided into a number of blocks, the boundaries of which could be major roads, railroad lines, rivers, etc. Figure 5.1 is a schematic diagram of a city, showing 16 blocks and 5 households in each block. The blocks form the groups (the strata) of this illustration.

Suppose that a stratified sample of 16 households is desired. Since there are 16 groups (blocks), a simple random sample of one household will be selected from *each* block. The selected households *could be* those shown in Figure 5.2.

In order to implement this stratified sample, it is necessary to have a list of the households in every block. Creating such lists may not be difficult or expensive in

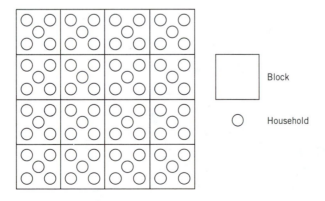

Figure 5.1   City blocks and households

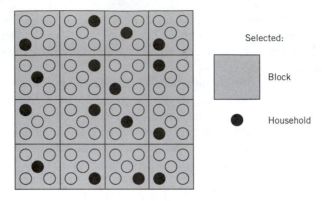

Figure 5.2 Stratified sample of 16 households

this illustration, but in a large city, with perhaps hundreds of blocks and hundreds of households in each block, the task would be formidable and the expense high.

It would certainly be much less difficult and expensive if the sample of households could be selected in two stages as follows: first, a number of blocks would be selected at random; second, a number of households would be selected at random from each *selected* block.

For example, in our illustration, we could decide to select at random four of the sixteen city blocks, and then four households from each of the four selected blocks. This procedure would yield a sample of sixteen households. The selected blocks and households *could be* those shown in Figure 5.3.

In order to select the random sample of four blocks, a list of all the blocks in the city is needed. But in order to select the households, it is necessary only to

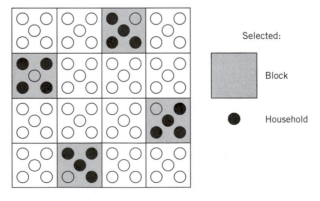

Figure 5.3 Two-stage sample of 16 households

construct or obtain lists of the households *in the selected blocks*, and not in all the blocks as required by the stratified sample.

A two-stage sample may have another cost advantage. The time spent by interviewers traveling from one household to another is probably less in the case of Figure 5.3 than in Figure 5.2, where the selected households tend to lie farther apart from one another.

As this example illustrates, a two-stage sample frequently tends to be more convenient and less expensive than a stratified sample of the same size. It is obviously a random sample, since the groups in the first stage as well as the elements from each selected group are drawn at random. It is reasonable then to ask whether or not a two-stage sample is effectively a simple or stratified sample of the same size, albeit randomly selected in a peculiar way. If the answer to this question is affirmative, no new theory or estimators are needed, since those in one of the preceding two chapters would be suitable.

As we shall soon demonstrate, the answer to this question is negative—two-stage random sampling is not the same as simple or stratified random sampling. However, we shall show that an unbiased estimator of the population mean of a variable Y is

$$\bar{Y}_{ts} = \frac{M}{m}(w_1\bar{Y}_1 + w_2\bar{Y}_2 + \cdots + w_m\bar{Y}_m),$$

and an unbiased estimator of the proportion of population elements in a category is

$$P_{ts} = \frac{M}{m}(w_1P_1 + w_2P_2 + \cdots + w_mP_m).$$

In the above expressions, M is the number of groups in the population, m the number of groups selected in the first stage, $\bar{Y}_i$ and P_i the sample mean and the sample proportion respectively in the ith selected group. Finally, $w_i = N_i/N$ is the fraction of the population elements in the ith selected group.

The sums in the expressions for $\bar{Y}_{ts}$ and P_{ts}, it should be noted, involve the *selected* groups only.

An unbiased estimator of the population total of Y is $T_{ts} = N\bar{Y}_{ts}$, and one of the number of elements in the category is $T'_{ts} = NP_{ts}$.

The subscript ts in $\bar{Y}_{ts}$, P_{ts}, T_{ts} and T'_{ts} stands for "two-stage." We shall call these the *two-stage estimators*.

Before we describe the properties of these estimators, it will be useful to illustrate their calculation.

◪ EXAMPLE 5.1   Once each year, the Consumer Survey Bureau (CSB) conducts an extensive survey of household expenditures and attitudes in a city. The Office of the Census divides the city into "Enumeration Areas (EAs)" the size of a large city block. Detailed maps of each EA are available, as well as additional information from the latest

census. For its survey, CSB selects at random a number of EAs, and then, also at random, a number of households from each selected EA.

Each selected household is visited, and the head of the household is asked to complete a questionnaire with the assistance of the interviewer. We illustrate the calculation of the estimates for two questions in the questionnaire: "How much did your household spend on clothing last month? ($___)" and "Do you own a tape recorder? (___ Yes ___ No)."

The city is divided into $M = 200$ EAs. The latest census shows that there are $N = 60,000$ households in the city. Let us say $m = 4$ EAs are selected in the first stage (this is to illustrate simply the calculations; in practice, many more EAs would have to be selected). The relevant data are shown in Table 5.1.

In Table 5.1, N_i is the number of households (as given in the most recent census) in the ith selected block, $\bar{Y}_i$ is the average household expenditure on clothing (in $), and P_i the proportion of households owning tape recorders in the sample from the ith selected EA.

The two-stage estimate of the average expenditure on clothing per household in the city is

$$\bar{Y}_{ts} = \frac{M}{m}(w_1\bar{Y}_1 + w_2\bar{Y}_2 + w_3\bar{Y}_3 + w_4\bar{Y}_4)$$

$$= \frac{200}{4}[(0.0042)(95.0) + \cdots + (0.0047)(90.3)]$$

$$= \frac{200}{4}(1.6790)$$

$$= 83.95.$$

TABLE 5.1   Estimation in Two-Stage Sample, Example 5.1

EA, i	N_i	$w_i = N_i/N$	n_i	$\bar{Y}_i$	P_i	$w_i\bar{Y}_i$	w_iP_i
29	250	0.0042	50	95.0	0.75	0.3958	0.003125
67	310	0.0052	60	84.0	0.92	0.4340	0.004753
102	340	0.0057	70	75.5	0.83	0.4278	0.004703
143	280	0.0047	55	90.3	0.95	0.4214	0.004433
						1.6790	0.017014

The two-stage estimate of the proportion of households in the city owning a tape recorder is

$$P_{ts} = \frac{M}{m}(w_1 P_1 + w_2 P_2 + w_3 P_3 + w_4 P_4)$$

$$= \frac{200}{4}[(0.0042)(0.75) + \cdots + (0.0047)(0.95)]$$

$$= \frac{200}{4}(0.017014)$$

$$= 0.851.$$

The estimated total household expenditure on clothing in the city is $N\bar{Y}_{ts} = (60,000)(83.95)$ or $\$5,037,000$. The estimate of the number of households in the city that own a tape recorder is $NP_{ts} = (60,000)(0.851)$ or $51,060$. ◼

The two-stage estimators are not arbitrary, but can be interpreted in intuitively appealing terms. To see this, suppose that the objective is to estimate the population mean of a variable Y. The simple estimator of the total of Y in the ith *selected* group is

$$N_i \bar{Y}_i.$$

Therefore, a reasonable estimator of the total of Y over all *selected* groups is

$$N_1 \bar{Y}_1 + N_2 \bar{Y}_2 + \cdots + N_m \bar{Y}_m.$$

It follows that an estimator of the average value of Y *per group* is

$$\frac{1}{m}(N_1 \bar{Y}_1 + N_2 \bar{Y}_2 + \cdots + N_m \bar{Y}_m),$$

and, since there are M groups in total, a reasonable estimator of the population total of Y is

$$M \frac{1}{m}(N_1 \bar{Y}_1 + N_2 \bar{Y}_2 + \cdots + N_m \bar{Y}_m).$$

This *is* the two-stage estimator T_{ts}, since

$$T_{ts} = N\bar{Y}_{ts}$$

$$= N\frac{M}{m}(w_1 \bar{Y}_1 + w_2 \bar{Y}_2 + \cdots + w_m \bar{Y}_m)$$

$$= N\frac{M}{m}(\frac{N_1}{N} \bar{Y}_1 + \frac{N_2}{N} \bar{Y}_2 + \cdots + \frac{N_m}{N} \bar{Y}_m)$$

$$= \frac{M}{m}(N_1 \bar{Y}_1 + N_2 \bar{Y}_2 + \cdots + N_m \bar{Y}_m).$$

A reasonable estimator of the population mean of Y, then, is T_{ts}/N, which is $\bar{Y}_{ts}$, the two-stage estimator of μ. A similar interpretation can be given to P_{ts} and T'_{ts}.

A final remark before we describe the properties of the two-stage estimators. In the event only one group is selected in the first stage, $m = 1$ and the expressions for the two-stage estimators simplify to

$$\bar{Y}_{ts} = Mw\bar{Y},$$

and

$$P_{ts} = MwP,$$

where $\bar{Y}$ and P are respectively the mean of the variable and the proportion in the category in the sample from the one and only selected group, while w is the proportion of population elements in that selected group. (We could have written, for example, $\bar{Y}_{ts} = Mw_1\bar{Y}_1$, but the subscript is redundant in this case.)

5.3 PROPERTIES OF TWO-STAGE ESTIMATORS

Let us consider once again the population of six firms in the Ackroyd case divided into two groups according to geographic location, as shown in Table 4.3 and repeated here for convenience as Table 5.2.

For a stratified sample of size 2 we selected one firm at random from each group. We now examine what will happen if we were to draw from this population

---

TABLE 5.2 Ackroyd Population and Its Characteristics, by Groups Formed According to Geographic Location

	Group 1 (East)			Group 2 (West)		
	Firm	Employees	Hiring?	Firm	Employees	Hiring?
	A	9	Y	D	2	N
	B	8	Y	E	1	N
	C	6	N			
	F	5	N			
Group size	$N_1 = 4$			$N_2 = 2$		
Group mean[a]	$\mu_1 = 7$			$\mu_2 = 1.5$		
Group variance[a]	$\sigma_1^2 = 2.5$			$\sigma_2^2 = 0.25$		
Group proportion[b]	$\pi_1 = 0.5$			$\pi_2 = 0.0$		

[a]Number of employees.
[b]Proportion of firms intending to hire.

a random sample of 2 firms in two stages, as follows: in the first stage, we shall select at random one of the two groups; then, in the second stage, we shall select at random two firms from the group selected in the first stage. (Obviously, if we happen to select the second group in the first stage, the sample will consist of the only two firms in that group.) The sample outcomes and their probabilities are shown in Figure 5.4.

For example, the probability that the sample will consist of firms A and F is equal to the probability (1/2) that group 1 will be selected in the first stage, times the probability (1/6) that firms A and F will be drawn in the second stage, given that group 1 is selected in the first stage; this is (1/2)(1/6) or 1/12. The probability that the sample will consist of firms D and E is the probability that group 2 will be drawn in the first stage, times the probability that firms D and E will be selected in the second stage, given that group 2 is selected in the first stage; this is (1/2)(1) = 1/2.

The list of possible outcomes and their probabilities in Table 5.3 are different from those of a simple random sample or a stratified random sample of the same size shown in Tables 3.2 and 4.5. We note in particular that several outcomes in the latter lists cannot occur when the sample is selected in two stages. We conclude that a two-stage sample is not the same as a simple or stratified random sample of the same size.

We can also verify easily that the ordinary sample average and proportion (the simple estimators) are not unbiased when used in two-stage sampling. The simple estimators ($\bar{Y}$ and P) are shown in the fifth and seventh columns of Table 5.3. Their expected values are

$$E(\bar{Y}) = (8.5)(1/12) + \cdots + (5.5)(1/12) + (1.5)(6/12) = 4.25 \neq 5.167,$$

and

$$E(P) = (1.0)(1/12) + \cdots + (0.0)(1/12) + (0.0)(6/12) = 0.25 \neq 0.333.$$

How, then, should one estimate μ or π? In this illustration, only one group is selected in the first stage. We would like to show that $\bar{Y}_{ts} = Mw\bar{Y}$, the two-stage

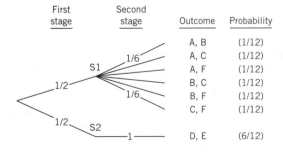

Figure 5.4 Two-stage random sample

TABLE 5.3 Two-Stage Random Sample without Replacement from Ackroyd Population Stratified According to Geographic Location (Table 5.2)

Outcome							
First Stage	Second Stage	w	Proba-bility	Sample Mean, $\bar{Y}$	$\bar{Y}_{ts} = 2w\bar{Y}$	Sample Proportion, P	$P_{ts} = 2wP$
S1	A,B	4/6	1/12	8.5	11.333	1.0	1.333
	A,C	4/6	1/12	7.5	10.000	0.5	0.667
	A,F	4/6	1/12	7.0	9.333	0.5	0.667
	B,C	4/6	1/12	7.0	9.333	0.5	0.667
	B,F	4/6	1/12	6.5	8.667	0.5	0.667
	C,F	4/6	1/12	5.5	7.333	0.0	0.000
S2	D,E	2/6	6/12	1.5	1.000	0.0	0.000
			12/12				

$$E(\bar{Y}_{ts}) = 5.167, \qquad Var(\bar{Y}_{ts}) = 18.102$$
$$E(P_{ts}) = 0.333, \qquad Var(P_{ts}) = 0.185$$

estimator when $m = 1$, is indeed an unbiased estimator of the population mean, μ. In the above expression, $M = 2$, $\bar{Y}$ the average number of employees of the two firms in the sample, and w the fraction of population elements in the group from which the sample of two firms is selected. Thus, if the sample is from group 1, w is 4/6; if the sample is from group 2, w is 2/6.

We would also like to show that an unbiased estimator of the population proportion, π, is $P_{ts} = MwP$. P is the proportion of firms intending to hire in the sample, and w is either 4/6 or 2/6, depending on which group is selected in the first stage.

Table 5.3 shows the values of these estimators for each outcome of the two-stage sample.

The expected values of the two-stage estimators are

$$E(\bar{Y}_{ts}) = (1/12)(11.333) + \cdots + (6/12)(1) = 5.167,$$

and

$$E(P_{ts}) = (1/12)(1.333) + \cdots + (6/12)(0) = 0.333,$$

respectively, as shown in Table 5.3. Therefore, $\bar{Y}_{ts}$ is indeed an unbiased estimator of the population mean number of employees, and P_{ts} is an unbiased estimator of the population proportion of firms planning to hire.

Summary 5.1 lists the unbiased estimators under two-stage sampling and their variances. Not surprisingly, the variances of the two-stage estimators are considerably more complicated than those of simpler estimators.

Observe that the sums in Eqs. 5.1 and 5.3 extend over the *selected* groups, while those in Eqs. 5.2 and 5.4 extend over *all* groups.

SUMMARY 5.1 Two-stage random sampling, unbiased estimators

- *Method of sample selection:* Two-stage random sampling.
- *An unbiased estimator of the population mean of a variable Y is the two-stage estimator of μ:*

$$\bar{Y}_{ts} = \frac{M}{m}(w_1 \bar{Y}_1 + w_2 \bar{Y}_2 + \cdots + w_m \bar{Y}_m) = \frac{M}{m} \sum_{i=1}^{m} w_i \bar{Y}_i, \qquad (5.1)$$

and its variance is given by

$$Var(\bar{Y}_{ts}) = \left(\frac{M}{N}\right)^2 \frac{\sigma_{01}^2}{m} \frac{M-m}{M-1} + \frac{M}{m} \sum_{i=1}^{M} w_i^2 \frac{\sigma_i^2}{n_i} \frac{N_i - n_i}{N_i - 1}, \qquad (5.2)$$

- *An unbiased estimator of the proportion of elements in the population belonging to a category C is the two-stage estimator of π:*

$$P_{ts} = \frac{M}{m}(w_1 P_1 + w_2 P_2 + \cdots + w_m P_m) = \frac{M}{m} \sum_{i=1}^{m} w_i P_i, \qquad (5.3)$$

and its variance is given by

$$Var(P_{ts}) = \left(\frac{M}{N}\right)^2 \frac{\sigma_{02}^2}{m} \frac{M-m}{M-1} + \frac{M}{m} \sum_{i=1}^{M} w_i^2 \frac{\pi_i(1-\pi_i)}{n_i} \frac{N_i - n_i}{N_i - 1}. \qquad (5.4)$$

In the above expressions, $w_i = N_i/N$, and

$$M = \text{number of groups in the population}$$
$$m = \text{number of groups selected in the first stage}$$
$$N = \text{number of elements in the population}$$
$$N_i = \text{number of elements in the } i\text{th group}$$
$$n_i = \text{number of elements selected from the } i\text{th group}$$
$$\bar{Y}_i = \text{sample mean in the } i\text{th group}$$
$$P_i = \text{sample proportion in the } i\text{th group}$$

In Eqs. 5.2 and 5.4,

$$\sigma_{01}^2 = \frac{1}{M} \sum_{i=1}^{M} \left(N_i \mu_i - \frac{N\mu}{M}\right)^2,$$

and

$$\sigma_{02}^2 = \frac{1}{M} \sum_{i=1}^{M} \left(N_i \pi_i - \frac{N\pi}{M}\right)^2.$$

Observe also that when $m = M$, that is, when all the groups are selected in the first stage, the two-stage estimators become in effect the stratified estimators, in which case, $Var(\bar{Y}_{ts}) = Var(\bar{Y}_s)$ and $Var(P_{ts}) = Var(P_s)$.

Finally, as we noted earlier, observe that when $m = 1$ the two-stage estimators simplify to $\bar{Y}_{ts} = Mw\bar{Y}$ and $P_{ts} = MwP$.

Before we close this section, let us verify the expressions of the variances of the two-stage estimators given in Summary 5.1, using for this purpose the results of Table 5.3. Readers who do not wish to question the Summary may skip the remainder of this section.

From Table 5.2, $N_1\mu_1 = 28$, $N_2\mu_2 = 3$, and $N\mu/M = 15.5$. Therefore,

$$\sigma_{01}^2 = \frac{1}{2}[(28 - 15.5)^2 + (3 - 15.5)^2] = 156.25.$$

It follows that

$$\left(\frac{M}{N}\right)^2 \frac{\sigma_{01}^2}{m} \frac{M - m}{M - 1} = \left(\frac{2}{6}\right)^2 \frac{156.25}{1} \frac{2 - 1}{2 - 1} = 17.361.$$

Also,

$$\left(\frac{M}{m}\right) \sum_{i=1}^{M} w_i^2 \frac{\sigma_i^2}{n_i} \frac{N_i - n_i}{N_i - 1} = \left(\frac{2}{1}\right)\left[\left(\frac{4}{6}\right)^2 \frac{2.5}{2} \frac{4 - 2}{4 - 1} + \left(\frac{2}{6}\right)^2 \frac{0.25}{2} \frac{2 - 2}{2 - 1}\right] = 0.741.$$

Therefore, according to Eq. 5.2, we should have

$$Var(\bar{Y}_{ts}) = 17.361 + 0.741 = 18.102.$$

Now, according to Table 5.3,

$$Var(\bar{Y}_{ts}) = (11.333)(1/12) + \cdots + (1.000)(1/12) - (5.167)^2$$
$$= 18.102.$$

The direct calculations based on Table 5.3, therefore, confirm Eq. 5.2.

The verification of the variance of P_{ts} proceeds along similar lines. Again referring to Table 5.2, $N_1\pi_1 = 2$, $N_2\pi_2 = 0$, and $N\pi/M = 1$. Therefore,

$$\sigma_{02}^2 = \frac{1}{2}[(2 - 1)^2 + (0 - 1)^2] = 1.$$

Also,

$$\left(\frac{M}{N}\right)^2 \frac{\sigma_{02}^2}{m} \frac{M - m}{M - 1} = \left(\frac{2}{6}\right)^2 \frac{1}{1} \frac{2 - 1}{2 - 1} = 0.111,$$

and

$$\left(\frac{M}{m}\right) \sum_{i=1}^{M} w_i^2 \frac{\pi_i(1 - \pi_i)}{n_i} \frac{N_i - n_i}{N_i - 1} = \left(\frac{2}{1}\right)\left[\left(\frac{4}{6}\right)^2 \frac{(0.5)(0.5)}{2} \frac{4 - 2}{4 - 1} + \left(\frac{2}{6}\right)^2 \frac{(0)(1)}{2} \frac{2 - 2}{2 - 2}\right]$$
$$= 0.074.$$

Therefore, according to Eq. 5.4, we should have

$$Var(P_{ts}) = 0.111 + 0.074 = 0.185.$$

Referring to Table 5.3, we calculate

$$Var(P_{ts}) = (1.333)(1/12) + \cdots + (0.000)(1/12) - (1/3)^2$$
$$= 0.185,$$

which is in agreement with Eq. 5.4.

5.4 SPECIAL CASE: CLUSTER SAMPLING

In the event that *all* the elements of each group drawn in the first stage are selected, we say the sample is a *cluster random sample*.[1] It should be clear that such a sample is a special type of a two-stage sample with all $n_i = N_i$. The cluster estimators, $\bar{Y}_{ts}$ and P_{ts}, are, of course, calculated exactly as in the general case.

■ EXAMPLE 5.2 In order to monitor on a monthly basis the quality of service provided to its customers, Gamma Airlines plans to question each month a number of passengers to be selected at random from among all the passengers flying Gamma that month.

A problem facing the designers of next month's sample is that the airline does not know who its passengers will be. It does, however, know all its scheduled flights next month. Although some changes will undoubtedly take place (some flights will be canceled, others will be added), the number of such changes is usually relatively small in relation to the total number of scheduled flights and the airline believes they can be ignored.

The plan, therefore, is for the airline to select a random sample *of flights* without replacement from the known list of all scheduled flights next month. All the passengers in each selected flight will be given a short questionnaire. The questionnaire will be distributed during the flight by flight attendants, who will also collect the completed questionnaires.

The main advantage of this cluster sample is, of course, that it does not require knowledge of next month's passengers.

For the sake of illustration, let us suppose that 4 of the 140 scheduled flights in March were selected using a random number generator in the usual way. By the end of March, the passenger responses to two of the questions in the questionnaire are known. These questions are

---

[1]The terms *single-stage sample* and *single-stage cluster sample* are also used to refer to this kind of sample.

5. Overall, how do you rate the quality of Gamma's service? Excellent ___ Very good ___ Average ___ Poor ___

8. How many days did/will the present flight keep you away from home? ___ days

The responses to these two questions are summarized in Table 5.4.

By the end of March, Gamma also knows the total number of passengers carried in all its flights that month. Again for the purpose of this illustration, let us suppose this number is 10,700. In the notation of this chapter, therefore, $N = 10,700$, $M = 140$, and $m = 4$. Also, $w_1 = N_1/N = 60/10,700 = 0.0056$, $w_2 = 45/10,700 = 0.0042$, $w_3 = 128/10,700 = 0.0119$, and $w_4 = 80/10,700 = 0.0075$. Therefore, the estimate of the proportion of all Gamma Airline passengers who consider the service excellent is

$$P_{ts} = \frac{140}{4}[(0.0056)(0.24) + \cdots + (0.0075)(0.18)] = 0.182,$$

and the estimate of the overall average trip length is

$$\bar{Y}_{ts} = \frac{140}{4}[(0.0056)(2.4) + \cdots + (0.0075)(2.7)] = 2.952.$$

The estimate of the total number of Gamma passengers in March who consider the service excellent is $(10,700)(0.182)$, or 1,948, and the estimate of the total number of days Gamma passengers were away from home in March is $(10,700)(2.952)$ or 31,587. ◪

In the special case of cluster sampling, the variances of the two-stage estimators, Eqs. 5.2 and 5.4, reduce to

$$Var(\bar{Y}_{ts}) = \left(\frac{M}{N}\right)^2 \frac{\sigma_{01}^2}{m} \frac{M-m}{M-1}, \tag{5.5}$$

and

$$Var(P_{ts}) = \left(\frac{M}{N}\right)^2 \frac{\sigma_{02}^2}{m} \frac{M-m}{M-1}. \tag{5.6}$$

TABLE 5.4 Gamma Airlines Questionnaire Responses

i	Date/Flight	Number of Passengers, $n_i = N_i$	Proportion Rating Excellent, P_i	Average Trip Length (days), $\bar{Y}_i$
1	03-08/417	60	0.24	2.4
2	03-15/200	45	0.17	2.1
3	03-16/315	128	0.15	3.5
4	03-25/167	80	0.18	2.7

We can explore these expressions to gain some insight into the conditions that recommend the use of cluster sampling. For present purposes, Eq. 5.5 can be written as

$$Var(\bar{Y}_{ts}) = (\text{Constant})\sigma_{01}^2 = (\text{Constant})\frac{1}{M}\sum_{i=1}^{M}(N_i\mu_i - \frac{N\mu}{M})^2.$$

If the group means are approximately equal, then $\mu_i \approx \mu$, and

$$Var(\bar{Y}_{ts}) \approx (\text{Constant})\frac{\mu^2}{M}\sum_{i=1}^{M}(N_i - \frac{N}{M})^2,$$

where N/M is the average number of elements per group. If, in addition, the groups are of approximately the same size, then $N_i \approx N/M$ and the sum of the squared deviations of the group sizes N_i from their average N/M—hence also $Var(\bar{Y}_{ts})$—will be small. We conclude that, *for estimating the population mean of a variable, cluster sampling is desirable when the group means μ_i and group sizes N_i are approximately equal.*

The same reasoning applied to Eq. 5.6 suggests that, *for estimating the population proportion in a category, cluster sampling is desirable when the group sizes and proportions π_i are approximately equal.*

5.5 SPECIAL CASE: SYSTEMATIC SAMPLING

Suppose that a population has 15 elements, numbered serially from 01 to 15, and that a random sample of 3 elements is desired. Consider the following method of sample selection.

Select at random *one* of the first five elements, 01 to 05, and then every 5th element in the sequence. For example, if the first element is 03, then the sample will consist of elements 03, 08, and 13; if the first element is 01, the sample will consist of elements 01, 06, and 11. Figure 5.5 illustrates.

This type of sample is called a *systematic random sample.* In general, to select a systematic random sample of size n from a population of size N when the ratio $k = N/n$ is an integer, select at random one of the first k elements and then every

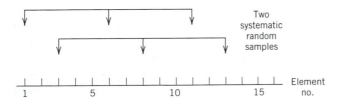

Figure 5.5 Systematic sampling illustrated

*k*th element in the sequence thereafter. (We consider below the case where the ratio N/n is not an integer.)

It should be clear that *a systematic sample is a cluster sample* with $m = 1$. The systematic procedure described above essentially divides the population elements into k groups, one of which with all its elements is selected with equal probability.

Group No.	Elements		
1	01	06	11
2	02	07	12
3	03	08	13
4	04	09	14
5	05	10	15

In the above illustration, the first group consists of elements 01, 06, and 11; the second, of elements 02, 07, and 12; and so on. Each group has the same probability of selection.

Exactly the same general selection procedure is used when the ratio N/n is not an integer: if k is the largest integer less than N/n, select at random one of the first k elements, and then every kth element thereafter. The actual sample size will differ from the desired by one, depending on which element is selected first, but this difference is negligible when the sample size is reasonably large.

To illustrate, suppose that the population size is $N = 17$ and that a sample of size $n = 3$ is initially desired. $N/n = 17/3 = 5\frac{2}{3}$, therefore $k = 5$. Imagine the population elements divided into five groups, as follows:

Group No.	Elements			
1	01	06	11	16
2	02	07	12	17
3	03	08	13	
4	04	09	14	
5	05	10	15	

Following the general selection procedure, if the first selected element is 01 or 02, the sample will be of size 4 (and will consist of the elements in group 1 or 2); if the first element is 03, 04, or 05, the sample will be of size 3 (and will consist of the elements in group 3, 4, or 5). Again, such a selection procedure yields a cluster sample.

It should also be clear that a systematic sample is not a simple random sample. This follows from the fact that it is a special case of a two-stage sample, but also by a direct argument. In simple random sampling, it will be recalled from Section 3.4, any two distinct population elements have the same probability of appearing together in the sample. By contrast, consider the preceding example. The probability that elements 01 and 02 will appear in the sample is zero, while the probability that elements 01 and 06 will appear in the sample is 1/5, the probability that group 1 will be selected.

Nevertheless, a systematic sample is often used in practice as if it were a simple random sample, when it can be assumed that the order in which the population elements are listed is itself random.

For example, let us suppose that a random sample of telephone numbers is desired from those listed in the city telephone book. The numbers are listed in alphabetical order of the subscriber's name. It could be argued that this list can be treated, for all practical purposes, *like* one list produced by copying each number on a tag, putting the tags in a hat, and then drawing at random and without replacement the tags, one after the other, until none is left in the hat. If this argument is accepted, a systematic sample consisting of, say, the first name listed at the top of each page of the telephone book could be regarded essentially as a simple random sample. (In fact, *any* sample, whether systematic or not, would qualify as a simple random sample in this case.) However, if the first entry on each page of the directory is always that of a firm, a systematic sample may consist entirely of firms at the exclusion of residences.

If then a systematic sample is essentially a simple random sample, the simple estimators (the ordinary sample mean, $\bar{Y}$, and proportion, P) are unbiased and could be used. If it is not, appropriate are the unbiased two-stage estimators $\bar{Y}_{ts} = Mw\bar{Y}$ and $P_{ts} = MwP$.

■ EXAMPLE 5.2
(Continued)

It is interesting to examine how the Consumer Survey Bureau sample of households in each Enumeration Area (EA) is actually selected. We had implied that these samples were random, but, in fact, they are not strictly so.

CSB provides its interviewers with a map of the EA to which they are assigned; one such map is shown in Figure 5.6. The instructions to the interviewer read as follows:

> Starting at the designated point, and moving in the direction shown—FOR HOUSES: Select the *5*th house for your first interview. Then skip *4* houses, selecting every *5*th house thereafter.—FOR APARTMENTS: Take the elevator to the *top* floor, turn *right*, and select every *5*th apartment going *down* one floor at a time .... Continue until you have selected *20* houses or apartments.

The words in *italics* vary for each EA. The starting point, the initial direction, and—in the case of apartments—the first floor visited and the direction of travel are all chosen at random. The selection interval—5 in this illustration—is calculated for each EA so that (Number of households in EA) ÷ (Selection interval) ≈ (Desired sample size in EA).

CSB treats the resulting sample of households as if it were a random one. ■

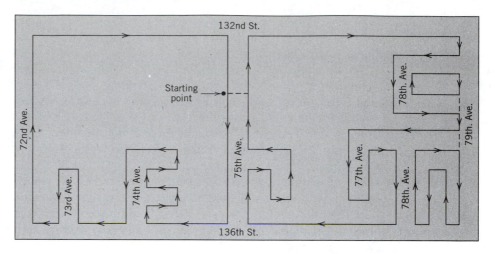

Figure 5.6 Enumeration area map, Example 5.1

5.6 STRATIFIED OR TWO-STAGE SAMPLING?

As we observed at the beginning of this chapter, two-stage sampling is often preferred because it is convenient and inexpensive. In this section we consider stratified and two-stage sampling and the conditions under which one of these methods may be preferable to the other when cost and convenience are the same.

It is clear that both sampling methods yield unbiased estimators. But it can be demonstrated that in some populations the stratified estimators have lower variances than the two-stage estimators based on samples of the same size, while in other populations the reverse is true.

For example, a comparison of the variances shown in Tables 4.5 and 5.3 shows that stratified sampling is better than two-stage sampling; in this case, $Var(\bar{Y}_s) = 1.139$ while $Var(\bar{Y}_{ts}) = 18.102$, and $Var(P_s) = 0.111$ while $Var(P_{ts}) = 0.185$.

Consider, however, the same population of six firms in Ackroyd, but this time stratified according to the nature of the firms' operations (manufacturing and non-manufacturing, Table 4.6). Suppose we draw a two-stage sample by selecting at random first one group, and then two firms from the group selected in the first stage. Table 5.5 shows the possible outcomes of this two-stage sample and the means and variances of $\bar{Y}_{ts}$ and P_{ts}. Compare these results with those of a stratified sample of the same size (Table 4.7). Clearly the two-stage sample would be preferred in this situation, since $Var(\bar{Y}_{ts}) = 2.139 < Var(\bar{Y}_s) = 4.222$, and $Var(P_{ts}) = 0.055 < Var(P_s) = 0.111$.

For a general two-stage sample, suppose that the total sample size (the total number of elements to be selected) has been decided, perhaps on the basis of the available budget. Just as was done to determine the "optimal" stratified sample, it is interesting to examine how the total sample size should be allocated among the

TABLE 5.5   Two-Stage Random Sample without Replacement from Ackroyd Population Stratified According to the Nature of Operations

Outcome					
First Stage	Second Stage	w	Proba- bility	$\bar{Y}_{ts}$	P_{ts}
S1	A,F	3/6	1/6	7.0	0.5
S1	A,D	3/6	1/6	5.5	0.5
S1	F,D	3/6	1/6	3.5	0.0
S2	B,C	3/6	1/6	7.0	0.5
S2	B,E	3/6	1/6	4.5	0.5
S2	C,E	3/6	1/6	3.5	0.0
			6/6		

$$E(\bar{Y}_{ts}) = 5.167, \qquad Var(\bar{Y}_{ts}) = 2.139$$
$$E(P_{ts}) = 0.333, \qquad Var(P_{ts}) = 0.055$$

groups so as to make the two-stage estimator as precise as possible, that is, so as to minimize $Var(\bar{Y}_{ts})$.

There is, however, a difficulty in attempting to follow the same approach in the case of a two-stage sample. The plan for such a sample calls for selecting n_i elements from the ith group *if that group is selected in the first stage*. The total sample size, therefore, cannot be known in advance with certainty, as was possible under stratified sampling.

One way out of this difficulty is to consider only the case where all the n_i are equal, that is, all $n_i = n_0$, where n_0 is the common sample size in the second stage. The total sample size, therefore, *is* known in advance to be $n = mn_0$. (We assume that $n_0 < N_i$ for all i, and $1 \leq m \leq M$.) The problem then is to determine m and n_0 which minimize $Var(\bar{Y}_{ts})$, subject to the constraint that $mn_0 = n$, a constant.

This is not a difficult problem to solve. Suppose that the purpose of the sample is to estimate the population mean of a given variable, and consider $Var(\bar{Y}_{ts})$, Eq. 5.2, when all $n_i = n_0$:

$$Var(\bar{Y}_{ts}) = (\frac{M}{N})^2 \frac{\sigma_{01}^2}{m} \frac{M-m}{M-1} + (\frac{M}{m}) \sum_{i=1}^{M} w_i^2 \frac{\sigma_i^2}{n_0} \frac{N_i - n_0}{N_i - 1}$$

$$= \underbrace{[(\frac{M}{N})^2 \frac{\sigma_{01}^2}{M-1}]}_{a}(\frac{M}{m} - 1) + (\frac{M}{m}) \sum_{i=1}^{M} \underbrace{[w_i^2 \frac{\sigma_i^2}{N_i - 1}]}_{b_i}(\frac{N_i}{n_0} - 1)$$

$$= a(\frac{M}{m} - 1) + (\frac{M}{m}) \sum_{i=1}^{M} b_i(\frac{N_i}{n_0} - 1),$$

where $a \geq 0$ and $b_i \geq 0$ do not depend on m or n_0 and can be considered as constants for present purposes. Therefore,

$$Var(\bar{Y}_{ts}) = a(\frac{M}{m}) - a + \frac{M}{mn_0} \sum_{i=1}^{M} b_i N_i - (\frac{M}{m}) \sum_{i=1}^{M} b_i$$

$$= -a + (\frac{M}{m})(a - \sum_{i=1}^{M} b_i) + \frac{M}{mn_0} \sum_{i=1}^{M} b_i N_i.$$

Since $mn_0 = n =$ constant, the last term in the above expression is also a constant and can be ignored.

It should be clear, therefore, that if $(a - \sum_{i=1}^{M} b_i) > 0$, $Var(\bar{Y}_{ts})$ is minimized by making m as large as possible, that is, equal to M, in which case, $n_0 = n/M$. On the other hand, if $(a - \sum_{i=1}^{M} b_i) < 0$, $Var(\bar{Y}_{ts})$ is minimized by making m as small as possible, that is, with $m = 1$, in which case $n_0 = n$.

Note that setting $m = M$ means effectively taking a stratified sample, whereas $m = 1$ means selecting only one group in the first stage.

We conclude that, *if the total sample size is treated as a constant and if the samples from each selected group are to be of the same size, the optimal strategy is to either take a stratified sample with n/M elements sampled from each group, or to take a two-stage sample with only one group selected in the first stage.* The choice is determined by the sign of

$$a - \sum_{i=1}^{M} b_i = (\frac{M}{N})^2 \frac{\sigma_{01}^2}{M - 1} - \sum_{i=1}^{M} w_i^2 \frac{\sigma_i^2}{N_i - 1}.$$

This quantity, of course, depends on the particular variable that is being considered and its sign may be different for different variables. Therefore, even in this restricted case, the optimal strategy for estimating the population mean of one variable may be different from that for another.

Very similar conclusions apply when it is desired to make $Var(P_{ts})$ as small as possible under the same constraints.

A final comment on the comparison between stratified and two-stage sampling: It will be recalled from Section 2.10 that a stratified sample is indicated if the groups are approximately homogeneous with respect to the variable or category of interest. In Section 2.11, it was suggested that a two-stage sample is indicated in the event the groups are approximately identical—miniatures, we said, of the entire population. These guidelines apply for all types of samples, including random ones, and can be defended mathematically.

As a practical rule, *for estimating the population mean of a variable, the smaller the differences among the group means (μ_i) and the group variances (σ_i^2), the better is a two-stage sample in relation to a stratified sample of the same size; for estimating a population proportion, the smaller the differences among the group*

proportions (π_i), *the better is a two-stage sample relative to a stratified sample of the same size.* Conversely, *the greater the differences among the group means or proportions and the smaller the group variances, the better is a stratified sample in relation to a two-stage sample of the same size.*

If, as a result of having obtained an earlier, similar or pilot sample, estimates are available of the μ_i, σ_i^2, and π_i, then $Var(\bar{Y}_{ts})$ and $Var(P_{ts})$ may be estimated using Eqs. 5.2 and 5.4, and these, in turn, may be compared to the estimated variances under stratified or simple random sampling. The calculations and comparisons, of course, provide guidelines only, as they are based on estimates.

5.7 INTERVAL ESTIMATES

The manner in which confidence intervals are constructed should be clear by now. The general principles remain the same in two-stage sampling, but the formulas presented in Summary 5.2 *are* complicated. Mercifully, computers can be easily programmed to replace long and tedious calculations by hand.

■ EXAMPLE 5.2
(Continued)

Table 5.6 summarizes the available information after the sample was taken by CSB. It is in effect Table 5.1 with the addition of S_i^2, the group sample variances of the household expenditures on clothing, which were not needed when this example was first discussed. It will be recalled that CSB selected $m = 4$ enumeration areas (EAs) from the $M = 200$ into which the city is divided, and that there are $N = 60,000$ households in the city. Earlier, we had calculated $\bar{Y}_{ts} = 83.95$ and $P_{ts} = 0.851$. In this example, of course, m is not large, but we pretend it is strictly in order to illustrate the calculations.

TABLE 5.6 Interval Estimation, Example 5.1

EA	N_i	n_i	$\bar{Y}_i$	S_i^2	P_i	$N_i\bar{Y}_i$	N_iP_i
29	250	50	95.0	110	0.75	23,750	187.5
67	310	60	84.0	80	0.92	26,040	285.2
102	340	70	75.5	124	0.83	25,670	282.2
143	280	55	90.3	105	0.95	25,284	266.0
						100,744	1,020.9

SUMMARY 5.2   Two-stage sampling, confidence intervals

- When m, $M - m$, and all n_i and $N_i - n_i$ are large, an approximate $100(1 - \alpha)\%$ confidence interval for μ, the population mean of a variable Y, is

$$\bar{Y}_{ts} \pm Z_{\alpha/2}\sqrt{\widehat{Var(\bar{Y}_{ts})}}, \tag{5.7}$$

and that for π, the population proportion in a category, is

$$P_{ts} \pm Z_{\alpha/2}\sqrt{\widehat{Var(P_{ts})}}. \tag{5.8}$$

In Eq. 5.7, $\widehat{Var(\bar{Y}_{ts})}$ is an unbiased estimator of $Var(\bar{Y}_{ts})$:

$$\widehat{Var(\bar{Y}_{ts})} = \left(\frac{M}{N}\right)^2 \frac{S_{01}^2}{m-1}\frac{M-m}{M} + \frac{M}{m}\sum_{i=1}^{m} w_i^2 \frac{S_i^2}{n_i - 1}\frac{N_i - n_i}{N_i},$$

and

$$S_{01}^2 = \frac{1}{m}\sum_{i=1}^{m}(N_i\bar{Y}_i - \bar{Y}_0)^2,$$

where $\bar{Y}_0 = (\sum_{i=1}^{m} N_i\bar{Y}_i)/m$. In Eq. 5.8, $\widehat{Var(P_{ts})}$ is an unbiased estimator of $Var(P_{ts})$:

$$\widehat{Var(P_{ts})} = \left(\frac{M}{N}\right)^2 \frac{S_{02}^2}{m-1}\frac{M-m}{M} + \frac{M}{m}\sum_{i=1}^{m} w_i^2 \frac{P_i(1 - P_i)}{n_i - 1}\frac{N_i - n_i}{N_i}.$$

and

$$S_{02}^2 = \frac{1}{m}\sum_{i=1}^{m}(N_i P_i - P_0)^2,$$

where $P_0 = (\sum_{i=1}^{m} N_i P_i)/m$.
- $100(1 - \alpha)\%$ confidence intervals for the totals $\tau = N\mu$ and $\tau' = N\pi$ are calculated by multiplying the limits of Eqs. 5.7 and 5.8, respectively, by N.
- $Z_{\alpha/2}$ for selected $(1 - \alpha)$ are given in Table 3.6.

We calculate $\bar{Y}_0 = 100{,}744/4 = 25{,}186$ and $P_0 = 1{,}020.9/4 = 255.225$. We need also

$$S_{01}^2 = \frac{1}{4}[(23{,}750 - 25{,}186)^2 + \cdots + (25{,}284 - 25{,}186)^2] = 758{,}818,$$

and

$$S_{02}^2 = \frac{1}{4}[(187.5 - 255.225)^2 + \cdots + (266.0 - 255.225)^2] = 1{,}582.225.$$

Substituting in Eq. 5.2, we get

$$
\widehat{Var}(\bar{Y}_{ts}) = (\frac{200}{60,000})^2 \frac{758,818}{4-1} \frac{200-4}{200} + (\frac{200}{4})[(\frac{250}{60,000})^2 \frac{110}{50-1} \frac{250-50}{250} + \cdots
$$
$$
+ (\frac{280}{60,000})^2 \frac{105}{55-1} \frac{280-55}{280}]
$$
$$
= 2.7612.
$$

Substituting in Eq. 5.4, we get

$$
\widehat{Var}(P_{ts}) = (\frac{200}{60,000})^2 \frac{1,582.225}{4-1} \frac{200-4}{200}
$$
$$
+ (\frac{200}{4})[(\frac{250}{60,000})^2 \frac{(0.75)(1-0.75)}{50-1} \frac{250-50}{250} + \cdots
$$
$$
+ (\frac{280}{60,000})^2 \frac{(0.95)(1-0.95)}{55-1} \frac{280-55}{280}]
$$
$$
= 0.00575.
$$

A, say, 95% confidence interval for the average household expenditure on clothing in the city is

$$
83.95 \pm (1.96)\sqrt{2.7612},
$$

or from $80.69 to $87.21.

A 95% confidence interval for the total household expenditure on clothing in the city is calculated by multiplying the above limits by $N = 60,000$; this gives the interval from $4,841,400 to $5,232,600.

A 95% confidence interval for the proportion of households owning a tape recorder is

$$
0.851 \pm (1.96)\sqrt{0.00575},
$$

or from 70.24 to 99.96%. The 95% confidence interval for the number of households in the city owning a tape recorder is from 42,144 to 59,976. ◪

A special note should be made for the case of a systematic sample. If a systematic sample can be treated as a simple random sample, confidence intervals may be constructed using the simple estimators and the prescriptions of Summary 3.3. If not, Summary 5.2 may not be used on the grounds that a systematic sample is a cluster sample with $m = 1$. Not only is m not large, as required by Summary 5.2, but the first terms of $\widehat{Var}(\bar{Y}_{ts})$ and $\widehat{Var}(P_{ts})$ are undefined since their denominator is zero. If it is anticipated that confidence intervals will be required, the proper strategy is to plan at the design stage to select a large number of systematic samples, thereby ensuring that m and $M - m$ are large for the prescriptions of Summary 5.2 to apply.

5.8 COMBINED STRATIFIED/TWO-STAGE SAMPLING

Imagine a city divided into a number of districts. Each district contains a number of residences, and in each residence there live a number of persons. The situation is outlined in Figure 5.7.

A survey is planned to estimate, let us say, the average age of and the proportion of males among the city residents. The elements of this population are individuals, who are grouped into residences, which in turn are grouped into districts.

One possible sampling plan is to first select a simple random sample of districts, then a simple random sample of residences from each selected district, and, finally, a simple random sample of individuals from each selected residence. This type of sample is a *three-stage random sample*, which we examine briefly in Section 8.4. If, instead of selecting residences from a *sample* of districts, we select residences from *each* district, the sample is a *combined stratified/two-stage random sample*. It is on this sampling method that we shall concentrate in this section.

Let Y be a variable of interest, and μ its mean in the population; for example, the variable could be age and μ the average age of all individuals in the city. Let μ_i be the mean age, and $w_i = N_i/N$ the proportion of individuals in the ith district. There is a relationship between μ and the μ_i's (see Section 4.5); this is

$$\mu = w_1\mu_1 + w_2\mu_2 + \cdots + w_M\mu_M = \sum_1^M w_i\mu_i,$$

where M is the number of districts.

It can be shown that if $\bar{Y}'_1, \bar{Y}'_2, \ldots, \bar{Y}'_M$ are *any* unbiased estimators of $\mu_1, \mu_2, \ldots, \mu_M$, respectively, then

$$\bar{Y}' = w_1\bar{Y}'_1 + w_2\bar{Y}'_2 + \cdots + w_M\bar{Y}'_M = \sum_1^M w_i\bar{Y}'_i,$$

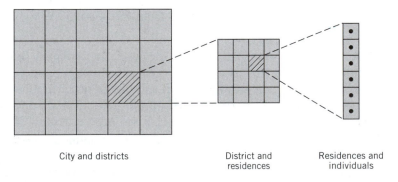

City and districts   District and residences   Residences and individuals

Figure 5.7 City, districts, residences and individuals

is an unbiased estimator of μ.

In particular, $\bar{Y}_i'$ could be the two-stage estimator of μ_i, which we know to be unbiased. (Think of each district as a separate population from which a two-stage sample of residences and individuals is selected.) An unbiased estimator of the population mean is then the weighted average of these two-stage estimators.

The same reasoning can show that an unbiased estimator of the proportion π of population elements in a category (for example, the proportion of males in the city) is

$$R' = w_1 R_1' + w_2 R_2' + \cdots + w_M R_M' = \sum_1^M w_i R_i',$$

where R_i' is the two-stage estimator of the proportion in the category in the ith district. The following simple example should explain the calculations.

◢ EXAMPLE 5.3   The city consists of $M = 2$ districts. The residences in each district and the individuals in each residence are as shown in cols. (1) to (3) of Table 5.7.

Suppose that the plan calls for selecting a simple random sample of 2 residences from each district, and 1 individual from each selected home with 1 or 2 residents or 2 individuals from each selected home with more than 2 residents. The actual selections are indicated with an asterisk in Table 5.7.

Suppose further that the selected individuals were asked for their age and gender, and that their responses were as shown in cols. (4) and (5) of Table 5.7. Columns (6) and (8) show the estimated average age and proportion of male individuals in each selected residence; these are, of course, the ordinary sample average and proportion, respectively.

Now, District 1 has a population of 9 individuals, grouped into 3 residences, from which 2 are selected at random. The two-stage estimate of the average age of all individuals in District 1 is

$$(\frac{3}{2})[(\frac{3}{9})(20) + (\frac{4}{9})(51)] = 44.$$

District 2 has a population of 10 individuals, grouped into 4 residences, from which 2 are selected at random. The two-stage estimate of the average age of all individuals in District 2 is

$$(\frac{4}{2})[(\frac{3}{10})(16) + (\frac{2}{10})(64)] = 35.2.$$

Therefore, the combined estimate of the average age of all individuals in the city is

$$(\frac{9}{19})(44) + (\frac{10}{19})(35.2) = 39.37.$$

TABLE 5.7 A Combined Stratified/Two-Stage Sample

District (1)	Residence (2)	Individual (3)	Age (4)	Gender[a] (5)	Estimated Average Age in Res. (6)	Dist. (7)	Estimated Proportion of Males in Res. (8)	Dist. (9)
1*	1*	1*	35	0	20	34.0	0.5	0.583
		2						
		3*	5	1				
	2	1						
		2						
	3*	1*	53	1	51		0.5	
		2*	49	0				
		3						
		4						
2*	1	1				25.6		0.700
	2*	1			16		0.5	
		2*	20	0				
		3*	12	1				
	3*	1			64		1.0	
		2*	64	1				
	4	1						
		2						
		3						
		4						

[a] 1 = Male, 0 = Female.
*Selected in sample.

We let the reader verify that the two-stage estimates of the proportion of males in Districts 1 and 2 are 0.583 and 0.700, respectively, and that the combined estimate of the proportion of all males in the city is 0.644. ◾

5.9 MULTISTAGE SAMPLING

Two-stage sampling can be extended to *multistage sampling*, that is, sampling in three, four, or more stages. We shall describe briefly, but will not study thoroughly, such complicated samples.

Many national samples, for example, are selected in four stages, which are roughly as follows.

(a) In the first stage, a sample of urban and rural areas (counties, metropolitan areas, etc.) is selected from a list of such areas covering the entire nation.

(b) Using city or area maps, each rural or urban area selected in the first stage is divided into a number of subareas: city blocks in cities, or subareas formed by highways, railway lines, rivers, and other easily identified landmarks in rural areas. From each area selected in the first stage, a sample of subareas is selected in the second stage of sampling.

(c) A list of all households is constructed for each subarea selected in the second stage, and a sample of households is selected from each such list.

(d) Interviewers visit the selected households and obtain from each a list of household members. Using these lists, a sample of individuals is drawn from each household selected in the third stage. The individuals thus selected in this fourth and final stage constitute the national sample.

Multistage sampling is sometimes combined with stratified sampling in a process that may be called *combination sampling*. For example, the urban and rural areas in the first stage of a national sample may be stratified into groups according to some criterion: by geographic location into provinces or states, by mode of life into rural and urban areas, by population into densely and sparsely populated areas, and so on. Instead of selecting a sample of areas from the entire list, a sample may be selected from *each* group separately. The remaining three stages of the national sample could be carried out as described earlier.

5.10 TO SUM UP

- A two-stage random sample is selected in two stages. In the first stage, a simple random sample of groups is drawn from the list of all groups. In the second stage, a simple random sample of elements is drawn from every group selected in the first stage.
- Two-stage is not the same as simple or stratified random sampling, and the simple and stratified estimators are not unbiased under two-stage sampling. The two-stage estimators defined in this chapter are unbiased and their variances can be determined.
- Two-stage random sampling requires lists of the groups and of the elements in each selected group. It is often cheaper to implement than a simple or stratified sample of the same size.
- Cluster sampling is a special case of two-stage sampling in which all the elements of each selected group are selected in the second stage. Cluster sampling is desirable when the group sizes and the group means or proportions are approximately equal.
- Another special case is systematic sampling, in effect cluster sampling, but often treated in practice as simple random sampling.
- Other things being equal, stratified sampling should be preferred to two-stage sampling when the population consists of approximately homogeneous groups with respect to the variables and categories of interest; two-stage should be the

preferred method when the population consists of groups that are similar to one another and to the population at large.

• Multistage sampling is sampling in three, four, or more stages. Combination sampling refers to mixtures of multistage and stratified sampling.

PROBLEMS

5.1 *Computing exercise:* Using the program SCALC, verify the numerical results presented in this chapter. Note any deviations, and determine if these are the result of rounding at intermediate stages.

5.2 For a survey of household expenditures in a certain city, a two-stage sample was used. The city was divided into 118 blocks, of which 5 were selected at random and without replacement in the first stage. From each selected block, 4 households were selected at random without replacement. An interviewer was sent to each selected household to inquire about a number of aspects of the financing of the household. A summary of some of the survey results is given in Table 5.8.

The latest census shows that there were 9,600 households in the city. Estimate the average grocery bill in the previous month of all the households in the city. Estimate the proportion of all households in the city that own a color TV set.

5.3 In order to estimate the condition of highways under its jurisdiction and the cost of urgent repairs, the state Department of Transportation selected a number of "highway miles" in two stages. In the first stage, a number of highways were selected at random and without replacement from the list of all highways maintained by the Department. In the second stage, a number of one-mile segments were selected at random and without replacement from the total length of each selected highway; for example, if the length of highway 101 is 73 miles, it is seen as consisting of 73 one-mile segments ("highway miles"), from which a number are

TABLE 5.8   Data for Problem 5.2

Block No.	Number of Households in Block	Average Grocery Bill in Previous Month of Selected Households ($)	Percentage of Selected Households Owning a Color TV Set
21	68	85	32
45	115	110	15
63	84	88	22
89	144	92	21
92	70	122	25

selected at random. Highway engineers then visit the selected segments, inspect the pavement condition, rate the condition of the segment, and estimate the cost of urgently needed repairs.

For the purpose of this exercise, assume there are 352 highways in the state, with a total length of 28,950 miles. A simple random sample of five highways was selected without replacement. From each selected highway, approximately 10% of its one-mile segments were then selected. The inspection results were as follows:

Highway No.	Length (miles)	Selected One-mile Segments	Number Rated Excellent	Cost of Urgent Repairs ($000)
155	85	10	1	80
489	120	10	0	100
283	47	5	0	50
698	98	10	1	90
311	34	5	1	30

For example, Highway 155 has a length of 85 miles. Ten of its 85 one-mile segments were selected and inspected. One of these segments was rated Excellent. The total cost of urgent repairs on the 10 selected segments was $80,000.

(a) Estimate the proportion and number of state highway miles that are in Excellent condition.

(b) Estimate the average cost per highway mile and the total cost of urgently needed repairs.

5.4 In the manner of Section 5.2, show that the two-stage estimator P_{ts} of a population proportion can be interpreted in intuitively appealing terms.

5.5 Suppose that the six firms in the town of Ackroyd are classified alphabetically into three groups: group 1 consists of firms A and B; group 2 of firms C and D; and group 3 of firms E and F. A two-stage sample of size 2 will be taken by selecting at random and without replacement *two* groups and then, also at random, *one* firm from each group selected in the first stage.

(a) Determine the probability distributions of $\bar{Y}_{ts}$ and P_{ts}.

(b) Calculate the expected values and variances of these estimators.

(c) Are $\bar{Y}_{ts}$ and P_{ts} unbiased in this case?

(d) Confirm that the variances of $\bar{Y}_{ts}$ and P_{ts} are given by Eqs. 5.2 and 5.4, respectively.

5.6 A box contains 10 items in two compartments, A and B. These items form the population of this exercise. They are inspected for quality and declared to be Good or Defective, as follows:

Compartment	Number of Good Items	Number of Defective Items	Total
A	3	1	4
B	1	5	6
Total	4	6	10

Obviously, the proportion of defective items (π) in the box is 0.6.

You plan to select a random sample of *two* items in two stages, as follows. First, you select at random one compartment; then, from the selected compartment, you select at random two items without replacement.

The two-stage estimator of π is $P_{ts} = 2wP$, where w is the proportion of items in the selected compartment (0.4 or 0.6), and P is the proportion of defectives in the sample of two items.

(a) With the help of a probability tree, determine the probability distribution of P_{ts}. Express probabilities to at least five decimal places.

(b) Show that P_{ts} is an unbiased estimator of π. (If you cannot show this, you have made a mistake in (a). Find it before proceeding further.)

(c) Calculate the variance of P_{ts}.

(d) For this particular population, is this two-stage sample better than a simple random sample of the same size?

5.7 Consider the Ackroyd population of six firms stratified according to the nature of their operations as shown in Table 4.6. Suppose that a two-stage sample will be selected by drawing at random one of the two groups (manufacturing and nonmanufacturing) and then, also at random, two firms from the selected group.

(a) Determine the probability distribution of the simple estimators $\bar{Y}$ (the average number of employees of the two selected firms) and P (the proportion of selected firms that intend to hire). Are these unbiased estimators of μ and π, respectively?

(b) Confirm the calculations shown in Table 5.5.

(c) Confirm that the variances of $\bar{Y}_{ts}$ and P_{ts} shown in Table 5.5 agree with the theoretical expressions given by Eqs. 5.2 and 5.4, respectively.

5.8 In order to determine the general pattern of railway freight movements and the changing trends of railway traffic in the country, the Transport Commission annually selects a "one percent continuous sample" of waybills. (*Webster's 10th Ed.* defines a waybill as a document prepared by the carrier of a shipment of goods and containing details of the shipment, route, and charges.) The participating railways are requested to forward copies of 1% of all waybills they issue. This 1% sample is drawn by selecting all waybills bearing serial numbers ending in "01."

Comment on this method of selection. Can this 1% sample be considered a simple random sample without replacement from the population of all waybills?

5.9 A population has 10,000 elements, which are divided into 50 groups according to a certain criterion. A two-stage random sample of four groups and fifty elements from each selected group was taken. The following table summarizes the sample results.

Group, i	N_i	n_i	$\bar{Y}_i$	S_i^2	P_i
1	150	50	142	540	0.42
2	250	50	154	482	0.38
3	200	50	130	623	0.36
4	180	50	165	508	0.32

$\bar{Y}_i$ is the sample mean and S_i^2 the sample variance of a variable Y, and P_i is the sample proportion in a category.

(a) Calculate the two-stage estimators $\bar{Y}_{ts}$ and P_{ts}.

(b) Calculate 95% confidence intervals for the population mean and total of Y, and the proportion and number of elements in the category in the population. Pretend that the number of selected groups is large.

5.10 A population has 50,000 elements, which are divided into 200 groups according to a certain criterion. A two-stage random sample of three groups and fifty elements from each selected group was taken. The following table summarizes the sample results.

Group, i	N_i	n_i	$\bar{Y}_i$	S_i^2	P_i
1	300	50	−10	240	0.025
2	250	50	−12	160	0.018
3	200	50	−15	180	0.036

$\bar{Y}_i$ is the sample mean and S_i^2 the sample variance of a variable Y, and P_i is the sample proportion in a category.

(a) Calculate the two-stage estimators $\bar{Y}_{ts}$ and P_{ts}.

(b) Calculate 95% confidence intervals for the population mean and total of Y, and the proportion and number of elements in the category in the population. Pretend that the number of selected groups is large.

5.11 Suppose the main purpose of a two-stage sample is to estimate the population proportion (π) in a category. Assume that the same number of elements will be selected from each group selected in the first stage (that is, all $n_i = n_0$) and that the total sample size (n) is given. In the manner of Section 5.6, develop a sampling strategy for minimizing $Var(P_{ts})$.

5.12 "Stratified sampling can be considered a special case of two-stage sampling." Comment.

5.13 Examine Table 5.3 and observe that for some sample outcomes the two-stage estimates do not make sense. For example, in four cases $\bar{Y}_{ts} > 9$, yet we know that in the Ackroyd population $1 \leq Y \leq 9$; also, for one outcome $P_{ts} > 1$, when obviously π cannot be greater than 1. Can such implausible estimates occur in other situations? Under other sampling methods? What, in your opinion, should be done about them?

5.14 Assume that the same number of elements will be selected from each group selected in the first stage of a two-stage sample (that is, all $n_i = n_0$). Assume further that the total sampling cost is given by

$$C = c_1 m + c_2 m n_0,$$

where c_1 is the cost per sampled group and c_2 the cost per sampled element. Suppose the main purpose of the sample is to estimate the population mean of a variable Y and that the sampling budget C is fixed.

(a) Show that $m = C/(c_1 + c_2 n_0)$.

(b) Show that the variance of the two-stage estimator $\bar{Y}_{ts}$ can be written as follows in the notation of Section 5.6:

$$Var(\bar{Y}_{ts}) = -a + \frac{M}{C}(a - \sum_{i=1}^{M} b_i)(c_1 + c_2 n_0) + \frac{M}{Cn_0}(\sum_{i=1}^{M} N_i b_i)(c_1 + c_2 n_0).$$

(c) Describe a procedure for determining the values of m and n_0 which minimize $Var(\bar{Y}_{ts})$.

(d) Modify **(a)** to **(c)** and describe a procedure for determining the values of m and n_0 which minimize $Var(P_{ts})$.

5.15 When the ratio $k = N/n$ is an integer in systematic sampling, the population elements can be thought of as divided into k groups of n elements each. A systematic sample, we noted in Section 5.5, is a cluster sample, in turn a special case of a two-stage sample with $M = k$, $m = 1$, and all $N_i = n_i = n$.

(a) Show that the two-stage estimator $(\bar{Y}_{ts})$ under systematic random sampling is the ordinary sample average $(\bar{Y})$.

(b) In view of **(a)**, explain why the ordinary sample average is an unbiased estimator of μ under systematic random sampling.

(c) Making use of Eq. 4.8, show that the variance of $\bar{Y}$ under systematic random sampling is

$$\underbrace{Var(\bar{Y})}_{system} = \frac{1}{k}\sum_{i=1}^{k}(\mu_i - \mu)^2 = \sigma^2 - \frac{1}{k}\sum_{i=1}^{k}\sigma_i^2.$$

(d) Show that under the assumptions of this problem the variance of $\bar{Y}$ under simple random sampling is

$$\underbrace{Var(\bar{Y})}_{simple} = \sigma^2 \frac{k-1}{kn-1}.$$

(e) Show that a systematic random sample is better than a simple random sample of the same size if

$$\sigma^2 < \frac{kn-1}{k^2(n-1)} \sum_{i=1}^{k} \sigma_i^2.$$

(f) Interpret the result in **(e)**.

5.16 (a) Determine the variance of the two-stage estimator $\bar{Y}_{ts}$ in the event only one group is selected with all its elements (that is, in the event $m = 1$ and all $n_i = N_i$).

(b) In Section 2.11, we argued that when each group is a "miniature" of the population with respect to a variable Y of interest, we should select just one group with all its elements, since the sample average is the correct estimate of the population mean μ of Y. This would appear to suggest that the variance in **(a)** ought to be zero ("correct $\bar{Y}_{ts}$" means $\bar{Y}_{ts} = \mu$ always, hence $Var(\bar{Y}_{ts}) = 0$). Is this impression justified? If not, how do you explain the apparent inconsistency?

5.17 (a) Determine the variance of the two-stage estimator P_{ts} in the event only one group is selected with all its elements (that is, in the event $m = 1$ and all $n_i = N_i$).

(b) In Section 2.11, we argued that when each group is a "miniature" of the population with respect to a category of interest, we should select just one group with all its elements, since the sample proportion is the correct estimate of the population proportion π in the category. This would appear to suggest that the variance in **(a)** ought to be zero ("correct P_{ts}" means $P_{ts} = \pi$ always, hence $Var(P_{ts}) = 0$). Is this impression justified? If not, how do you explain the apparent inconsistency?

5.18 In Section 2.10, we argued that in the event a population consists of homogeneous groups with respect to a variable Y of interest, it is sufficient to select just one element from each group, since the estimate of μ calculated along the lines of Table 2.6 is correct (that is, always equal to μ). This type of stratified sample, then, ought to be better than any type of two-stage sample in this situation.

(a) Explain why, when all groups are homogeneous with respect to a variable Y, all $\sigma_i^2 = 0$.

(b) In view of **(a)**, explain why $Var(\bar{Y}_s) = 0$ for any stratified random sample and why a stratified random sample with all $n_i = 1$ is best in this situation.

(c) Show that any stratified sample is better than a two-stage sample of the same size in this situation.

5.19 In Section 2.10, we argued that in the event a population consists of homogeneous groups with respect to a category of interest, it is sufficient to select just one element from each group, since the estimate of π calculated along the lines of Table 2.7 is correct (that is, always equal to π). This type of stratified sample, then, ought to be better than any type of two-stage sample in this situation.

(a) Explain why, when all groups are homogeneous with respect to a variable Y, all π_i are equal to either 1 or 0.

(b) In view of (a), explain why $Var(P_s) = 0$ for any stratified random sample and why a stratified random sample with all $n_i = 1$ is best in this situation.

(c) Show that any stratified sample is better than a two-stage sample of the same size in this situation.

5.20 Verify the combined estimate of the proportion of males in the city (0.644) stated in Example 5.3.

5.21 A survey was conducted in a town to estimate the number of townspeople who visited a physician and their expenditures on drugs in the month immediately preceding the survey.

The survey made use of an available list of all residential units in the town. A number of residential units were selected from this list at random and without replacement, and all the residents of the unit interviewed (in the case of infants and children, the questions were answered by an adult).

For the purpose of this exercise, assume that the town population is 1,248 and the number of residential units 643.

(a) Using a table of random numbers show how a simple random sample of five units and all its residents could be selected.

(b) Suppose that a simple random sample of five residences was selected and that interviews with all the residents in the selected units gave the following results:

Residence No.	Number of Residents	Number who Visited a Physician	Expenditures on Drugs in Residence ($)
45	3	0	0, 0, 65
131	2	1	400, 350
207	1	1	105
398	2	0	0, 0
519	2	1	50, 0

For example, unit no. 131 had two residents, one of whom had visited a physician; in the preceding month, the expenditure on drugs by the first resident was $400 and by the second $350.

Estimate the proportion and number of residents in the town who visited a physician in the preceding month. Estimate the average expenditure on drugs per person and the total expenditure on drugs in the town in the preceding month.

(c) Calculate 90% confidence intervals for all the population characteristics estimated in **(b)**. Pretend that the sample sizes are large enough.

(d) Is there an altogether different and possibly better method for estimating the population characteristics in **(b)**?

5.22 TORELEC is a company selling electrical equipment and supplies to electricians, contractors, and other wholesale customers. It carries a huge inventory of items, ranging from electrical motors to cables and connectors. The items are stored in stacks, which are like large bookcases except that they have bins (compartments, drawers, or dividers) instead of shelves. The arrangement is illustrated in Figure 5.8.

Each bin holds one item. The bins are adjustable, so that items of varying size can be accommodated.

The company uses a "grid" classification system to locate items. For example, when a customer requests a Honeywell D4 house thermostat, the clerk leafs through the product catalog to thermostats, and, within that, to manufacturer Honeywell, then down to model D4, which may have a storage code 107A2. The clerk then

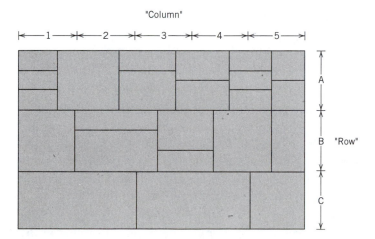

Figure 5.8   Stack layout, Problem 5.22

looks for the requested item in stack 107, approximately in "row" A and "column" 2 (see Figure 5.8).

The company's accounting system does show the quantity and value of each stored item (for example, the number and total value of Honeywell D4 thermostats), but mistakes do occur quite often. In addition, management is always concerned about theft and pilferage.

At irregular intervals, the warehouse manager arranges to have a number of items "audited," in the following manner. First, some stacks are selected at random. The bins in each selected stack are counted (remember, the number of bins varies from month to month), and some of them are selected at random. The actual quantity and value of the items in the selected bin are ascertained, and compared to the book quantity and value. Any discrepancy found is investigated carefully, but the information collected is also used to provide estimates of the total discrepancy for all items in the warehouse.

TORELEC's warehouse contains 500 stacks of the same shape and size, and about 15,000 bins.

(a) Using a table of random numbers, illustrate clearly how you would have selected at random five of these stacks. Having found that a stack has 30 bins, illustrate how you would have selected three of them at random. All selections are, of course, without replacement.

(b) Suppose that five stacks are selected at random. The relevant information is shown in the following table.

Selected Stack	Total Number of Bins	Number of Selected Bins and Items	Number of Items with a Discrepancy	Total Discrepancy ($)
051	30	3	3	142
105	20	2	0	0
252	25	2	2	−85
343	40	4	1	30
417	32	3	2	−210

Total discrepancy is the sum of the differences (in dollars) between actual and book values of the selected items. For example, the actual values of the three selected items in stack 051 differed from their book values by 85, −100, and 157 respectively; in this stack, all selected items were found to have a discrepancy.

Estimate the aggregate discrepancy (for all items in the warehouse). Briefly explain your calculations.

(c) Using the data in (b), estimate the total number of items in the warehouse with a discrepancy. If your procedure differs from that in (b), briefly explain the calculations and why your estimator is unbiased.

(d) Given TORELEC's concern with pilferage, do you think aggregate discrepancy should be measured as in **(b)**? If so, explain why. If not, explain clearly your recommendation and why it should be preferred.

(e) Suppose TORELEC stores 15,000 items, and a computer file of these items (with identification numbers, product description, and storage location) is available. Would there be any *statistical* advantage to selecting instead a *simple* random sample of the same size as the current sampling method?

(f) A consultant hired to review TORELEC's sampling method claimed that "the method has a fundamental logical—not statistical—flaw, requiring that it be replaced as soon as possible by a simple random sample." Do you agree? If so, name and explain clearly the *one* most important reason for this opinion.

CHAPTER 6

Ratio and Regression Estimators

6.1 INTRODUCTION AND SUMMARY

The auxiliary information about the population may include a known variable to which the variable of interest is approximately related. As observed in Chapter 2, it may be possible in such a situation to take advantage of this relationship and form more accurate estimates than when the relationship is ignored. In this chapter, we investigate some simple relationships and the manner in which they may be utilized.

We shall consider first the case where the variable of interest, Y, can be assumed to be approximately proportional to a known auxiliary variable; X, that is, $Y \approx bX$, where b is the constant of proportionality. For example, Y may be the sales of a given book title at a bookstore and X the size of that bookstore.

We shall show that there are at least three ways in which this constant of proportionality and the population mean or total of Y can be estimated on the basis of a sample. One of these is the widely used "ratio estimator." We show that the ratio estimator is not unbiased under simple random sampling, but becomes unbiased if the selection method is modified so that the first element is selected with a probability proportional to its X value.

The proportional relationship $Y \approx bX$ is a special case of the linear relationship $Y \approx a + bX$, the parameters of which can be estimated by the method of least squares. The resulting "regression estimator," like the ratio estimator, is not

unbiased under simple random sampling but becomes unbiased when the selection method is further modified.

These modifications of simple random sampling are not necessary when the sample size and the population remainder are large, because in such a case both the ratio and regression estimators are approximately unbiased; in addition, the variances of these estimators can be approximated by relatively simple expressions. These approximate expressions, in turn, will show that the regression estimator should nearly always be preferred to either the ratio or the simple estimator when the sample is simple random and the sample size and remainder are large.

This last conclusion has implications for stratified and two-stage sampling, suggesting better estimators than the standard stratified and two-stage estimators described in the preceding chapters.

Finally, in this chapter we shall describe how to calculate the sample size needed to meet the usual accuracy requirements and how to form confidence intervals using ratio and regression estimators with simple random sampling.

6.2 A SIMPLE LINEAR RELATIONSHIP

In Chapter 2, we examined some situations in which with each population element there is associated a known value of an auxiliary variable, exactly or approximately related to the variable of interest. We had concluded that in order to estimate the population mean or total of the variable of interest, it may be better to exploit the relationship than ignore it. We now intend to investigate further such relationships and the manner in which they may be utilized.

◪ EXAMPLE 6.1 Consider a market research company planning to estimate the total purchases by hospitals of each of about 3,200 pharmaceutical products. There are 1,158 hospitals in the country; a list of these hospitals is available. We concentrate on one pharmaceutical product, Product Y, and on the problem of estimating the total purchases by hospitals of this product over a given period of time (e.g., one month). Table 6.1 shows in outline what is known and not known about this population.

The number of beds in each hospital is known, as is the total number of beds in all hospitals (186,030).

It is reasonable to assume that there is a relationship between the purchases of Product Y by a given hospital and the number of beds in that hospital, as larger hospitals (those with more beds) will tend to use more of the product. It is not anticipated, of course, that this relationship is exact, but approximate, and perhaps of the form

$$Y_i \approx bX_i,$$

where b is some constant.

TABLE 6.1 Population of Hospitals, Example 6.1		
Hospital No., i	Number of Beds, X_i	Purchases of Product Y, ($000), Y_i
1	675	?
2	450	?
$\vdots$	$\vdots$	$\vdots$
$N = 1,158$	1,500	?
Total	186,030	?

If the value of b providing the best approximation to this relationship was known, the estimate of the purchases of Product Y at hospital i would be $\hat{Y}_i = bX_i$. The estimate (T_r) of the total purchases of Product Y by all $N = 1,158$ hospitals would be:

$$\begin{aligned} T_r &= \hat{Y}_1 + \hat{Y}_2 + \cdots + \hat{Y}_N \\ &= bX_1 + bX_2 + \cdots + bX_N \\ &= b(X_1 + X_2 + \cdots + X_N) \\ &= b\sum_{i=1}^{N} X_i, \end{aligned}$$

that is,

$$T_r = b \text{ (Total number of beds in all hospitals)}.$$

The estimate of the average purchases of Product Y per hospital would be:

$$\bar{Y}_r = \frac{1}{N}T_r = b\left(\frac{1}{N}\sum_{i=1}^{N} X_i\right),$$

that is,

$$\bar{Y}_r = b \text{ (Average number of beds per hospital)}.$$

The total number of beds and the average number of beds per hospital are known: they are 186,030 and 186,030/1,158 (or 160.648), respectively. The value of b is unknown, and must be estimated from a sample.

Suppose that a simple random sample of three hospitals is selected, and that this sample happens to consist of hospitals 1, 2, and 1,158. The information in col. (3) of Table 6.2 will be available after the selected hospitals are questioned concerning their purchases of Product Y. Column (2) shows the number of beds of

TABLE 6.2   Sample of Hospitals and Calculations

Sampled Hospital, i (1)	Number of Beds, X_i (2)	Purchases of Product Y, Y_i ($000) (3)	Y_i/X_i (4)	X_iY_i (5)	X_i^2 (6)
1	675	500	0.7407	337,500	455,625
2	450	350	0.7778	157,500	202,500
1,158	1,500	1,100	0.7333	1,650,000	2,250,000
	2,625	1,950	2.2518	2,145,000	2,908,125

the selected hospitals, known even before the sample is taken. Colunms (4) to (6) will be explained shortly.

The relationship $Y_i \approx bX_i$ implies $b \approx Y_i/X_i$ or, in words, that the ratio of purchases to number of beds is approximately constant. One reasonable estimate of b, therefore, is *the average of the ratios Y_i/X_i in the sample*. These ratios are shown in col. (4) of Table 6.2, and the estimate of b is:

$$b = \frac{1}{n} \sum_{i=1}^{n} \frac{Y_i}{X_i} = \frac{2.2518}{3} = 0.7506.$$

The estimate of the total hospital purchases of Product Y is

$$T_r = (0.7506)(186,030) = 139,634 \ (\$000),$$

and the estimate of the average purchases per hospital is

$$\bar{Y}_r = \frac{T_r}{N} = \frac{139,634}{1,158} = 120.582.$$

The relationship $Y_i \approx bX_i$ also implies $(\sum Y_i) \approx b(\sum X_i)$ and $b \approx (\sum Y_i)/(\sum X_i)$. Therefore, another reasonable estimate of b is *the ratio of the totals* or, what amounts to the same thing, *the ratio of the averages* of Y and X in the sample:

$$b = \frac{\sum Y_i}{\sum X_i} = \frac{n\bar{Y}}{n\bar{X}} = \frac{\bar{Y}}{\bar{X}} = \frac{1,950/3}{2,625/3} = \frac{650}{875} = 0.7428.$$

With this method, the estimate of the total hospital purchases is

$$T_r = (0.7428)(186,030) = 138,183 \ (\$000),$$

and that of the average purchases per hospital $\bar{Y}_r = 138,183/1,158 = 119.329.$

There is yet another reasonable method for estimating b. We could determine that value of b which minimizes the sum of the squared deviations between actual and estimated Y values in the sample:

$$\sum_{i=1}^{n}(Y_i - \hat{Y}_i)^2 = \sum_{i=1}^{n}(Y_i - bX_i)^2.$$

This "best-fitting" value of b is the ordinary *least squares estimate* of regression analysis (see Appendix A.7):

$$b = \frac{\sum_{i=1}^{n} X_i Y_i}{\sum_{i=1}^{n} X_i^2}.$$

The numerator and denominator of this last expression are calculated in cols. (5) and (6) of Table 6.2. The least squares estimate is given by:

$$b = \frac{2,145,000}{2,908,125} = 0.7376.$$

The estimate of the total hospital purchases of Product Y is now

$$T_r = (0.7376)(186,030) = 137,216\ (\$000),$$

and that of the average purchases per hospital is $\bar{Y}_r = 137,216/1,158 = 118.494$.

The three estimates of b are quite close and, as shown in Figure 6.1, the estimated relationships $Y \approx bX$ are practically indistinguishable from one another.

All three methods for estimating the constant b appear reasonable. Which one should be used? ◼

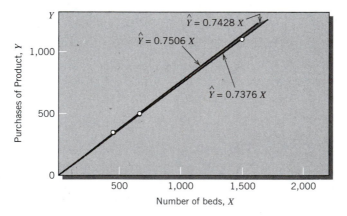

Figure 6.1 Data and estimates, Example 6.1

The second method has been extensively studied in the statistical literature, where it is referred to as *the* ratio estimation method, and it is on this method that we shall focus first.

6.3 RATIO ESTIMATORS AND MODIFIED SIMPLE RANDOM SAMPLING

In general, *the ratio estimator* of the population mean of a variable Y, utilizing a known auxiliary variable X, is

$$\bar{Y}_r = \frac{\bar{Y}}{\bar{X}}\mu_x,$$ (6.1)

where μ_x is the known population mean of X.[1] The ratio estimator of the population total of Y is $T_r = N\bar{Y}_r$. The subscript r in $\bar{Y}_r$ and T_r stands for "ratio."

Thus, in Example 6.1, the ratio estimate of the average purchases of Product Y by hospitals is $\bar{Y}_r = 119.329$, and that of the total hospital purchases is $T_r = 138,183$ ($000).

It can be shown that (1) *the ratio estimator is not unbiased under simple random sampling;* (2) *the ratio estimator is unbiased if* (a) *the first element is selected with probability proportional to its X-value, while* (b) *the other $n-1$ elements are selected from the remaining $N-1$ elements at random and without replacement in the usual manner.*

This last selection method is a form of *modified simple random sampling* due to D. B. Lahiri. It is our first encounter with a method of sampling under which elements are purposely selected with unequal probabilities. In this case, the purpose is to ensure the unbiasedness of the ratio estimator; other purposes are discussed in the next chapter.

Before we confirm these properties of the ratio estimator, let us explain how modified simple random sampling can be implemented.

■ EXAMPLE 6.2   Suppose it is desired to select a modified simple random sample of three hospitals from the population of hospitals outlined in Table 6.1.

The first hospital should be selected with probability proportional to the number of its beds. Hospital 1 has 625 beds; the probability of it being the first selected should be 625/186,030. Likewise, the probability of hospital 2 being the first selected hospital should be 450/186,030. And so on, down to 1,500/186,030 for the probability that hospital 1,158 will be the first selected hospital.

---

[1]It is now necessary to distinguish two population means: μ_y is that of the variable Y of interest and is unknown; μ_x is that of the auxiliary variable X and is assumed to be known.

Consider Table 6.3, in which 675 six-digit numbers are assigned to hospital 1, 450 to no. 2, down to 1,500 to no. 1,158.

Using Table 3.5 or a computer program, generate a six-digit random number. If the random number is 000000 or greater than 186030, it is ignored and another six-digit number generated. If the random number is between 000001 and 000675, inclusive, no. 1 is the first selected hospital. If it is between 000676 and 001125, no. 2 is the first selected hospital. And so on.

The assignments shown in Table 6.3 are not the only ones satisfying the selection requirements. *Any* 675 six-digit numbers between 000001 and 186030 can be assigned to hospital no. 1, any 450 of the remaining ones to no. 2, and so on. All such assignments make the probability of selecting a hospital proportional to the number of its beds, but the assignment based on the cumulative number of beds and illustrated in Table 6.3 is the simplest and easiest to use.

Having selected the first hospital with probability proportional to the number of its beds, the selection of the other two hospitals from the remaining 1,157 proceeds in the usual manner. Generate a four-digit random number; if it is 0000, greater than 1,158, *or the number of the first selected hospital*, discard it and generate another; otherwise, the four-digit random number identifies the second hospital to be selected. Another four-digit random number is then generated; if it is 0000, greater than 1,158, *or the number of a previously selected hospital*, it is discarded; otherwise, the four-digit random number identifies the third hospital to be selected.

For example, if the generated random numbers are 456213, 000795, 0002, 1158, and 0001, hospitals 2, 1,158, and 1 will be selected. ◼

6.4 PROPERTIES OF THE RATIO ESTIMATOR

We would like to demonstrate first that the ratio estimator, Eq. 6.1, is not unbiased under simple random sampling.

TABLE 6.3	Assignment of Random Numbers to Hospitals		
Hospital No., i	Number of Beds	Cumulative Number of Beds	Assigned Numbers
1	675	675	000001–000675
2	450	1,125	000676–001125
⋮	⋮	⋮	
$N = 1,158$	1,500	186,030	184531–186030
Total	186,030		

◪ EXAMPLE 6.3 Suppose that last year's sales of the six firms in the town of Ackroyd are as shown in the second column of Table 6.4.

Sales, X, are the known auxiliary variable of this illustration. The total sales of the six firms in Ackroyd are 45.

Let us consider what would happen if we were to select a simple random sample of two firms without replacement, and use Eq. 6.1 to estimate μ_y, the average number of employees per firm in Ackroyd.

To begin, suppose that the two selected firms happen to be firms A and B. Table 6.5 shows the information that will be available after these two firms are interviewed.

The sample average number of employees is 8.5. (This would have been the simple estimate of μ_y.) The ratio estimate of μ_y is

$$\bar{Y}_r = \frac{\bar{Y}}{\bar{X}}\mu_x = \frac{8.5}{12.5}7.5 = 5.1.$$

TABLE 6.4   Ackroyd Population and Its Characteristics

Firm	Annual Sales, X ($00,000)	Number of Employees, Y
A	13	9
B	12	8
C	9	6
D	3	2
E	1	1
F	7	5
	45	

Number of firms, $N = 6$
Average number of employees, $\mu_y = 31/6 \approx 5.167$
Average annual sales, $\mu_x = 45/6 = 7.5$

TABLE 6.5   Sample Consisting of Firms A and B

Firm	Sales, X	Number of Employees, Y
A	13	9
B	12	8
Sample totals:	25	17
Sample averages:	$\bar{X} = 12.5$	$\bar{Y} = 8.5$

		TABLE 6.6 Simple Random Sample of Size $n = 2$ without Replacement, Ratio Estimator					
Outcome (1)	Sales, X (2)	Number of Employees, Y (3)	Probability (4)	$\bar{X}$ (5)	$\bar{Y}$ (6)	$\bar{Y}_r = (\bar{Y}/\bar{X})(7.5)$ (7)	
A,B	13,12	9,8	1/15	12.5	8.5	5.100	
A,C	13,9	9,6	1/15	11.0	7.5	5.114	
A,D	13,3	9,2	1/15	8.0	5.5	5.156	
A,E	13,1	9,1	1/15	7.0	5.0	5.357	
A,F	13,7	9,5	1/15	10.0	7.0	5.250	
B,C	12,9	8,6	1/15	10.5	7.0	5.000	
B,D	12,3	8,2	1/15	7.5	5.0	5.000	
B,E	12,1	8,1	1/15	6.5	4.5	5.192	
B,F	12,7	8,5	1/15	9.5	6.5	5.132	
C,D	9,3	6,2	1/15	6.0	4.0	5.000	
C,E	9,1	6,1	1/15	5.0	3.5	5.250	
C,F	9,7	6,5	1/15	8.0	5.5	5.156	
D,E	3,1	2,1	1/15	2.0	1.5	5.625	
D,F	3,7	2,5	1/15	5.0	3.5	5.250	
E,F	1,7	1,5	1/15	4.0	3.0	5.625	

The possible sample outcomes, their probabilities, and the corresponding values of the ratio estimator are shown in Table 6.6.

Columns (1), (3), (4), and (6) of Table 6.6 are identical to those in Table 3.2, since, in both instances, the sample selection method is the same. Column (2) shows the sales, and col. (5) the average sales ($\bar{X}$) of the two selected firms. The calculation of the ratio estimate, $\bar{Y}_r$, in col. (7) was illustrated in Table 6.5 for the case where the sample consists of firms A and B (the first line in Table 6.6).

The expected value of $\bar{Y}_r$ is

$$E(\bar{Y}_r) = (5.100)(1/15) + \cdots + (5.625)(1/15) = 5.214.$$

The population average number of employees is 5.167. We conclude that the ratio estimator is not unbiased. ◪

Let us now confirm that the ratio estimator is unbiased under modified simple random sampling. We continue to use the Ackroyd population for our illustration.

◪ EXAMPLE 6.3
(Continued)

We examine what would happen if a random sample of two firms is selected from the Ackroyd population according to the modified selection method described in the last section. Figure 6.2 outlines the possible sample outcomes and their probabilities.

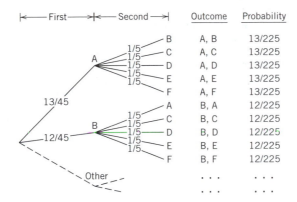

First	Second	Outcome	Probability

Figure 6.2 Probability tree, modified simple random sampling

There are altogether 30 possible outcomes, but, as noted in Section 3.4, the order in which the firms appear in the sample does not affect the calculation of the ratio estimates; thus, (A,B) can be combined with (B,A), (A,C) with (C,A), and so on. The probability that the sample will consist of, say, firms A and B, regardless of order, is $(13/225) + (12/225) = 25/225$, or 0.111; and so on. The combined sample outcomes and their probabilities are shown in cols. (1) and (2) of Table 6.7.

For each sample outcome, the ratio estimator is calculated exactly as in Table 6.6; col. (3) of Table 6.7 is col. (7) of Table 6.6.

The expected value of the ratio estimator is

$$E(\bar{Y}_r) = (5.100)(0.111) + \cdots + (5.625)(0.036) = 5.167.$$

Since $E(\bar{Y}_r)$ equals the population average number of employees, $\mu_y = 5.167$, we conclude that $\bar{Y}_r$ is unbiased. ◼

6.5 THE REGRESSION ESTIMATOR

There are situations in which the variable Y of interest is approximately linearly related to a known auxiliary variable X, but Y does not tend to zero when X is zero, as implied by the simple linear model of the preceding sections. In such cases, it may be reasonable to assume that

$$Y_i \approx a + bX_i. \tag{6.2}$$

If the values of a and b providing the best approximation to this relationship were known, an estimate of the Y value of every population element would be:

$$\hat{Y}_i = a + bX_i. \tag{6.3}$$

TABLE 6.7 Modified Random Sample of Size n = 2 without Replacement, Ratio Estimator

Outcome (1)	Probability (2)	$\bar{Y}_r = (\bar{Y}/\bar{X})(7.5)$ (3)
A,B	25/225=0.111	5.100
A,C	22/225=0.098	5.114
A,D	16/225=0.071	5.156
A,E	14/225=0.062	5.357
A,F	20/225=0.089	5.250
B,C	21/225=0.093	5.000
B,D	15/225=0.067	5.000
B,E	13/225=0.058	5.192
B,F	19/225=0.084	5.132
C,D	12/225=0.053	5.000
C,E	10/225=0.044	5.250
C,F	16/225=0.071	5.156
D,E	4/225=0.018	5.625
D,F	10/225=0.044	5.250
E,F	8/225=0.036	5.625
	225/225=1.000	

The estimated population total would be

$$T_{lr} = \sum_{i=1}^{N} \hat{Y}_i = \sum_{i=1}^{N}(a + bX_i) = Na + b\sum_{i=1}^{N} X_i = Na + bN\mu_x, \qquad (6.4)$$

since, by definition, $\mu_x = (\sum_{i=1}^{N} X_i)/N$. The estimate of the population mean of the variable Y would be

$$\bar{Y}_{lr} = \frac{T}{N} = a + b\mu_x.$$

Since μ_x, the population mean of the auxiliary variable X, is known, the main problem is how to estimate a and b on the basis of n pairs of sample observations (X_i, Y_i). There are not as many choices as in Section 6.2 for the simple linear relationship, but a very reasonable method is that of least squares. The least squares estimates are the values of a and b which minimize the sum of squared deviations between actual and estimated Y values in the sample

$$\sum_{i=1}^{n}(Y_i - \hat{Y}_i)^2 = \sum_{i=1}^{n}(Y_i - a - bX_i)^2.$$

The objective, in other words, is to find the best-fitting values of a and b, with least squares serving as the criterion of fit. It is well-known from regression theory that these *least squares estimates* are given by

$$b = \frac{\sum_{i=1}^{n}(X_i - \bar{X})(Y_i - \bar{Y})}{\sum_{i=1}^{n}(X_i - \bar{X})^2} = \frac{\sum_{i=1}^{n} X_i Y_i - n\bar{X}\bar{Y}}{\sum_{i=1}^{n} X_i^2 - n\bar{X}^2}, \tag{6.5}$$

and

$$a = \bar{Y} - b\bar{X}. \tag{6.6}$$

Substituting Eq. 6.6 into Eq. 6.4, we arrive at *the regression estimator of the population total of Y*:

$$T_{lr} = N(\bar{Y} - b\bar{X}) + bN\mu_x = N\left[\bar{Y} + b(\mu_x - \bar{X})\right], \tag{6.7}$$

where, of course, b is calculated according to Eq. 6.5. Dividing Eq. 6.7 by N, we obtain the *regression estimator of the population mean of Y*,

$$\bar{Y}_{lr} = \bar{Y} + b(\mu_x - \bar{X}). \tag{6.8}$$

The subscript lr in T_{lr} and $\bar{Y}_{lr}$ stands for "linear regression."

■ EXAMPLE 6.4   Consider again the sample of three hospitals and the data of Table 6.2. Table 6.2 shows that $\bar{X} = 2{,}625/3 = 875$ and $\bar{Y} = 1{,}950/3 = 650$. Making use of the preliminary calculations of Table 6.2 we find

$$b = \frac{\sum_{i=1}^{n} X_i Y_i - n\bar{X}\bar{Y}}{\sum_{i=1}^{n} X_i^2 - n\bar{X}^2} = \frac{(2{,}145{,}000) - (3)(875)(650)}{(2{,}908{,}125) - (3)(875)^2} = 0.7178.$$

and

$$a = \bar{Y} - b\bar{X} = (650) - (0.7178)(875) = 21.932.$$

The least squares estimates can also be calculated using widely available computer programs for regression. The results of regressing Y against X using one such program and the data in Table 6.2 are shown in Figure 6.3.

Of interest are the numbers in the column labeled ESTIMATE, which agree with the manual calculations. The other items of the computer output need not concern us at this stage.

The estimated relationship between purchases and number of beds is given by

$$\hat{Y} = (21.932) + (0.7178)X,$$

and is plotted in Figure 6.4.

```
Dependent Variable: Y

                             STANDARD
    VARIABLE     ESTIMATE      ERROR      T-RATIO     P-VALUE

    INTERCEPT     21.9325     10.4621       2.096      0.2834
          X        0.7178      0.0106      67.550      0.0094

    S= 8.3077          R-SQUARED= 0.9998
```

Figure 6.3   Computer output, Example 6.4

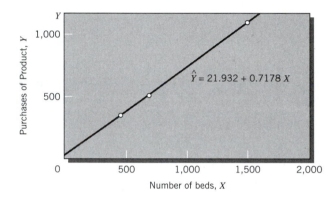

Figure 6.4   Estimated linear relationship, Example 6.4

There are $N = 1,158$ hospitals in the population, with a total of 186,030 beds. The average number of beds per hospital is $\mu_x = 186,030/1,158$ or about 160.65. The regression estimate of the total hospital purchases of Product Y is

$$T_{lr} = N[\bar{Y} + b(\mu_x - \bar{X})] = (1,158)[650 + (0.7178)(160.65 - 875)]$$
$$= (1,158)(137.245)$$
$$= 158,929 \ (\$000).$$

The regression estimate of the average purchases per hospital is $\bar{Y}_{lr} = T_{lr}/N = 158,929/1,158 = 137.245 \ (\$000)$.   ◼

6.6 PROPERTIES OF THE REGRESSION ESTIMATOR

The regression estimator has properties similar to those of the ratio estimator. It can be shown that (1) *the regression estimator is not unbiased under simple random sampling;* (2) *the regression estimator is unbiased if* (a) *the first two elements are*

selected with probability proportional to the square of the difference between their X values, and (b) *the other n − 2 elements are selected from the remaining N − 2 elements at random and without replacement in the usual manner.*

This selection procedure is another form of *modified simple random sampling*, similar but not identical to that for the ratio estimator.

To explain and confirm the properties of the regression estimator, we turn, one more time, to the Ackroyd population.

◪ EXAMPLE 6.5 Let us suppose that a modified random sample of $n = 2$ firms will be selected from the Ackroyd population described in Table 6.3. The first two firms will be selected with probability proportional to the squared difference of their annual sales, X. That is, the probability that firms i and j will be the first two selected should be proportional to $(X_i - X_j)^2$. In this simple illustration, of course, the first two selected firms will form the entire sample. Had the sample size been larger, the remaining firms in the sample would have been selected at random and without replacement from the remainder of the population.

The first three columns of Table 6.8 are identical to those of Table 6.6 and show the possible pairs of firms and the associated X and Y values.

TABLE 6.8 Modified Simple Random Sample of Size $n = 2$ without Replacement, Regression Estimator

Outcome (1)	Sales, X (2)	Number of Employees, Y (3)	$(X_i - X_j)^2$ (4)	Probability (5)	b (6)	$\bar{Y}_{lr}$ (7)
A,B	13,12	9,8	1	0.001443	1.00000	3.50000
A,C	13,9	9,6	16	0.023088	0.75000	4.87500
A,D	13,3	9,2	100	0.144300	0.70000	5.15000
A,E	13,1	9,1	144	0.207792	0.66667	5.33333
A,F	13,7	9,5	36	0.051948	0.66667	5.33333
B,C	12,9	8,6	9	0.012987	0.66667	5.00000
B,D	12,3	8,2	81	0.116883	0.66667	5.00000
B,E	12,1	8,1	121	0.174603	0.63636	5.13636
B,F	12,7	8,5	25	0.036075	0.60000	5.30000
C,D	9,3	6,2	36	0.051948	0.66667	5.00000
C,E	9,1	6,1	64	0.092352	0.62500	5.06250
C,F	9,7	6,5	4	0.005772	0.50000	5.25000
D,E	3,1	2,1	4	0.005772	0.50000	4.25000
D,F	3,7	2,5	16	0.023088	0.75000	5.37500
E,F	1,7	1,5	36	0.051948	0.66667	5.33333
			693	1		

For each pair of firms in col. (1), col. (4) shows the square of the difference of their X values, and col. (5) the desired probability of selection. For example, for the pair of firms (A,B), the quantity $(X_i - X_j)^2$ is $(13 - 12)^2$, or 1. The sum of the figures in col. (4) is 693. Therefore, the pair (A,B) should be selected with probability 1/693 or 0.001443. The remaining entries in cols. (4) and (5) are similarly calculated.

Column (6) shows the least squares estimate b, and col. (7) the regression estimate of the population mean number of employees, μ_y. The calculations are long and tedious by hand, but very easy with the help of a computer program.

Consider, for example, the first line of Table 6.8. If the sample consists of firms A and B, then

$$b = \frac{\sum_{i=1}^{n} X_i Y_i - n \bar{X} \bar{Y}}{\sum_{i=1}^{n} X_i^2 - n \bar{X}^2} = \frac{(13)(9) + (12)(8) - (2)(12.5)(8.5)}{(13)^2 + (12)^2 - (2)(12.5)^2} = \frac{0.5}{0.5} = 1,$$

as shown in col. (6), and the regression estimate is

$$\bar{Y}_{lr} = \bar{Y} + b(\mu_x - \bar{X}) = (8.5) + (1)(7.5 - 12.5) = 3.5,$$

as shown in col. (7).

For this modified random sample, therefore, the possible regression estimates are listed in col. (7) and their probabilities in col. (5). The expected value of $\bar{Y}_{lr}$ is

$$E(\bar{Y}_{lr}) = (3.5)(0.001443) + \cdots + (5.3333)(0.05195) = 5.167,$$

which is the value of μ_y for this case. We have confirmed that $\bar{Y}_{lr}$ is an unbiased estimator of μ_y.

To implement this modified selection procedure, begin by assigning five-digit numbers to each possible pair of firms using the cumulative probabilities as a guide. Table 6.9 shows a part of this assignment.

It can be observed that 144 five-digit numbers are assigned to the pair (A,B), 2,309 to the pair (A,C), and so on. Using a table or computer program, a five-digit random number will be generated, and the pair corresponding to that number

TABLE 6.9 **Assignment of Random Numbers, Example 6.5**

Pair	Desired Probability	Cumulative Probability	Assigned Numbers
A,B	0.00144	0.00144	00001–00144
A,C	0.02309	0.02453	00145–02453
A,D	0.14430	0.16883	02454–16883
⋮	⋮	⋮	⋮

selected. For example, if the generated five-digit random number is 01852, firms A and C will be the first two selected.

To demonstrate that the regression estimator is not unbiased under simple random sampling, we need only combine its possible values in col. (7) of Table 6.8 and the probabilities in col. (4) of Table 6.6, to calculate

$$E(\bar{Y}_{lr}) = (3.5)(1/15) + \cdots + (5.33333)(1/15) = 4.993.$$

We know that the population average number of employees is 5.167. Therefore, the regression estimator is not unbiased under simple random sampling. ◪

6.7 RATIO AND REGRESSION ESTIMATORS IN LARGE SAMPLES

We have shown that the ratio and regression estimators are not unbiased under simple random sampling. The variances of these estimators, whether under simple or modified simple random sampling, are very complicated and will not be written here. However, when the sample size (n) and the population remainder ($N - n$) are large, it can be shown that

- the ratio and regression estimators are approximately unbiased under simple random sampling
- the variances of the ratio and regression estimators can be approximated by simple expressions
- the regression estimator is better than either the ratio or the simple estimator.

Put simply, *if the planned sample is large enough, it is not necessary to apply the modified selection procedures to ensure unbiasedness; simple random sampling is sufficient and regression estimates are best.*

The proof of the first two results can be found in mathematically advanced texts. These proofs provide no guidance as to how large n and $N - n$ must be for the approximations to be satisfactory. The consensus appears to require n and $N - n$ to be greater than 50. This rule of thumb is comfortably satisfied in many practical situations. The proof of the third result is quite easy, as will soon be seen.

We begin with the ratio estimator. Its large-sample approximate variance is given by Eqs. 6.10 and 6.11 in Summary 6.1.

◪ EXAMPLE 6.6   We use the Ackroyd data in Table 6.3 to clarify the calculation of $Var(\bar{Y}_r)$. The calculations are for illustration only, since a sample of size 2 is much too small for a good approximation.

We have already calculated $\mu_y = 31/6$, $\mu_x = 45/6$, and $\sigma_y^2 = 8.4722$ (see Table 3.1). Therefore, $\mu_r = \mu_y/\mu_x = 31/45 = 0.689$.

SUMMARY 6.1   Properties of the ratio estimator

- The *ratio estimator* of the population mean of a variable Y is

$$\bar{Y}_r = \frac{\bar{Y}}{\bar{X}}\mu_x, \tag{6.9}$$

where μ_x is the known population mean of an auxiliary variable X.
- The ratio estimator is unbiased under the modified simple random sampling described in Section 6.3.
- The ratio estimator is approximately unbiased under simple random sampling when n and $N - n$ are large.
- For large n and $N - n$, under either simple or modified simple random sampling, the approximate variance of $\bar{Y}_r$ is

$$Var(\bar{Y}_r) \approx \frac{1}{n}\frac{N-n}{N-1}\sigma_r^2, \tag{6.10}$$

where

$$\sigma_r^2 = \sigma_y^2 + \mu_r^2\sigma_x^2 - 2\mu_r\rho_{xy}\sigma_x\sigma_y. \tag{6.11}$$

In Eq. 6.11, $\mu_r = \mu_y/\mu_x$ is the ratio of the population means, σ_x^2 and σ_y^2 are the population variances, σ_x and σ_y the standard deviations, and ρ_{xy} is the population correlation coefficient of X and Y.

To calculate σ_r^2, we need the population variance of X, which is

$$\sigma_x^2 = \frac{1}{N}\sum_{i=1}^{N} X_i^2 - \mu_x^2$$
$$= \frac{1}{6}[13^2 + \cdots + 7^2] - (45/6)^2$$
$$= 19.25.$$

The population correlation coefficient of X and Y, by definition (see Appendix A.6), is

$$\rho_{xy} = \frac{\sigma_{xy}}{\sigma_x\sigma_y},$$

where σ_{xy} is the population covariance of X and Y. Therefore, Eq. 6.11 can be written more simply as $\sigma_r^2 = \sigma_y^2 + \mu_r^2\sigma_x^2 - 2\mu_r\sigma_{xy}$.

The population covariance of X and Y (see Appendix A.6) is

$$\sigma_{xy} = \frac{1}{N} \sum_{i=1}^{N} X_i Y_i - \mu_x \mu_y$$

$$= \frac{1}{6}\left[(13)(9) + \cdots + (7)(5)\right] - (31/6)(45/6)$$

$$= 12.75.$$

It follows that

$$\sigma_r^2 = \sigma_y^2 + \mu_r^2 \sigma_x^2 - 2\mu_r \sigma_{xy}$$

$$= 8.4722 + \left(\frac{31}{45}\right)^2(19.25) - (2)\left(\frac{31}{45}\right)(12.75)$$

$$= 0.040988.$$

Thus, with $n = 2$, Eq. 6.10 yields

$$Var(\bar{Y}_r) \approx \frac{1}{n}\frac{N-n}{N-1}\sigma_r^2$$

$$= \frac{1}{2}\frac{6-2}{6-1}(0.040988)$$

$$= 0.0164. \quad \blacksquare$$

Let us see what can be learned from the approximate variance of $\bar{Y}_r$. Under simple random sampling, the ordinary sample average is an unbiased estimator of the population mean of the variable Y. The ratio estimator is also unbiased under modified simple random sampling, and approximately unbiased under simple random sampling when the sample size and the population remainder are large. Of two unbiased estimators, the one with the smaller variance is preferable. The ratio estimator is preferable if

$$Var(\bar{Y}_r) < Var(\bar{Y}).$$

For large n and $N - n$, this means

$$\frac{1}{n}\frac{N-n}{N-1}[\sigma_y^2 + \mu_r^2 \sigma_x^2 - 2\mu_r \rho_{xy}\sigma_x\sigma_y] < \frac{1}{n}\frac{N-n}{N-1}\sigma_y^2,$$

or

$$\mu_r^2 \sigma_x^2 - 2\mu_r \rho_{xy}\sigma_x\sigma_y < 0.$$

The ratio estimator is not preferable when $\mu_r < 0$ and $\rho_{xy} > 0$ or $\mu_r > 0$ and $\rho_{xy} < 0$, because then the left-hand side of the above expression is positive. If

$\mu_r > 0$, as is frequently the case in practice, the ratio estimator is preferable to the simple estimator if

$$\rho_{xy} > \frac{1}{2}\frac{(\sigma_x/\mu_x)}{(\sigma_y/\mu_y)}.$$

The ratio of the standard deviation (σ) to the mean (μ) is called the *coefficient of variation* of the variable. Therefore, the ratio is preferable to the simple estimator if the population correlation coefficient of X and Y is greater than half the ratio of the coefficients of variation of these two variables.

◢ EXAMPLE 6.7   For the Ackroyd case we have $\rho_{xy} = \sigma_{xy}/(\sigma_x\sigma_y) = 0.998$, $\sigma_x/\mu_x = \sqrt{19.25}/7.5 = 0.585$, $\sigma_y/\mu_y = \sqrt{8.472}/5.167 = 0.563$. Therefore,

$$\frac{1}{2}\frac{(\sigma_x/\mu_x)}{(\sigma_y/\mu_y)} = \frac{1}{2}\frac{0.585}{0.563} = 0.519.$$

Since $\mu_r > 0$ and $\rho_{xy} > 0.519$, the ratio estimator is preferable. Of course, in this case this result has no practical significance because the population size is small and does not permit a large sample without replacement.   ◢

In general, to determine which of the two estimators (simple, ratio) is preferable, we need estimates of the correlation coefficient and the coefficients of variation, perhaps on the basis of a pilot or similar earlier sample. Evidently, when there are many variables of interest and several auxiliary variables, the best method for estimating the characteristics of one variable need not be best for estimating those of another. As we shall show below, however, the regression estimator *is* better than either the ratio or the simple estimator for estimating the population mean or total of *any* variable of interest when the sample size and population remainder are large.

The large-sample variance of the regression estimator is given by Eqs. 6.14 and 6.15 in Summary 6.2.

◢ EXAMPLE 6.7
(Continued)   We clarify the calculation of the approximate variance of the regression estimator using the data for the Ackroyd population. Again, these calculations are for illustration only, as neither the sample size ($n = 2$) nor the population remainder ($N - n = 4$) is large.

We established earlier that $\sigma_y^2 = 8.472$ and $\rho_{xy} = 0.998$. According to 6.14 and 6.15,

$$Var(\bar{Y}_{lr}) \approx \frac{1}{n}\frac{N-n}{N-1}\sigma_y^2(1 - \rho_{xy}^2) = \frac{1}{2}\frac{6-2}{6-1}(8.472)[1 - (0.998)^2] = 0.0135. \quad ◢$$

SUMMARY 6.2   Properties of the regression estimator

- The *regression estimator* of the population mean of a variable Y is

$$\bar{Y}_{lr} = \bar{Y} + b(\mu_x - \bar{X}),\qquad(6.12)$$

where μ_x is the known population mean of an auxiliary variable X, and

$$b = \frac{\sum_{i=1}^{n}(X_i - \bar{X})(Y_i - \bar{Y})}{\sum_{i=1}^{n}(X_i - \bar{X})^2}$$

$$= \frac{\sum_{i=1}^{n} X_i Y_i - n\bar{X}\bar{Y}}{\sum_{i=1}^{n} X_i^2 - n\bar{X}^2}.\qquad(6.13)$$

- The regression estimator is unbiased under the modified simple random sampling described in Section 6.5.
- The regression estimator is approximately unbiased under simple random sampling when n and $N - n$ is large.
- For large n and $N - n$, under either simple or modified simple random sampling, the approximate variance of $\bar{Y}_{lr}$ is

$$Var(\bar{Y}_{lr}) \approx \frac{1}{n}\frac{N-n}{N-1}\sigma_{lr}^2,\qquad(6.14)$$

where

$$\sigma_{lr}^2 = \sigma_y^2(1 - \rho_{xy}^2).\qquad(6.15)$$

In Eq. 6.15, σ_y^2 is the population variance of Y, and ρ_{xy} the population correlation coefficient of X and Y.

The simple estimator (the ordinary sample average of the variable of interest) is unbiased under simple random sampling. The ratio and regression estimators are approximately unbiased for large simple random samples. Among unbiased estimators, best is the one with the smallest variance.

When n and $N - n$ are large, the variances of the three estimators are approximately as follows:

$$\text{Simple:}\quad Var(\bar{Y}) = \frac{1}{n}\frac{N-n}{N-1}\sigma_y^2$$

$$\text{Ratio:}\quad Var(\bar{Y}_r) = \frac{1}{n}\frac{N-n}{N-1}[\sigma_y^2 + \mu_r^2\sigma_x^2 - 2\mu_r\rho_{xy}\sigma_x\sigma_y]$$

$$\text{Regression:}\quad Var(\bar{Y}_{lr}) = \frac{1}{n}\frac{N-n}{N-1}[1 - \rho_{xy}^2]\sigma_y^2$$

Therefore, for large n and $N - n$,

$$Var(\bar{Y}) - Var(\bar{Y}_{lr}) = \frac{1}{n}\frac{N-n}{N-1}[\sigma_y^2 - (\sigma_y^2 - \rho_{xy}^2\sigma_y^2)]$$

$$= \frac{1}{n}\frac{N-n}{N-1}\rho_{xy}^2\sigma_y^2 \geq 0,$$

because both ρ_{xy}^2 and σ_y^2 are nonnegative. Therefore, $Var(\bar{Y}) \geq Var(\bar{Y}_{lr})$. The greater the correlation coefficient in absolute value, the greater the difference of these variances. The two estimators have equal variances if the correlation coefficient of X and Y is exactly zero. Since this is unlikely to happen in practice, we conclude that the regression estimator is better than the simple estimator.

Let us now compare the variances of the ratio and regression estimators. Again for large n and $N - n$,

$$Var(\bar{Y}_r) - Var(\bar{Y}_{lr}) = \frac{1}{n}\frac{N-n}{N-1}[(\sigma_y^2 + \mu_r^2\sigma_x^2 - 2\mu_r\rho_{xy}\sigma_x\sigma_y) - (\sigma_y^2 - \rho_{xy}^2\sigma_y^2)]$$

$$= \frac{1}{n}\frac{N-n}{N-1}[\mu_r^2\sigma_x^2 - 2\mu_r\rho_{xy}\sigma_x\sigma_y + \rho_{xy}^2\sigma_y^2]$$

$$= \frac{1}{n}\frac{N-n}{N-1}[\mu_r\sigma_x - \rho_{xy}\sigma_y]^2 \geq 0.$$

Therefore, the regression estimator is better than the ratio estimator, unless the quantity in brackets happens to be equal to zero, in which case the two estimators have the same variance.

This exception is also unlikely to hold in practice. We conclude that, when the sample is simple random and the sample size and population remainder are large, the regression estimator is better than either the ratio or the simple estimator.

6.8 SAMPLE SIZE

The large-sample variance of the regression estimator, Eq. 6.14, is very similar in form to that of the simple estimator, $\bar{Y}$. It is not surprising, therefore, that the prescription for determining the sample size shown in Summary 6.3 is also similar.

As for earlier sampling methods, c must be small enough and $1 - \alpha$ large enough so that n calculated from Eq. 6.16 and $N - n$ are large—greater than 50, according to the prevailing rule of thumb. For estimating $\tau_y = N\mu_y$ within $\pm c'$, apply Eq. 6.16 with $c = c'/N$. In either case, Eq. 6.16 requires estimates of σ_{lr}^2 (or σ_y^2 and ρ_{xy}), which may be based on pilot, similar, or earlier samples.

■ **EXAMPLE 6.8**   A population has 10,000 elements. It is believed that the population variance of a variable of interest is about 25, and that its correlation coefficient with a known auxiliary variable is about 0.7. How large a simple random sample should be taken

---

SUMMARY 6.3 Regression estimator, sample size

Provided the accuracy requirements are stringent enough, the size of a simple or modified simple random sample needed for the regression estimate to be in the interval $\mu_y \pm c$ with probability $1 - \alpha$ is given by

$$n = \frac{N\sigma_{lr}^2}{(N-1)D^2 + \sigma_{lr}^2},\tag{6.16}$$

where $D = (c/Z_{\alpha/2})$ and $\sigma_{lr}^2 = \sigma_y^2(1 - \rho_{xy}^2)$. $Z_{\alpha/2}$ for selected values of $1 - \alpha$ are listed in Table 3.6.

---

so that the regression estimate of the population mean of the variable of interest is within ± 0.5 with probability 95%?

We calculate first $\sigma_{lr}^2 = (25)(1 - 0.7^2) = 12.75$. For $1 - \alpha = 0.95$, $Z_{\alpha/2} = 1.96$ and $D = (0.5/1.96) = 0.255$. Applying Eq. 6.16, we find

$$n = \frac{(10{,}000)(12.75)}{(10{,}000 - 1)(0.255)^2 + 12.75} = 192,$$

or, about 200 in round figures.   ◼

As we now know, the regression estimator is better than the ratio estimator for large n and $N - n$. If, for some reason, the ratio estimator is used instead of the regression estimator, Eq. 6.16 may still be applied, except that σ_{lr}^2 should be replaced by σ_r^2, Eq. 6.11, and appropriate estimates of its components.

6.9 INTERVAL ESTIMATES

Large-sample confidence intervals for the population mean or total of a variable may be constructed using either the ratio or the regression estimator. Interval estimates based on the regression estimator are described in Summary 6.4.

◼ EXAMPLE 6.9   We illustrate the calculations using the data of Example 6.1, which are reproduced in the first three columns of Table 6.10.

The least squares estimates were $a = 21.932$ and $b = 0.7178$, and the regression estimate of the average purchases of Product Y per hospital was $\bar{Y}_{lr} = 137.245$. As shown in the last three columns of Table 6.10, we get

$$S_{lr}^2 = \frac{1}{n-2}\sum_{i=1}^{n}(Y_i - a - bX_i)^2 = \frac{1}{3-2}(69.0185) = 69.0185,$$

SUMMARY 6.4 Confidence intervals, regression estimator

For large n and $N - n$, an approximate $100(1 - \alpha)\%$ confidence interval for μ_y, the population mean of a variable Y, based on the regression estimator under either simple or modified simple random sampling, is

$$\bar{Y}_{lr} \pm Z_{\alpha/2}\sqrt{\widehat{Var(\bar{Y}_{lr})}} \qquad (6.17)$$

where

$$\widehat{Var(\bar{Y}_{lr})} = \frac{S_{lr}^2}{n}\frac{N - n}{N},$$

and

$$S_{lr}^2 = \frac{1}{n-2}\sum_{i=1}^{n}(Y_i - \hat{Y}_i)^2 = \frac{1}{n-2}\sum_{i=1}^{n}(Y_i - a - bX_i)^2. \qquad (6.18)$$

A $100(1 - \alpha)\%$ confidence interval for $\tau_y = N\mu_y$, the population total of Y, is calculated by multiplying the limits in (6.17) by N. $Z_{\alpha/2}$ for selected values of $1 - \alpha$ can be found in Table 3.6.

TABLE 6.10	**Sample of Hospitals, Regression Estimator**				
i	X_i	Y_i	$\hat{Y}_i = a + bX_i$	$(Y_i - \hat{Y}_i)$	$(Y_i - \hat{Y}_i)^2$
1	675	500	506.45	−6.44702	41.5641
2	450	350	344.94	5.05798	25.5832
3	1,500	1,100	1,098.63	1.36792	1.8712
					69.0185

from which we calculate

$$\begin{aligned}
\widehat{Var(\bar{Y}_{lr})} &= \frac{S_{lr}^2}{n}\frac{N - n}{N} \\
&= \frac{69.0185}{3}\frac{1,158 - 3}{1,158} \\
&= 22.9466.
\end{aligned}$$

In computer programs for regression, S_{lr}, the square root of S_{lr}^2, is the "standard deviation of residuals." See Figure 6.3, where S_{lr} is displayed as $S=8.3077$.

If these results were based on a large simple or modified simple random sample (which, of course, they are not), a, say, 90% confidence interval for the average purchases of Product Y per hospital would be

$$(137.245) \pm (1.645)\sqrt{22.9466},$$

which is the interval from 129.365 to 145.125 ($000).

A 90% confidence interval for the total hospital purchases of Product Y would be from $(1,158)(129.365)$ to $(1,158)(145.125)$, or from about 149,805 to 168,055 ($000). ◪

Large-sample confidence intervals based on the ratio estimator are shown in Summary 6.5.

◪ **EXAMPLE 6.10** We use the data for the same sample of three hospitals to illustrate the calculation of these confidence intervals. The data are reproduced once again in the first three columns of Table 6.11.

We had earlier calculated $\bar{Y}/\bar{X} = (1,950/3)/(2,625/3) = 0.7428$; also, the ratio estimate of the average purchases of Product Y per hospital, $\bar{Y}_r = (\bar{Y}/\bar{X})\mu_x =$

SUMMARY 6.5 Confidence intervals, ratio estimator

For large n and $N - n$, an approximate $100(1 - \alpha)\%$ confidence interval for μ_y, the population mean of a variable Y, based on the ratio estimator under either simple or modified simple random sampling, is

$$\bar{Y}_r \pm Z_{\alpha/2}\sqrt{\widehat{Var(\bar{Y}_r)}} \qquad (6.19)$$

where

$$\widehat{Var(\bar{Y}_r)} = \frac{S_r^2}{n}\frac{N - n}{N},$$

and

$$S_r^2 = \frac{1}{n - 1}\sum_{i=1}^{n}(Y_i - \hat{Y}_i)^2 = \frac{1}{n - 1}\sum_{i=1}^{n}\left(Y_i - \frac{\bar{Y}}{\bar{X}}X_i\right)^2. \qquad (6.20)$$

A $100(1 - \alpha)\%$ confidence interval for $\tau_y = N\mu_y$, the population total of Y, is calculated by multiplying the limits in (6.19) by N. $Z_{\alpha/2}$ for selected values of $1 - \alpha$ can be found in Table 3.6.

TABLE 6.11	Sample of Hospitals, Ratio Estimator				
i	X_i	Y_i	$\hat{Y}_i = 0.7428X_i$	$(Y_i - \hat{Y}_i)$	$(Y_i - \hat{Y}_i)^2$
1	675	500	501.43	-1.4286	2.041
2	450	350	334.29	15.7143	246.938
3	1,500	1,100	1,114.29	-14.2858	204.083
	2,625	1,950			453.062

$(0.7428)(160.648) = 119.329$. The estimated purchases by hospital i are $\hat{Y}_i = 0.7428X_i$, and, as shown in the last column of Table 6.11,

$$S_r^2 = \frac{1}{n-1} \sum_{i=1}^{n}(Y_i - \hat{Y}_i)^2 = \frac{1}{3-1}(453.062) = 226.531.$$

Therefore,

$$\widehat{Var(\bar{Y}_r)} = \frac{S_r^2}{n}\frac{N-n}{N}$$
$$= \frac{226.531}{3}\frac{1,158-3}{1,158}$$
$$= 75.3147.$$

If the sample was simple or modified simple random and the sample size and population remainder large (which, in this illustration, they are not), a 90% confidence interval for the average purchases of Product Y per hospital would be

$$(119.329) \pm (1.645)\sqrt{75.3147},$$

or from 105.053 to 133.605 ($000). A 90% confidence interval for the total hospital purchases would be calculated by multiplying the above limits by $N = 1,158$, to give, in this illustration, the interval from 121,651 to 154,715 ($000). ◪

6.10 REGRESSION AND RATIO ESTIMATORS IN STRATIFIED SAMPLING

A stratified sample, it will be recalled, is made up of a number of simple random samples, one from each group into which the population elements are classified. The stratified estimator of the population mean of a variable is a weighted average of the simple estimators of the group means of the variable. It stands to reason that the stratified estimator can be improved if the simple estimator of any group mean is replaced by a better estimator. That is indeed the case, as we shall show below. In particular, *a better stratified estimator can be constructed if, in any group*

in which the number of sampled elements and the group remainder are large, the simple estimator of the group mean is replaced by the regression estimator.

Let us explain. Stratified random sampling requires that the elements of the population be divided into M groups, from each of which a simple random sample is selected. The stratified estimator of the population mean (μ_y) of a variable Y is

$$\bar{Y}_s = w_1 \bar{Y}_1 + \cdots + w_M \bar{Y}_M = \sum_{i=1}^{M} w_i \bar{Y}_i,$$

where $w_i = N_i/N$ and $\bar{Y}_i$ is the simple estimator of the group mean, μ_i, of variable Y. $\bar{Y}_i$ is an unbiased estimator of μ_i, and $\bar{Y}_s$ an unbiased estimator of μ. The variance of $\bar{Y}_s$ is

$$Var(\bar{Y}_s) = w_1^2 \frac{\sigma_1^2}{n_1} \frac{N_1 - n_1}{N_1 - 1} + \cdots + w_M^2 \frac{\sigma_M^2}{n_M} \frac{N_M - n_M}{N_M - 1}$$
$$= w_1^2 Var(\bar{Y}_1) + \cdots + w_M^2 Var(\bar{Y}_M).$$

It can be shown that $\bar{Y}_s$ remains unbiased if any $\bar{Y}_i$ is replaced by another unbiased estimator—say, $\bar{Y}_i'$—of μ_i. This was also noted in connection with combined stratified/two-stage sampling in Section 5.8.

The variance of the resulting new stratified estimator is still given by the last expression, except that $Var(\bar{Y}_i)$ is replaced by $Var(\bar{Y}_i')$. Since all $w_i^2 = (N_i/N)^2 > 0$, it is clear that the new stratified estimator will have smaller variance if $Var(\bar{Y}_i') < Var(\bar{Y}_i)$.

Earlier in this chapter, we established that the regression estimator is approximately unbiased and has smaller variance than the simple estimator when the sample size n and the population remainder $N - n$ are large. Therefore, the stratified estimator $\bar{Y}_s$ can be improved if, in any group for which n_i and $N_i - n_i$ are large, $\bar{Y}_i$ is replaced by the regression estimator of μ_i.

◪ EXAMPLE 6.11   In Example 4.1, we examined the case of Books Research, which provides estimates of the retail sales of book titles. Books Research classifies the 10,000 bookstores in the country into three groups (Large, Medium, and Small) on the basis of their floor area. A simple random sample of bookstores is selected from each group. Table 6.12 repeats the relevant data first presented in Table 4.1. The last column will be explained shortly.

Let us consider the sample of Large bookstores. First, observe that the sample size (500) and the group remainder (1,000 − 500) are large. Second, note that Books Research does know the floor area of every large bookstore in the country since it employs this auxiliary variable to classify bookstores into the three size groups. Third, it is reasonable to assume that the sales of a given title are correlated to the size of the bookstore as measured by floor area. These three observations suggest

		Population		Sample	
		Size,	**Percent of Elements**	**Size,**	**Regression Estimate**
i	**Group**	N_i	$w_i = N_i/N$	n_i	$\bar{Y}'_i$
1	Large	1,000	0.1	500	231.4
2	Medium	3,000	0.3	200	125.5
3	Small	6,000	0.6	100	42.3
	Total	$N = 10,000$	1.0	$n = 800$	

TABLE 6.12   Improved Estimates of Book Sales, Example 6.11

that the regression estimator is better than the simple estimator for estimating the average sales of any title per large bookstore. The information outlined in Table 6.13 is available after the 500 selected large bookstores are questioned concerning their sales of the given title.

Using this information and the data in Table 6.12, it is straightforward to calculate the regression estimate of the average sales per *large* bookstore; for the sake of this illustration, let us suppose that this is $\bar{Y}_{lr} = 231.4$.

Exactly the same reasoning suggests that the regression estimator is better than the simple estimator for estimating the average sales per bookstore in the other two groups. Assume that the regression estimates are as shown in the last column of Table 6.12.

The improved stratified estimate of the average sales per bookstore in the country is

$$\bar{Y}'_s = w_1\bar{Y}'_1 + w_2\bar{Y}'_2 + w_3\bar{Y}'_3 = (0.1)(231.4) + (0.3)(125.5) + (0.6)(42.3) = 86.17.$$

The corresponding estimate of the total number of copies sold in the country is $N\bar{Y}'_s = (10,000)(86.17) = 861,700.$ ◼

TABLE 6.13   Sample of Large Bookstores

Sampled Bookstore	Floor area, X (000 sq ft)	Sales, Y (number of copies)
1	1.75	340
2	1.30	120
⋮	⋮	⋮
500	1.40	250

The ratio estimator, it will be recalled, is also approximately unbiased for large n and $N - n$, but does not necessarily have smaller variance than the simple estimator (see Section 6.7). However, if in any group i the ratio estimator ($\bar{Y}_i'$) of the group mean μ_i has smaller variance than the simple estimator ($\bar{Y}_i$), a better estimator of the population mean is $\bar{Y}_s'$, which is the stratified estimator $\bar{Y}_s$ but with $\bar{Y}_i$ replaced by $\bar{Y}_i'$.

6.11 REGRESSION AND RATIO ESTIMATORS IN TWO-STAGE SAMPLING

The two-stage estimator, it will be recalled, is

$$\bar{Y}_{ts} = \frac{M}{m} \sum_{i=1}^{m} w_i \bar{Y}_i, \tag{6.21}$$

where M is the number of groups, m the number of selected groups, and $w_i = N_i/N$ the proportion of population elements in selected group i.

Reasoning similar to that of Section 6.10 can show that if for any group i there is another unbiased estimator $\bar{Y}_i'$ of the group mean μ_i with smaller variance than $\bar{Y}_i$, a better two-stage estimator is given by Eq. 6.21 but with $\bar{Y}_i$ replaced by $\bar{Y}_i'$. In particular, the regression estimator of μ_i should replace $\bar{Y}_i$ in any selected group i in which n_i and $N_i - n_i$ are large.

◪ EXAMPLE 6.12 In Example 5.1, we described the two-stage sample used by the Consumer Survey Bureau (CSB) to estimate household expenditures in a certain city. A simple random sample of enumeration areas (EAs) is selected in the first stage; in the second stage, a simple random sample of households is selected from each EA selected in the first stage.

Table 6.14 shows the number of households (N_i), the number of selected households (n_i), and the sample average expenditure on clothing per selected houshold ($\bar{Y}_i$) in each of $m = 4$ selected EAs. There are $M = 200$ EAs in the city. The last column of Table 6.14 will be explained shortly.

CSB can obtain from public property tax records the assessed value of each selected household's dwelling. Table 6.15 outlines the available data from the sample of 50 households for the first selected EA. Similar data are available for the other selected EAs. The only other information needed to calculate the regression estimate of the average expenditure on clothing in a given EA is the average assessed value of all dwellings in the EA. This figure is also known from the property tax records.

For the purpose of this example, assume that the regression estimates are as shown in the last column of Table 6.14 under the heading $\bar{Y}_i'$.

TABLE 6.14	**CSB Sample, Example 6.12**				
EA, i	N_i	$w_i = N_i/N$	n_i	$\bar{Y}_i$	$\bar{Y}_i'$
1	250	0.0042	50	95.0	97
2	310	0.0052	60	84.0	81
3	340	0.0057	70	75.5	76
4	280	0.0047	55	90.3	102

TABLE 6.15 Sample of Households, First Selected EA		
Sampled Household	Assessed Value of Dwelling	Expenditure on Clothing
1	10,700	125
2	4,800	60
$\vdots$	$\vdots$	$\vdots$
50	8,500	90

Although some of the n_i are close to the recommended minimum of 50, all n_i and $N_i - n_i$ can be considered large enough for claiming that the regression estimators are better than the simple estimators. Therefore, a better estimate of the average expenditure on clothing by all households in the city is

$$
\begin{aligned}
\bar{Y}_{ts}' &= \frac{M}{m}[w_1\bar{Y}_1' + w_2\bar{Y}_2' + w_3\bar{Y}_3' + w_4\bar{Y}_4'] \\
&= \frac{200}{4}[(0.0042)(97) + (0.0052)(81) + (0.0057)(76) + (0.0047)(102)] \\
&= 87.06.
\end{aligned}
$$

There are, it will be recalled, $N = 60,000$ households in the city. The estimate of the total expenditures on clothing by all households in the city, then, is $N\bar{Y}_{ts}' = (60,000)(87.06)$ or $5,223,600. ◪

The ratio estimator does not necessarily have smaller variance than the simple estimator when the sample size and group remainder are large, but, if it has, it may replace $\bar{Y}_i$ in any group i for which n_i and $N_i - n_i$ are large in order to form a two-stage estimator $\bar{Y}_{ts}'$ with smaller variance than $\bar{Y}_{ts}$.

6.12 TO SUM UP

- The ratio estimator is not unbiased under simple random sampling, but is unbiased if (a) the first element is selected with probability proportional to its X value, and (b) the other $n - 1$ elements are selected from the remaining $N - 1$ elements at random and without replacement.
- Likewise, the regression estimator is not unbiased under simple random sampling, but is unbiased if (a) the first two elements are selected with probability proportional to the square of the difference between their X values, (b) the other $n - 2$ elements are selected from the remaining $N - 2$ elements at random and without replacement.
- When the sample size (n) and the population remainder $(N - n)$ are large, (a) the ratio and regression estimators are approximately unbiased under simple random sampling; (b) the variances of the ratio and regression estimators can be approximated by simple expressions; and (c) the regression estimator is better than either the ratio or the simple estimator. Put simply, if the planned sample is large enough, it is not necessary to apply the modified selection procedures to ensure unbiasedness; simple random sampling is sufficient and regression estimates are best.
- In stratified or two-stage sampling, the stratified and two-stage estimators can be improved if, in any group in which the number of sampled elements and the group remainder are large, the simple estimator of the group mean is replaced by the regression estimator.

PROBLEMS

6.1 *Computing exercise:* Using the program SCALC, verify the numerical results presented in this chapter. Note any deviations, and determine if these are the result of rounding at intermediate stages.

6.2 Refer to Example 6.1, and suppose that a random sample of three hospitals yielded the data shown in Table 6.16.

(a) Assuming that $Y \approx bX$, estimate b using (i) the average of the ratios Y_i / X_i, (ii) the ratio of averages $\bar{Y} / \bar{X}$, and (iii) the method of least squares.

(b) Estimate (i) the total, and (ii) the average purchases of Product Y by all 1,158 hospitals for each of the three methods in part (a). Assume $\mu_x = 160.648$.

6.3 Same as Problem 6.2, except use the data of Table 6.17.

6.4 A population consists of five elements. These are listed in Table 6.18 together with the values of a known auxiliary variable X.

TABLE 6.16 Data for Problem 6.2		
Sampled Hospital, i	Number of Beds, X_i	Purchases of Product Y, Y_i ($000)
121	200	180
375	540	400
652	380	300

TABLE 6.17 Data for Problem 6.3		
Sampled Hospital, i	Number of Beds, X_i	Purchases of Product Y, Y_i ($000)
212	900	520
430	600	390
705	140	180

TABLE 6.18 Data for Problem 6.4	
Element No., i	X_i
1	110
2	10
3	25
4	80
5	150

Using Table 3.5 of random numbers, select three elements according to the modified simple random sampling procedure described in Section 6.3. Show the assigned numbers, the random numbers read, and their interpretation.

6.5 Suppose there are only three firms in Ackroyd, as shown in Table 6.19.

(a) Calculate μ_x and μ_y.

(b) In the manner of Table 6.6, show the possible outcomes of a simple random sample of two firms without replacement, the probabilities of these outcomes, and the associated values of the ratio estimator of μ_y, $\bar{Y}_r$.

TABLE 6.19 Half-of-Ackroyd Population

Firm	Sales, X	Employees, Y
A	13	9
B	12	8
C	9	6

(c) Calculate the expected value of $\bar{Y}_r$. Is $\bar{Y}_r$ an unbiased estimator of μ_y in this case?

(d) In the manner of Table 6.7, determine the possible values of $\bar{Y}_r$ and their probabilities if the sample of two firms is selected according to the modified sampling procedure described in Section 6.3.

(e) Calculate the expected value of $\bar{Y}_r$. Is $\bar{Y}_r$ unbiased in this case?

6.6 Same as Problem 6.5, except assume that the population is as shown in Table 6.20.

6.7 Refer to Table 6.16, showing the available information from a sample of three hospitals.

(a) Plot the three pairs of (X_i, Y_i) values in a scatter diagram. Which of the two assumptions, $Y \approx bX$ or $Y \approx a + bX$, appears to describe better the relationship between X and Y?

(b) Calculate the least squares estimates a and b of $Y \approx a + bX$.

(c) Confirm your calculations using a computer program for regression.

(d) Calculate the regression estimate, $\bar{Y}_{lr}$, of the average purchases of Product Y per hospital. Assume $\mu_x = 160.65$.

(e) Calculate the regression estimate, T_{lr}, of the total purchases of Product Y by all hospitals. Assume $N = 1{,}158$.

6.8 Same as Problem 6.7, except use the data in Table 6.17.

TABLE 6.20 Other-Half-of-Ackroyd Population

Firm	Sales, X	Employees, Y
D	3	2
E	1	1
F	7	5

6.9 Refer to Table 6.19, describing the "Half-of-Ackroyd" population.

(a) In the manner of Table 6.8, list the possible outcomes of a sample of two firms selected according to the modified procedure described in Section 6.6, the probabilities of these outcomes, and the associated regression estimates $\bar{Y}_{lr}$. Calculate the expected value of $\bar{Y}_{lr}$. Is $\bar{Y}_{lr}$ an unbiased estimator in this case?

(b) Suppose the sample of two firms is instead simple random. List the possible values of $\bar{Y}_{lr}$ and their probabilities. Calculate the expected value of $\bar{Y}_{lr}$. Is $\bar{Y}_{lr}$ an unbiased estimator in this case?

6.10 Same as Problem 6.9, except refer to Table 6.20, describing the "Other-half-of-Ackroyd" population.

6.11 Refer to the population listed in Table 6.18. Using Table 3.5 of random numbers, select a modified simple random sample of three elements according to the procedure described in Section 6.6. Show the assigned numbers, the random numbers read, and their interpretation.

6.12 For a certain population, it is known that $N = 15{,}200$, $\mu_x = 123$, and $\sigma_x = 65$. In addition, it is believed that μ_y is in the range from 140 to 150, σ_y in the range from 80 to 100, and ρ_{xy} in the range from 0.4 to 0.6. A simple random sample of size $n = 200$ is planned.

(a) Calculate the variance of the simple, ratio, and regression estimators using the midpoints of the above ranges.

(b) Calculate the range of the variance of the simple, ratio, and regression estimators.

(c) Rank the three estimators in order of preference on the basis of the calculations in **(a)** and **(b)**.

6.13 Same as Problem 6.12, except that $N = 7{,}800$, $\mu_x = -10$, and $\sigma_x = 5$; μ_y is believed to be in the range from 30 to 40, σ_y in the range from 3 to 4, and ρ_{xy} in the range from -0.8 to -0.7. A simple random sample of size $n = 150$ is planned.

6.14 For a certain population, it is known that $N = 5{,}200$, $\mu_x = 30$, and $\sigma_x^2 = 180$. In addition, it is believed that $\mu_y = 45$, $\sigma_y^2 = 240$, and $\rho_{xy} = 0.6$.

(a) How large a simple random sample should be taken so that the regression estimate of μ_y will be within ± 2 of μ_y with probability 99%?

(b) How large a simple random sample should be taken so that the regression estimate of τ_y will be within $\pm 6{,}000$ of τ_y with probability 95%?

(c) How large a simple random sample should be taken so that the ratio estimate of μ_y will be within ± 1 of μ_y with probability 90%?

(d) How large a simple random sample should be taken so that the ratio estimate of τ_y will be within $\pm 5{,}500$ of τ_y with probability 99%?

6.15 Same as Problem 6.14, except that $N = 8,500$, $\mu_x = 30$, and $\sigma_x^2 = 200$. In addition, it is believed that $\mu_y = -45$, $\sigma_y^2 = 150$, and $\rho_{xy} = -0.75$.

6.16 A simple random sample of 5 elements from a population of 20 yielded the data in Table 6.21.

(a) Calculate S_r^2 and $\widehat{Var}(\bar{Y}_r)$, as defined in Summary 6.5.

(b) Calculate S_{lr}^2 and $\widehat{Var}(\bar{Y}_{lr})$, as defined in Summary 6.4.

(c) Pretending that n and $N - n$ are large, calculate 95% confidence intervals for μ_y and τ_y based on the regression estimator. Assume $\mu_x = 17.5$.

(d) Pretending that n and $N - n$ are large, calculate 95% confidence intervals for μ_y and τ_y based on the ratio estimator. Assume $\mu_x = 17.5$.

6.17 Same as Problem 6.16, except that the sample results are as shown in Table 6.22.

6.18 **(a)** Show that Eq. 6.20 can also be written as

$$S_r^2 = \frac{1}{n-1}[\sum_{i=1}^n Y_i^2 - 2\frac{\bar{Y}}{\bar{X}}\sum_{i=1}^n X_i Y_i + (\frac{\bar{Y}}{\bar{X}})^2 \sum_{i=1}^n X_i^2].$$

TABLE 6.21 Data for Problem 6.16

i	X_i	Y_i
1	13	4
2	15	5
3	10	1
4	17	6
5	12	3

TABLE 6.22 Data for Problem 6.17

i	X_i	Y_i
1	13	-2
2	15	-6
3	10	1
4	17	-8
5	12	0

(b) Calculate S_r^2 using the expression in **(a)** and the data in Table 6.21.

(c) Calculate S_r^2 using the expression in **(a)** and the data in Table 6.22.

6.19 (a) Using the data in Table 6.21, verify that

$$S_{lr}^2 = \frac{1}{n-2}[\sum_{i=1}^{n} Y_i^2 - a \sum_{i=1}^{n} Y_i - b \sum_{i=1}^{n} X_i Y_i].$$

(b) Same as **(a)**, except use the data in Table 6.22.

6.20 At any point in time, there is a stock of houses in a given region. The houses are of different type (e.g., detached, semidetached, townhouses, apartments), construction, location, size, age, style, layout, etc., and therefore of different value. The value of a house is defined as the price that house would fetch in a free market if offered for sale and sold. Only a few of the houses in a region are sold during a period of time. The characteristics of the sold houses need not be those of the entire stock of houses.

"Many factors prompt an owner to sell and a buyer to buy," says Mr. Jones, an experienced real estate agent. "These include financial need or desire to invest, relocation, change of income, standard of living, or family size, and a host of others. It is reasonable to assume that one house is just as likely to be offered for sale and sold as another. It is also reasonable to assume that whether or not a particular property is offered for sale and sold does not depend on whether or not any other property is offered for sale and sold. In other words," concludes Mr. Jones, "it is reasonable to treat the properties that change ownership in a period of time as a simple random sample with replacement from the population of all the properties in the region, even though no act of selection is involved."

(a) Comment on Mr. Jones's opinion.

(b) Five houses were sold last month in a certain city. Table 6.23 shows the assessed value and sale price of these houses.

TABLE 6.23 House Sale Prices and Appraised Values, Problem 6.20		
Property	Appraised Value ($)	Selling Price ($000)
1	4,750	166
2	5,000	195
3	6,300	320
4	5,500	249
5	10,100	650

The assessed value is determined by the City and is used for levying property taxes. The property tax equals the assessed value times a constant "mill rate." For example, if the mill rate is 3%, two city properties with assessed values of $10,000 and $5,000 are levied annual taxes of $300 and $150, respectively.

Public property tax records show that the city has 23,000 houses and that the average assessed value of these houses is $5,300.

Estimate the average value of a house, and the total value of all houses in the city. Justify your method of estimation. Pretending that the sample size is large, calculate 95% confidence intervals for the average and total value of all houses in the city.

(c) State and explain any reservations you may have about this approach to estimating the average and total of the house values in a region. Of what practical use can the estimates be?

6.21 Smith and Smith (S&S) is a firm of chartered accountants given the task of estimating the current value of the inventory of Reliable Supplies, a hardware wholesaler that went into receivership. Reliable regularly keeps in stock 1,100 different items. A complete stocktaking is not feasible.

S&S selected a random sample of 100 items, and determined the current inventory value of these items. In addition, S&S recorded the inventory value of these same items at the end of last fiscal year, with the results shown in Table 6.24.

The values shown in Table 6.24 equal the number of units times the cost per unit. For example, the current inventory found 1,500 units of Item 1, which, at a unit cost of $20.80, have a value of $31,200.

The total value of the last year-end inventory was $7,243,212. S&S's estimate of the current inventory value is

$$(\frac{675.8}{650.3})(7,243,212) = (1.039)(7,243,212) = \$7,527,238.$$

TABLE 6.24 Sample Results, Problem 6.21

Item No.	Current Value, Y ($000)	End-of-Year Value, X ($000)	Ratio, Y/X
1	31.2	30.1	1.036
2	18.4	16.0	1.150
⋮	⋮	⋮	⋮
100	5.3	6.4	0.828
Total	675.8	650.3	107.352

S&S argues that Reliable's hardware sales and inventories do not fluctuate seasonally. Changes in the inventory value of an item are due to random fluctuations of demand and to changes of unit values. The current value of the sampled items is 3.9% higher than at year-end. Therefore, it is reasonable to assume that the current value of *all* items is approximately 3.9% higher than at year-end.

(a) Estimate the total value of current inventory based on the theory of simple random samples.

(b) Another reasonable estimate of the total current inventory value is:

$$\frac{1}{100}(107.352)(7,243,212) = \$7,775,733.$$

Why is this reasonable? Is there any reason for preferring S&S's estimate to this one?

(c) Can you suggest a third estimate of the total current inventory value? If so, outline the information needed for its calculation.

(d) Would there be any advantage to a *stratified random sample* in this situation over a simple random sample of the same size?

(e) Determine if the ratio estimator of S&S is unbiased in the following simple version of the problem. Assume that Reliable regularly stocks only $N = 5$ items, whose current and year-end values are *known* to be as follows:

Item No.	Current Value, Y	Year-End Value, X
1	5	4
2	10	9
3	25	24
4	8	10
5	30	32
Total	78	79

Suppose that a random sample of $n = 2$ items will be selected and a ratio estimator formed as $T_r = (\sum Y / \sum X)(79)$, where $\sum Y$ is the sum of the current and $\sum X$ the sum of the year-end values of the selected items.

Determine the probability distribution of T_r. Calculate its mean. Is T_r indeed an unbiased estimator of the total current value of all items?

(f) Same as **(e)** but using the estimator described in **(b)**:

$$\frac{1}{n}\left(\sum \frac{Y}{X}\right)(79).$$

6.22 Voting takes place in voting stations. Each such station usually serves a small neighborhood, so there may be hundreds of stations in a medium-sized city. On

voting day, all stations in a region close at the same time, but the counting of the votes and the settlement of any disputes or protests take varying times. As a result, the official tallies are forwarded to the central election office at various times after the closing of the stations.

Radio and television stations have found significant interested audiences in the so-called "election projections." These are forecasts of the total and percentage vote for each party or candidate and of the winning candidate or party, which are made on the basis of tallies already in at the central office. The projections are updated frequently as new tallies arrive. The final estimate is, of course, the final count after all stations have reported.

(a) "In the absense of reasons to the contrary, it is reasonable to assume that the first n stations reporting to central are determined *as if* by a simple random sample without replacement from the population of N stations in the region. Imagine, if you will, that N identical tags, each bearing the identification number of one of the N stations in the region, are put into a hat. Select n tags, one at a time, at random, and without replacement. The 'process' that generates the arrival of station results at the central office can be thought of *like* the above process." Comment on this statement.

(b) By 9 P.M. of election night, five of the 300 voting stations in the region have reported to the central election office. There are 80,000 registered voters in the region. The reported vote counts, in the order in which they arrived, are as follows:

Voting Station No.	Number of Registered Voters	Number Voting for Party A
256	340	105
092	260	75
145	170	50
113	310	96
222	220	65

Assuming that the five stations that have reported can be considered a simple random sample without replacement from the population of 300 stations, estimate the total number of voters favoring Party A. You may pretend that the sample size is large. Explain and defend any other assumptions you are forced to make.

(c) Under the same assumptions as **(b)**, about how many voting stations should have reported before the estimate of the total vote for Party A is within ± 100 votes with probability 90%?

6.23 The commercial viability of new consumer products (for example, a new type of toothpaste, a new tool for the handyman, etc.) is frequently estimated by arranging with selected retail outlets to have the new product carried and displayed

on the shelves alongside similar products for a certain period of time. At the end of this trial period, the store sales of the new product are ascertained and an estimate is made of the total sales had the new product been carried by all retail outlets in the region under similar conditions.

(a) A directory is available listing all 1,500 retail outlets in a region. A simple random sample of five retail outlets is selected, and, for suitable compensation, all selected outlets agree to carry the new product and to display it with similar products for one month. At the end of the month, it is found that the sales (in $000) of the new product by the five selected outlets were 14, 9, 95, 65, and 80. Based on this information only, estimate the potential total sales of the new product in the region.

(b) Same as (a), except that there is information concerning the total sales (all products) of each selected outlet during the trial month, as shown below:

Selected Outlet	Sales of All Products ($000)	Sales of New Product ($000)
1	1,500	14
2	800	9
3	9,900	95
4	5,600	65
5	7,200	80

During the trial month, the total sales (all products) of all 1,500 retail outlets in the region were $8,700,000 (000). Explain your choice of method and justify any assumptions you are forced to make.

(c) What are the practical shortcomings of this method for estimating the potential sales of a new product? What can be done to alleviate these shortcomings?

CHAPTER 7

Some Special Topics

7.1 INTRODUCTION AND SUMMARY

Until now, we have concentrated on estimating the population mean or total of a variable, and the population proportion or number in a category. These are the population characteristics of greatest interest in practice, but obviously not the only ones. In preceding chapters, for instance, we provided estimators of the variance of a variable in the population or a group, although these are used mainly for constructing interval estimates of the principal characteristics.

We begin this chapter by considering how to estimate the ratio of the population totals or means of two variables. Examples of such ratios are the ratio of aggregate savings to income of households, and the ratio of aggregate profits to sales of businesses. The ratio of totals is of direct interest occasionally, but also indirectly for its special cases concerning subpopulations, cluster sampling, poststratification, and the related double sampling.

As will be shown in this chapter, the mean of a variable or the proportion in a category among elements of the population that belong to a group (the so-called subpopulation mean and proportion) can be considered a ratio of population totals of specially created variables, to which the theory of the general case conveniently applies.

Another special case involves an alternative estimator of a population mean or proportion under cluster sampling. Yet another special case involving subpopulation means or proportions concerns poststratification, that is, stratification of a simple random sample after it is selected. Poststratification is often used to take advantage of a relationship between the variable of interest and an auxiliary variable, but this relationship can be exploited only after a sample is selected.

Related to poststratification is double sampling. Again, it is desirable to exploit a relationship between a variable of interest and an auxiliary variable, but some of the information necessary to implement stratified sampling or ratio or regression estimation is lacking. With double sampling, two phases are distinguished. In the first phase, a simple random sample of elements is selected. This first sample provides estimates of the proportions of elements in groups formed according to the auxiliary variable or attribute, or of the population mean of an auxiliary variable. For the purpose of the second phase, this first sample is treated as a population with known characteristics, and a subsample of its elements is selected to calculate stratified, ratio, or regression estimates of characteristics of the original population.

7.2 ESTIMATING THE RATIO OF TOTALS OR MEANS OF TWO VARIABLES

Suppose that a survey of households in a certain town is planned to ascertain annual household expenditures for dental and health care. The population of interest can be imagined as outlined in Table 7.1.

Income earners are household members with positive annual income. In preceding chapters, we described methods for estimating a population mean (μ) or total (τ). But suppose it is also of interest to estimate the average dental expenditures per income earner (τ_2/τ_1) or the ratio of total dental to total health expenditures (τ_3/τ_2). These characteristics are ratios of population totals. Since, for any variable, $\tau = N\mu$, the ratio of totals is equal to the ratio of the corresponding population means; for example, $\tau_2/\tau_1 = N\mu_2/N\mu_1 = \mu_2/\mu_1$.

Common sense suggests that a reasonable estimator of the ratio of population totals of any two variables Y_1 and Y_2, τ_1/τ_2, is the ratio of the estimators of these totals $T_1/T_2 = N\bar{Y}_1/N\bar{Y}_2 = \bar{Y}_1/\bar{Y}_2$.

For example, suppose that a simple random sample of three households from a population of 3,000 households yielded the observations shown in Table 7.2.

TABLE 7.1 Population of Households			
Household No.	Number of Income Earners, Y_1	Dental Care Expenditures, Y_2	Health Care Expenditures, Y_3
1	?	?	?
2	?	?	?
⋮	⋮	⋮	⋮
N	?	?	?
Total	τ_1	τ_2	τ_3
Mean	μ_1	μ_2	μ_3

TABLE 7.2 Sample of Households

Household No.	Number of Income Earners, Y_1	Dental Care Expenditures, Y_2	Health Care Expenditures, Y_3
1	3	80	540
2	1	0	100
3	2	130	320
Total	6	210	960
Average	2	70	320

The estimate of the total number of income earners in the town is $T_1 = N\bar{Y}_1 = (3,000)(2) = 6,000$, that of total dental care expenditures $T_2 = (3,000)(70) = \$210,000$, and that of total health expenditures $T_3 = (3,000)(320) = \$960,000$. Thus, a reasonable estimate of average dental care expenditures per income earner is

$$\frac{T_2}{T_1} = \frac{N\bar{Y}_2}{N\bar{Y}_1} = \frac{\bar{Y}_2}{\bar{Y}_1} = \frac{70}{2} = \$35.$$

The estimate of the ratio of total dental to total health expenditures is

$$\frac{T_2}{T_3} = \frac{\bar{Y}_2}{\bar{Y}_3} = \frac{70}{320} = 0.219.$$

In general, under simple random sampling, the estimator $\bar{Y}_1/\bar{Y}_2$ is not an unbiased estimator of the corresponding ratio of population totals (τ_1/τ_2) or means (μ_1/μ_2) of two variables Y_1 and Y_2. However, it can be shown that the estimator is approximately unbiased when the sample size is large, in which case its variance is approximately as given by Eq. 7.2 in Summary 7.1. Summary 7.1 also lists approximate large-sample confidence intervals for the ratio of population totals or means.

◪ EXAMPLE 7.1

We illustrate the calculations using the data of Table 7.2, pretending that the sample size is large. We begin with a 95% confidence interval for health care expenditures per income earner. Note that $R = 320/2 = 160$. Preliminary calculations are shown in Table 7.3.

The estimate of the variance of R is

$$\widehat{Var}(R) = \frac{1}{2^2}\frac{1}{3}\frac{3,000 - 3}{3,000}\frac{1}{3 - 1}(7,200) = 299.70.$$

The required confidence interval, therefore, is of the form

$$160 \pm 1.96\sqrt{299.7},$$

or from about 126.07 to 193.93.

SUMMARY 7.1   Estimation of ratio of totals or means

- *Method of sample selection:* Simple random sampling.
- For large n, an unbiased estimator of the ratio of population totals (τ_1/τ_2) or means (μ_1/μ_2) of any two variables Y_1 and Y_2 is

$$R = \frac{T_1}{T_2} = \frac{\bar{Y}_1}{\bar{Y}_2}, \qquad (7.1)$$

and its approximate variance is given by

$$Var(R) \approx \frac{1}{\mu_2^2} \frac{1}{n} \frac{N-n}{N-1} \frac{1}{N} \sum_{i=1}^{N} \left(Y_{1i} - \frac{\mu_1}{\mu_2} Y_{2i}\right)^2. \qquad (7.2)$$

- For large n and $N - n$, an approximate $100(1 - \alpha)\%$ confidence interval for τ_1/τ_2 or μ_1/μ_2 is

$$R \pm Z_{\alpha/2} \sqrt{\widehat{Var(R)}}, \qquad (7.3)$$

where

$$\widehat{Var(R)} = \frac{1}{\bar{Y}_2^2} \frac{1}{n} \frac{N-n}{N} \frac{1}{n-1} \sum_{i=1}^{n} (Y_{1i} - RY_{2i})^2. \qquad (7.4)$$

- In the above expressions, it is assumed that $\mu_2 \neq 0$ and $\bar{Y}_2 \neq 0$.

TABLE 7.3   Preliminary Calculations, Example 7.1

Household No.	Number of Income Earners, Y_{1i}	Health Expenditures, Y_{3i}	$Y_{3i} - 160Y_{1i}$	$(Y_{3i} - 160Y_{1i})^2$
1	3	540	60	3,600
2	1	100	−60	3,600
3	2	320	0	0
Total	6	960		7,200

A 95% confidence interval for the ratio of total dental to health expenditures is constructed similarly. It can be confirmed that $R = 0.219$, $\sum(Y_{2i} - RY_{3i})^2 = 5{,}533.85$, and

$$\widehat{Var(R)} = 0.008998.$$

The desired interval, therefore, is from about 0.033 to 0.405.   ◼

Equations 7.2 and 7.4 can also be written in terms of the variances and covariance of Y_1 and Y_2. It can be shown that

$$\sum_{i=1}^{n}(Y_{1i} - RY_{2i})^2 = n[S_1^2 + R^2 S_2^2 - 2RS_{12}],$$

where S_1^2, S_2^2, and S_{12} are, respectively, the sample variances and covariance of Y_1 and Y_2. It can also be shown that

$$\sum_{i=1}^{N}(Y_{1i} - \frac{\mu_1}{\mu_2} Y_{2i})^2 = N[\sigma_1^2 + (\frac{\mu_1}{\mu_2})^2 \sigma_2^2 - 2(\frac{\mu_1}{\mu_2})\sigma_{12}],$$

where σ_1^2, σ_2^2, and σ_{12} are, respectively, the population variances and covariance of Y_1 and Y_2.

7.3 ESTIMATING SUBPOPULATION MEANS AND PROPORTIONS

Frequently, an estimate is desired of the average value of a variable or of the proportion of elements in a category *among the elements of the population belonging to a certain group*. Because such a group of elements forms a part of the population, these characteristics are often referred to as *subpopulation* or *domain* means and proportions.

Imagine, for example, a survey of students at a large university. A simple random sample of students will be taken, and the selected students will be questioned concerning their gender, income, etc. The theory of the preceding chapters suggests unbiased estimators of, say, the proportion of male students, or the average annual income of all students. But suppose it is desirable to estimate also the average income of *male* students. The population characteristic of interest is not a population proportion or the population average of a variable, but the average value of a variable in a part of the population—the group of male students.

Intuition suggests that a reasonable estimator of the average value of a variable in the subpopulation is the average value of the variable among the sampled elements that belong to the subpopulation. For example, if, in a simple random sample of 500 students, 240 turn out to be male, and the average income of these 240 male students is $5,700, then $5,700 is a reasonable estimate of the average income of all male students in the university.

Likewise, intuition suggests that a reasonable estimator of the proportion of elements in a subpopulation that belong to a given category is the corresponding sample proportion. For example, if, among the 240 males in the sample of 500 students, 60 are smokers, then a reasonable estimate of the proportion of all male students who smoke is 60/240 or 25%.

Unlike the ordinary sample mean or proportion, these estimators are ratios, the numerator and denominator of which vary from sample to sample. For example, the average income of male students in the sample is the ratio

$$\frac{\text{Total income of male students in the sample}}{\text{Number of male students in the sample}}.$$

Both the numerator and the denominator depend on the sample observations—on the particular students who happened to be selected in the sample. By contrast, the average income of the sampled students is

$$\frac{\text{Total income of sampled students}}{\text{Sample size}}.$$

The denominator of this last expression is fixed in advance.

The same comments apply to the estimate of the proportion in a subpopulation that fall into a certain category. For example,

$$\frac{\text{Number of male smokers in the sample}}{\text{Number of males in the sample}}$$

differs from, say, the proportion of smokers in the sample:

$$\frac{\text{Number of smokers in the sample}}{\text{Sample size}}.$$

The difference matters because, among other problems, the expected value of a ratio is not necessarily equal to the ratio of the expected values of its numerator and denominator.

We shall show that a subpopulation mean or proportion can be regarded as a ratio of population totals of specially created variables to which the results of Section 7.2 apply. We shall also show that the subpopulation total of a variable or number in a category are the population totals of these special variables, to which the results of Chapter 3 apply.

To see this, let us return to the example of this section and suppose that there are 25,400 students in the university. Imagine the population of students as is outlined in cols. (1) to (4) of Table 7.4.

In col. (5), Y_1 is a dummy (indicator) variable, representing a student's gender: $Y_1 = 1$, if male; $Y_1 = 0$, if female. Note that the population total of Y_1, τ_1, equals the number of male students in the university. The population mean μ_1 is the proportion of male students in the university.

In col. (6), variable Y_2 is equal to the product of variables Y_1 and Y. Clearly, Y_2 is equal to zero if the student is female or to the student's income if the student is male. Note that the total of Y_2, τ_2, is the total income of male students, and that the ratio τ_2/τ_1 is the average income of male students.

The dummy variable Y_3 in col. (7) takes the value 1 if the student is a smoker or 0 if a nonsmoker. The population total of this variable, τ_3, is the number of

TABLE 7.4   Population of University Students

Student No. (1)	Gender (2)	Annual Income, Y ($000) (3)	Smoker? (4)	Y_1 (5)	Created Variables $Y_2 = Y_1 Y$ (6)	Y_3 (7)	$Y_4 = Y_1 Y_3$ (8)
1	M	3.2	No	1	3.2	0	0
2	F	5.6	No	0	0	0	0
3	M	1.4	Yes	1	1.4	1	1
4	F	0.7	No	0	0	0	0
⋮	⋮	⋮	⋮	⋮	⋮	⋮	⋮
25,400	M	2.9	Yes	1	2.9	1	1
Total				τ_1	τ_2	τ_3	τ_4
Mean				μ_1	μ_2	μ_3	μ_4

students who smoke. The population mean μ_3 is the proportion of smokers among all students in the university.

Finally, variable Y_4 in col. (8) is the product of Y_1 and Y_3. It is equal to 1 if the student is male and a smoker or 0 if otherwise. Observe that the population total of this variable, τ_4, is the number of male students who smoke, and that the ratio τ_4/τ_1 equals the proportion of smokers among male students in the university.

We see that the subpopulation mean and proportion of interest in this illustration (the average income of male students, and the proportion of smokers among male students) are in effect ratios of population totals of created variables. Therefore, point and interval estimates of these population characteristics can be calculated as described in Section 7.2.

We see also that the subpopulation totals of interest in this illustration (the total income of male students, and the number of male student smokers) are the ordinary population totals of the created variables Y_2 and Y_4. Point and interval estimates for these totals can be calculated as shown in Chapter 3.

☑ EXAMPLE 7.2   Let us suppose that a simple random sample of four students yielded the observations shown in cols. (1) to (4) of Table 7.5. The variables Y_1 to Y_4, shown in cols. (5) to (8), have the same interpretation as in Table 7.4. The sample size, of course, is very small. We shall pretend it is large in order to explain the calculations as simply as possible.

The estimate of the total income of male students in the university is

$$T_2 = N\bar{Y}_2 = (25,400)(4.575) = 116,205 \ (\$000)$$

TABLE 7.5 Sample of University Students, Example 7.2

Student No. (1)	Gender (2)	Annual Income, Y ($000) (3)	Smoker? (4)	Created Variables Y_1 (5)	$Y_2 = Y_1 Y$ (6)	Y_3 (7)	$Y_4 = Y_1 Y_3$ (8)
1	F	2.3	Yes	0	0	1	0
2	M	7.1	No	1	7.1	0	0
3	M	6.4	Yes	1	6.4	1	1
4	M	4.8	No	1	4.8	0	0
Total				3	18.3	2	1
Average				0.75	4.575	0.5	0.25

and that of the number of male students who smoke is

$$T_4 = N\bar{Y}_4 = (25{,}400)(0.25) = 6{,}350.$$

The estimate of the average annual income of male students in the university (τ_2/τ_1) is

$$\frac{\bar{Y}_2}{\bar{Y}_1} = \frac{4.575}{0.75} = 6.1 \text{ (\$000)}.$$

This estimate can also be written as

$$\frac{n\bar{Y}_2}{n\bar{Y}_1} = \frac{(4)(4.575)}{(4)(0.75)} = \frac{18.3}{3} = 6.1.$$

Clearly, this estimate is simply the average income of male students in the sample.

The estimate of the proportion of smokers among male students in the university (τ_4/τ_1) is

$$\frac{\bar{Y}_4}{\bar{Y}_1} = \frac{0.25}{0.75} = 0.333.$$

This estimate can also be written as

$$\frac{n\bar{Y}_4}{n\bar{Y}_1} = \frac{(4)(0.25)}{(4)(0.75)} = \frac{1}{3} = 0.333,$$

and is simply the proportion of smokers among male students in the sample.

To form a confidence interval for the total income of male students and the number of male students who smoke, we calculate first the sample variances of the created variables Y_2 and Y_4. Applying Eq. (3.18), we find

$$S_2^2 = \frac{1}{4}(0^2 + 7.1^2 + 6.4^2 + 4.8^2) - (4.575)^2 = 7.672$$

and

$$S_4^2 = \frac{1}{4}(0^2 + 0^2 + 1^2 + 0^2) - (0.25)^2 = 0.1875.$$

Following the procedure described in Summary 3.3, we calculate next

$$\widehat{Var}(\bar{Y}_2) = \frac{S_2^2}{n-1}\frac{N-n}{N} = \frac{7.672}{4-1}\frac{25,400-4}{25,400} = 2.557$$

and

$$\widehat{Var}(\bar{Y}_4) = \frac{S_4^2}{n-1}\frac{N-n}{N} = \frac{0.1875}{4-1}\frac{25,400-4}{25,400} = 0.06249.$$

It follows that an approximate, say, 95% confidence interval for the total income of male students in the university is

$$N[\bar{Y}_2 \pm Z_{\alpha/2}\sqrt{\widehat{Var}(\bar{Y}_2)}] = (25,400)[4.575 \pm 1.96\sqrt{2.557}],$$

or from 36,597 to 195,813 ($000). Similarly, it can be verified that a 95% confidence interval for the number of male students who smoke is from −6,095 to 18,795. The negative lower limit, obviously, does not make sense, and reflects the poor quality of the approximations in such a small sample.

Preliminary calculations for forming confidence intervals regarding τ_2/τ_1 and τ_4/τ_1 are shown in Table 7.6, where $R_2 = \bar{Y}_2/\bar{Y}_1 = 6.1$ and $R_4 = \bar{Y}_4/\bar{Y}_1 = 0.333$.
Applying Eq. 7.4,

$$\widehat{Var}(R_2) = \frac{1}{\bar{Y}_1^2}\frac{N-n}{nN(n-1)}\sum_{i=1}^{n}(Y_{2i} - R_2 Y_{1i})^2$$

$$= \frac{1}{0.75^2}\frac{25,400-4}{(4)(25,400)(4-1)}(2.78)$$

$$= 0.4118.$$

TABLE 7.6   Preliminary Calculations, Example 7.2

i	Y_{1i}	Y_{2i}	$(Y_{2i} - R_2 Y_{1i})^2$	Y_{4i}	$(Y_{4i} - R_4 Y_{1i})^2$
1	0	0	0	0	0
2	1	7.1	1	0	0.1109
3	1	6.4	0.09	1	0.4449
4	1	4.8	1.69	0	0.1109
Total			2.78		0.6667

Similarly,

$$Var(\widehat{R_4}) = \frac{1}{0.75^2} \frac{25,400 - 4}{(4)(25,400)(4-1)}(0.6667) = 0.09875.$$

Therefore, an approximate 95% confidence interval for the average income of male students in the university (τ_2/τ_1) is

$$6.1 \pm 1.96\sqrt{0.4118},$$

or from about 4.842 to 7.358 ($000); that for the proportion of male students who smoke is

$$0.333 \pm 1.96\sqrt{0.09875},$$

or from about -0.283 to 0.949. The sample size is much too small, of course, for the approximations to be satisfactory or the lower limit to make sense.   ◪

We found it convenient to view a subpopulation mean or proportion as a ratio of population totals of created variables, and showed that the reasonable estimators (the average value of the variable or proportion in the category among the sampled elements belonging to the subpopulation) can be expressed in the form of estimators of these ratios. Although the latter are not unbiased for any size of simple random sample, they can be shown to be unbiased in this special case under certain conditions, as the following example will demonstrate.

◪ **EXAMPLE 7.3**   Consider the Ackroyd population of Table 3.1 and suppose that firms A, B, C, and F are located in the East side of the town, while D and E are located in the West side. Suppose we are interested in estimating the average number of employees of the East firms and the proportion of East firms that intend to hire.

For the purpose of this illustration, the firms in the East side make up the subpopulation of interest. As shown in Table 3.1, the East firms have 9, 8, 6, and 5 employees respectively; the average of these numbers, the subpopulation mean number of employees, is $\mu_d = 7$. Of the four firms in the East side, two (A and B) intend to hire; the proportion planning to hire in the subpopulation is therefore 2/4 or $\pi_d = 0.5$. The subscript d stands for "domain."

Let us pretend that these two subpopulation characteristics are not known, and examine the behavior of their estimators in a simple random sample of two firms selected without replacement. The calculations are laid out in Table 7.7.

The first four columns of Table 7.7 are identical to Table 3.2. The East firms are shown in boldface. Column (5) shows the number of East firms (n_d) in the sample, col. (6) the average number of employees of East firms in the sample $(\bar{Y}_d)$, and col. (7) the proportion of East firms intending to hire (P_d).

To illustrate, suppose the sample consists of firms B and C. Both firms are located in the East side, therefore $n_d = 2$; the average number of their employees is $(8 + 6)/2$ or 7, and the proportion intending to hire is 1/2 or 0.5.

TABLE 7.7 Estimation of Subpopulation Mean and Proportion, Ackroyd Case

Outcome (1)	Number of Employees (2)	Hiring? (Y:Yes, N:No) (3)	Proba-bility (4)	Number of East Firms, n_d (5)	$\bar{Y}_d$ (6)	P_d (7)
A,B	9,8	Y,Y	1/15	2	8.5	1.0
A,C	9,6	Y,N	1/15	2	7.5	0.5
A,D	9,2	Y,N	1/15	1	9.0	1.0
A,E	9,1	Y,N	1/15	1	9.0	1.0
A,F	9,5	Y,N	1/15	2	7.0	0.5
B,C	8,6	Y,N	1/15	2	7.0	0.5
B,D	8,2	Y,N	1/15	1	8.0	1.0
B,E	8,1	Y,N	1/15	1	8.0	1.0
B,F	8,5	Y,N	1/15	2	6.5	0.5
C,D	6,2	N,N	1/15	1	6.0	0.0
C,E	6,1	N,N	1/15	1	6.0	0.0
C,F	6,5	N,N	1/15	2	5.5	0.0
D,E	2,1	N,N	1/15	0	-	-
D,F	2,5	N,N	1/15	1	5.0	0.0
E,F	1,5	N,N	1/15	1	5.0	0.0
			─────			
			15/15			

col. (6): Average number of employees of East firms.
col. (7): Proportion of East firms intending to hire.

Observe that n_d, like $\bar{Y}_d$ and P_d, varies from sample to sample. In the event the sample consists of firms D and E, none of the selected firms is located in the East side, hence it is not possible to estimate μ_d and π_d.

We shall agree to make estimates only when it is possible, that is, when $n_d \geq 1$. The probability that the sample will consist of, say, B and C when estimates can be made is 1/14 (in the long run, B and C are expected to be selected once in every 14 samples that do not consist of D and E). Table 7.8 shows the probability distributions of $\bar{Y}_d$ and P_d.

It is now easy to calculate the expected values of $\bar{Y}_d$ and P_d, and to confirm that these equal the subpopulation characteristics of interest:

$$E(\bar{Y}_d) = (5.0)(2/14) + \cdots + (9.0)(2/14) = 7 = \mu_d,$$
$$E(P_d) = (0.0)(5/14) + \cdots + (1.0)(5/14) = 0.5 = \pi_d.$$

We conclude that $\bar{Y}_d$ and P_d are unbiased, provided that estimates will be made only when $n_d \geq 1$. ◼

Until now we considered only simple random sampling, but the approach of this section can be extended to stratified and two-stage sampling. In general, the

TABLE 7.8 Probability Distribution of $\bar{Y}_d$ and P_d

$\bar{Y}_d$	Probability	P_d	Probability
5.0	2/14	0.0	5/14
5.5	1/14	0.5	4/14
6.0	2/14	1.0	5/14
6.5	1/14		
7.0	2/14		1
7.5	1/14		
8.0	2/14		
8.5	1/14		
9.0	2/14		
	1		

subpopulation mean of a variable Y can be regarded as a ratio τ_2/τ_1 of two created variables Y_2 and Y_1, where

$$Y_1 = \begin{cases} 1 \text{ if the element belongs to the subpopulation} \\ 0 \text{ otherwise} \end{cases}$$

and

$$Y_2 = \begin{cases} Y \text{ if the element belongs to the subpopulation} \\ 0 \text{ otherwise} \end{cases}$$

The ratio τ_2/τ_1 can always be estimated by the ratio of unbiased estimators of these totals under the sampling method used.

Similarly, a subpopulation proportion is the ratio of the population totals τ_4/τ_1 of two created variables Y_1 and Y_4, where

$$Y_4 = \begin{cases} 1 \text{ if the element belongs to the category and the subpopulation} \\ 0 \text{ otherwise.} \end{cases}$$

τ_4/τ_1 can be estimated by the ratio of unbiased estimators of the population totals τ_4 and τ_1 under the sampling method.

When the sample sizes involved are large, the above estimators can be expected to be approximately unbiased. Large-sample confidence intervals can also be constructed, but they are not presented here as the formulas tend to be quite complicated. The following example illustrates the calculation of point estimates under stratified sampling.

■ EXAMPLE 7.4 Suppose that the university described earlier is one of three in a region. Simple random samples were selected also from the lists of students in the other two universities. The results of this stratified sample are summarized in Table 7.9.

TABLE 7.9   Stratified Random Sample of University Students

University, i	Number of Students, N_i	$w_i = N_i/N$	Sample Means		
			$\bar{Y}_{1i}$	$\bar{Y}_{2i}$	$\bar{Y}_{4i}$
1	25,400	0.508	0.75	4.575	0.25
2	6,700	0.134	0.60	5.120	0.30
3	17,900	0.358	0.70	4.225	0.28
	50,000	1.000			

$\bar{Y}_{1i}$, $\bar{Y}_{2i}$, and $\bar{Y}_{4i}$ are the means of the created variables in the sample from university i. The first set was calculated in Table 7.5. In the second and third lines are the results for the other two universities. In all other respects, the situation is as described in Example 7.2.

The stratified estimator of the population mean of the created variable Y_1 (that is, of the proportion of males among all university students in the region) is

$$\bar{Y}_{1s} = w_1 \bar{Y}_{11} + w_2 \bar{Y}_{12} + w_3 \bar{Y}_{13}$$
$$= (0.508)(0.75) + (0.134)(0.60) + (0.358)(0.70)$$
$$= 0.712.$$

In a similar fashion, we calculate $\bar{Y}_{2s} = 4.523$ and $\bar{Y}_{4s} = 0.267$. Thus, the estimate of the average income of male students in the region is

$$\frac{\bar{Y}_{2s}}{\bar{Y}_{1s}} = \frac{4.523}{0.712} = 6.352 \ (\$000).$$

The estimate of the proportion of smokers among male students in the region is

$$\frac{\bar{Y}_{4s}}{\bar{Y}_{1s}} = \frac{0.267}{0.712} = 0.375. \quad \blacksquare$$

7.4 CLUSTER SAMPLING— AN ALTERNATIVE VIEW

Cluster sampling, a special case of two-stage sampling, consists of selecting a random sample of groups and *all* their elements. In effect, then, cluster sampling is simple random sampling of groups.

Suppose that a sample of the 1,600 households in a town will be taken to estimate the average age of all town residents and the proportion of residents that are divorced. It is assumed that every town resident belongs to a household with

1, 2, or more members. Let us view the town as a population of *households* rather than of residents grouped into households, and let us imagine this population as it is outlined in Table 7.10.

For example, if household 2 consists of a divorced 35-year-old single parent with two children aged 10 and 5, then $Y_1 = 3$, $Y_2 = 50$, and $Y_3 = 1$.

Observe that the average age of all town residents is equal to the ratio τ_2/τ_1, and that the proportion of town residents who are divorced is equal to the ratio τ_3/τ_1. These population characteristics, therefore, may be estimated by the ratios of the sample means $\bar{Y}_2/\bar{Y}_1$ and $\bar{Y}_3/\bar{Y}_1$, rather than the cluster estimators of Section 5.4.

■ EXAMPLE 7.5   Suppose that a simple random sample of three households produced the observations shown in Table 7.11.

Note that the averages are per household. The estimate of the average age of town residents is $\bar{Y}_2/\bar{Y}_1 = 57/2$ or 28.5. The estimate of the proportion of divorced town residents is $\bar{Y}_3/\bar{Y}_1 = 0.333/2$ or 0.167. (The same results can be obtained as ratios of the sample totals 171/6 and 1/6, respectively.)   ■

TABLE 7.10   Population of Households

Household No.	Number of Members, Y_1	Total Age of Members, Y_2	Number of Divorced Members, Y_3
1	5	175	0
2	3	50	1
⋮	⋮	⋮	⋮
1,600	2	124	0
Total	τ_1	τ_2	τ_3

TABLE 7.11   Sample of Households, Example 7.5

Household No.	Number of Members, Y_1	Total Age of Members, Y_2	Number of Divorced Members, Y_3
1	2	85	0
2	1	26	1
3	3	60	0
Total	6	171	1
Average	2	57	0.333

In general, the alternative estimator of the population mean of a variable Y under cluster sampling can be written as

$$\frac{\text{Sum of all } Y \text{ values in the sample}}{\text{Number of elements in the sample}}.$$

The alternative estimator of the proportion of elements in a category of interest is

$$\frac{\text{Number of sampled elements in the category}}{\text{Number of elements in the sample}}.$$

If the number of sampled groups is large, the alternative estimators are approximately unbiased, and approximate confidence intervals can be constructed as described in Summary 7.1.

7.5 POSTSTRATIFICATION

Consider a situation in which a stratified random sample is desirable, but the elements cannot be assigned to groups until after the sample is taken.

Think, for example, of a survey of households in a city to estimate the average annual household expenditures for home repairs and improvements. Since these expenditures are expected to be correlated with the age of the household's residence, it would appear desirable to stratify the city residences into age groups. Assume there is one household per residence and one residence per household. A commercially compiled list of all residences in the city is available, giving the address and postal code—but not the age—of each residence. A random sample stratified according to age, therefore, cannot be implemented. It is possible, however, to determine from published reports based on the city property tax records how many residences fall into each of a number of age intervals. Consider selecting a *simple* random sample of residences, determining from each sampled household the age of its residence, classifying the sampled households into the age groups, and calculating the *poststratified estimator*

$$\bar{Y}_{ps} = w_1 \bar{Y}_1 + w_2 \bar{Y}_2 + \cdots + w_M \bar{Y}_M = \frac{N_1}{N} \bar{Y}_1 + \frac{N_2}{N} \bar{Y}_2 + \cdots + \frac{N_M}{N} \bar{Y}_M, \quad (7.5)$$

where M is the number of age groups, N_i the known number of residences in age group i, and $\bar{Y}_i$ the average expenditures of the sampled households in group i:

$$\bar{Y}_i = \frac{\text{Sum of expenditures of sampled households in group } i}{\text{Number of sampled households in group } i}.$$

The subscript ps stands for "poststratified."

Three features of the poststratified estimator 7.5 should be noted: (a) it is calculated exactly like the stratified estimator; (b) it is based on the results of a *simple*—not a stratified—random sample; (c) the weights $w_i = N_i/N$ are assumed known.

Note also that $\bar{Y}_i$ is the estimator of the subpopulation mean μ_i (the mean expenditures by all households in group i), and that $\bar{Y}_{ps}$ is a weighted average of these estimators.

When the size of the simple random sample, n, is large, it can be expected that the proportion of sampled elements falling into a given group will be approximately equal to the proportion of elements of that group in the population, that is,

$$\frac{n_i}{n} \approx \frac{N_i}{N}.$$

In other words, when n is large, it can be expected that the poststratified estimator based on a simple random sample will behave like the stratified estimator based on a proportional stratified sample.

Indeed, it can be shown that for large n, the poststratified estimator of the population mean of a variable Y is approximately unbiased, and that its approximate variance is given by

$$Var(\bar{Y}_{ps}) \approx \frac{1}{n}(1 - \frac{n}{N}) \sum_{i=1}^{M} w_i \sigma_i^2 + \frac{1}{n^2} \sum_{i=1}^{M}(1 - w_i)\sigma_i^2, \qquad (7.6)$$

where σ_i^2 is the variance of the Y values of all elements in the ith group.

The first term of Eq. 7.6 is the variance of a stratified estimator under proportional stratified sampling when the N_i are large (see Eq. 4.6 of Section 4.5). The second term is always positive, but can be expected to be small when n is large. We conclude that the accuracy of the poststratified estimator is always less than that of the stratified estimator based on a proportional stratified sample, but the difference can be expected to be small when the simple random sample upon which the poststratified estimator is based is large.

Again for large n, an approximate $100(1 - \alpha)\%$ confidence interval for the population mean of the variable Y is given by

$$\bar{Y}_{ps} \pm Z_{\alpha/2}\sqrt{\widehat{Var(\bar{Y}_{ps})}}, \qquad (7.7)$$

where $\widehat{Var(\bar{Y}_{ps})}$ is Eq. 7.6 with the σ_i^2 replaced by

$$\hat{S}_i^2 = \frac{n_i}{n_i - 1} \frac{N_i - 1}{N_i} S_i^2.$$

S_i^2 is the variance of the Y values of the sampled elements falling into group i. When n_i and N_i are large, $\hat{S}_i^2 \approx S_i^2$.

☑ EXAMPLE 7.6   A simple random sample of 1,200 households was selected from a city of 45,000 households. Two of the questions in the questionnaire were as follows:

5. When was your residence built?

5.a __ Less than 5 years ago

5.b __ 5 to 10 years ago

5.c __ More than 10 years ago

17. How much were your household's expenditures last year for repairs and improvements to your residence? $__ .

Table 7.12 summarizes the relevant information.

The numbers in cols. (2) and (3) are assumed known from property tax records. Columns (4) to (7) are based on a simple random sample of 1,200 households. For example, 140 of the 1,200 sampled households lived in residences less than 5 years old; their average expenditure for repair and improvements was $350, and the variance of these expenditures was 610.

The poststratified estimate of the average household expenditures for repairs and improvements in the city is

$$\bar{Y}_{ps} = (0.111)(350) + (0.333)(675) + (0.556)(920) = \$775.14.$$

The estimate of $Var(\bar{Y}_{ps})$ is

$$\widehat{Var(\bar{Y}_{ps})} = \frac{1}{1,200}(1 - \frac{1,200}{45,000})[(0.111)(614) + (0.333)(752) + (0.556)(941)]$$

$$+ \frac{1}{1,200^2}[(1 - 0.111)(614) + (1 - 0.333)(752) + (1 - 0.556)(941)]$$

$$= 0.682766 + 0.01017$$

$$= 0.684.$$

An approximate, say, 95% confidence interval for μ is

$$775.14 \pm 1.96\sqrt{0.684},$$

TABLE 7.12   Data for Example 7.6

Age of Residence (1)	Number of Households, N_i (2)	$w_i = N_i/N$ (3)	Sample of Households			
			n_i (4)	$\bar{Y}_i$ (5)	S_i^2 (6)	$\hat{S}_i^2$ (7)
Less than 5 years	5,000	0.111	140	350	610	614
5 to 10 years	15,000	0.333	420	675	750	752
More than 10 years	25,000	0.556	640	920	940	941
	45,000	1.000	1,200			

or from $ 773.52 to $776.77. The corresponding interval estimate for the total expenditures by households in the city is obtained by multiplying the above limits by 45,000. ☑

7.6 DOUBLE SAMPLING

Double or *two-phase sampling* is used in some situations where making use of an auxiliary variable is desirable, but the necessary information for stratification or for the application of ratio or regression estimators is lacking.

Consider for example the problem of estimating book sales in a region. It may be reasonable to assume that there is a positive correlation between the size of a bookstore and its sales of any book title. It appears desirable to use a random sample of book outlets stratified according to size into, say, Small, Medium, and Large. Suppose that a list of all retail book outlets in the region can be compiled, but this can only be a list of outlets and their addresses. There is no information on the size of the outlets. In other words, it is not possible to stratify the outlets according to their size.

A possible alternative to simple random sampling is to proceed in two phases. In the first phase, select a simple random sample of outlets, determine their sizes, and estimate the proportion of Small, Medium, and Large outlets in the population. In the second phase, select a simple random subsample of outlets from each set of Small, Medium, and Large outlets selected in the first phase. Determine their sales of the book title of interest and the mean sales per selected outlet in the group. Finally, calculate the stratified estimate using the group mean sales from the second phase and the estimated proportions from the first phase.

For example, suppose a simple random sample of 20 was selected from a population of 500 outlets. Table 7.13 shows the size of the sampled outlets (X, in square feet) and their classification into Small (less than 500 sq. ft.), Medium (500 to 1500 sq. ft.), and Large (over 1500 sq. ft.). The group size limits are, of course, arbitrary. The average size of the selected outlets is 930 sq. ft., as can be easily confirmed.

Since 10 of the 20 outlets are Small, 6 are Medium, and 4 Large, the estimated proportions of outlets in these three groups are $\hat{w}_1 = 0.50$, $\hat{w}_2 = 0.30$, and $\hat{w}_3 = 0.20$, respectively.

Suppose that a simple random subsample of size 4 is drawn from the 10 Small outlets selected in the first phase, and the number of copies of the given title sold by these outlets determined. Likewise, random subsamples of 3 and 2 outlets are selected from the Medium and Large groups, respectively. The results are shown in Table 7.14.

The double sample estimate of the average sales of the title per outlet in the region is

$$\hat{w}_1\bar{Y}_1 + \hat{w}_2\bar{Y}_2 + \hat{w}_3\bar{Y}_3 = (0.50)(37.5) + (0.30)(86.67) + (0.20)(265) = 97.75,$$

TABLE 7.13 Phase I Results

Outlet No.	Size, X	Group	Outlet No.	Size, X	Group
1	300	S	11	200	S
2	1,000	M	12	1,800	L
3	1,400	M	13	450	S
4	2,500	L	14	1,200	M
5	400	S	15	400	S
6	200	S	16	700	M
7	700	M	17	2,500	L
8	350	S	18	250	S
9	3,200	L	19	600	M
10	150	S	20	300	S

TABLE 7.14 Phase II Results

Small		Medium		Large	
Outlet No.	Sales	Outlet No.	Sales	Outlet No.	Sales
5	50	2	120	9	350
8	30	7	60	17	180
11	25	16	80		
15	45				
Mean	37.5	Mean	86.67	Mean	265

and the estimate of the total sales of this title in the region (500)(97.75) or 48,875 copies.

In general, under double sampling for stratification, a simple random sample of size n' is first selected from a population of N elements. The unknown proportion of elements in group i, $w_i = N_i/N$, is estimated by $\hat{w}_i = n'_i/n'$, where n'_i is the number of sampled elements falling into group i. Next, random subsamples are selected from the selected elements in each group, and the sample means of a variable of interest $\bar{Y}_i$ determined. The double sample stratified estimator of the population mean of a variable Y is

$$\hat{w}_1 \bar{Y}_1 + \hat{w}_2 \bar{Y}_2 + \cdots + \hat{w}_M \bar{Y}_M.$$

It resembles the familiar stratified estimator, $\bar{Y}_s$, except that the unknown w_i are replaced by the estimates $\hat{w}_i$.

If it is agreed that estimates will be made only when all $n'_i > 0$, a condition we expect to be satisfied when $n'w_i$ are large, then the double sampling estimator can be shown to be approximately unbiased.

Stratification, of course, is not the only method for exploiting a relationship between the variable of interest and an auxiliary variable. Ratio and regression estimation are alternatives, and double sampling may be applied if the required information is not initially available.

The ratio estimator, it will be recalled, is

$$\bar{Y}_r = \frac{\bar{Y}}{\bar{X}}\mu_x,$$

and the regression estimator is given by

$$\bar{Y}_{lr} = \bar{Y} + b(\mu_x - \bar{X}).$$

If the population mean of the auxiliary variable (μ_x) is not known, a simple random sample may be taken first to measure X and estimate μ_x, and then a subsample selected to measure Y and determine $\bar{X}$, $\bar{Y}$, and (in the case of the regression estimator) b. This second subsample may be simple, stratified, or of some other form.

To illustrate these applications of double sampling, consider again the estimation of book sales in a region. A simple random sample of 20 of the 500 outlets was selected, and the size of the selected outlets measured as shown in Table 7.13. The population average size of outlets (μ_x) can be estimated by the sample average size $\bar{X}' = 930$. A stratified subsample of these outlets was taken to measure the sales of the given title. The results were shown in Table 7.14 and are reproduced for present purposes in Table 7.15.

TABLE 7.15 Double Sampling for Ratio and Regression Estimation

Outlet No.	Size, X	Sales, Y	XY	X^2
5	400	50	20,000	160,000
8	350	30	10,500	122,500
11	200	25	5,000	40,000
15	400	45	18,000	160,000
2	1,000	120	120,000	1,000,000
7	700	60	42,000	490,000
16	700	80	56,000	490,000
9	3,200	350	1,120,000	10,240,000
17	2,500	180	450,000	6,250,000
Total	9,450	940	1,841,500	18,952,500

From this second sample we calculate $\bar{X} = 9{,}450/9 = 1{,}050$, $\bar{Y} = 940/9 = 104.4$, and

$$b = \frac{\sum XY - n\bar{X}\bar{Y}}{\sum X^2 - n\bar{X}^2} = \frac{(1{,}841{,}500) - (9)(1{,}050)(104.4)}{(18{,}952{,}500) - (9)(1{,}050)^2} = 0.0947.$$

The double sample ratio estimate of the average sales per retail book outlet in the region is

$$\frac{\bar{Y}}{\bar{X}}\bar{X}' = (\frac{104.4}{1{,}050})(930) = 92.47.$$

The double sample regression estimate is

$$\bar{Y} + b(\bar{X}' - \bar{X}) = (104.4) + (0.0947)(930 - 1{,}050) = 93.04.$$

The estimated total sales in the region would be calculated by multiplying the above figures by 500, the number of outlets in the region.

In general, the ratio and regression estimators based on double sampling are not unbiased, although their bias can be expected to be small when the sizes of the samples used in the first and second phase are large. Independent samples, rather than subsamples, may also be used in the second phase. The variances of all estimators based on double sampling depend on the manner in which the samples are selected and are not presented here because they tend to be quite complicated, as are their estimates needed for the calculation of confidence intervals.

7.7 TO SUM UP

- A reasonable estimator of the ratio of the population totals of two variables is the ratio of the estimators of these totals. In general, this estimator is not unbiased, but is approximately so when the sample is simple and large, in which case approximate confidence intervals can also be constructed.
- A reasonable estimator of the average value of a variable in a subpopulation is the average value of the variable among the sampled elements that belong to the subpopulation. Likewise, a reasonable estimator of the proportion of elements in a subpopulation that belong to a certain category is the corresponding sample proportion. These population characteristics can be regarded as ratios of population totals of specially created variables, and, if the sample is simple random, their reasonable estimators are the general estimators for such ratios. This conclusion also suggests using the ratio of estimates of the totals of the specially created variables when the sample is stratified, two-stage, or of some other form.

- An alternative estimator of the population mean of a variable Y under cluster sampling is the ratio of the sum of all Y values in the sample to the number of sampled elements. This estimator, and the similar one for a population proportion, is approximately unbiased in large cluster samples.
- Poststratification is sometimes practiced when stratification appears desirable, the proportions of elements in the groups are known, but the elements cannot be assigned to the groups until after a simple sample is selected. A poststratified estimator looks exactly like a stratified one, except that it is based on a simple—not stratified—random sample. When the size of such a sample is large, the poststratified estimator can be expected to have the properties of a stratified estimator under proportional stratified sampling.
- Double sampling, like poststratification, aims to make use of an auxiliary variable when the necessary information for stratification or ratio and regression estimation is not available. A simple random sample of elements provides estimates of the proportions of elements in the groups or of the population mean of the auxiliary variable. A subsample then provides the rest of the information needed to calculate stratified, ratio, or regression estimates of population characteristics.

PROBLEMS

7.1 *Computing exercise:* Using the program SCALC, verify the numerical results presented in Section 7.2. Note any deviations, and determine if these are the result of rounding at intermediate stages.

7.2 The ratio of population totals of two variables Y_2 and Y_1, τ_2/τ_1, can be interpreted as the average value of Y_2 per unit of Y_1. In Table 7.1, for example, the ratio τ_2/τ_1 is the average dental care expenditure per household income earner. A related average is the average of ratios

$$\frac{1}{N}\sum_{i=1}^{N}\frac{Y_{2i}}{Y_{1i}},$$

where Y_{2i} and Y_{1i} are the values of the variables Y_2 and Y_1 for population element i. In the following questions, assume simple random sampling will be used.

(a) Indicate an unbiased estimator of the average of ratios and describe how a large-sample confidence interval for this average may be constructed. Briefly but clearly explain your answer.

(b) Using the data of Table 7.2, estimate the average of the ratios Y_2/Y_1 (dental care expenditure per income earner) and, pretending that the sample size is large, construct a 95% confidence interval for this population characteristic.

(c) Using the data of Table 7.2, estimate the average of the ratios Y_3/Y_1 (health care expenditure per income earner) and, pretending that the sample size is large, construct a 95% confidence interval for this population characteristic.

(d) Using the data of Table 7.2, estimate the average of the ratios Y_3/Y_2 (health care expenditure per dollar of dental care expenditure) and, pretending that the sample size is large, construct a 95% confidence interval for this population characteristic.

7.3 Confirm the calculation in Example 7.1 of the 95% confidence interval (0.033 to 0.405) for the ratio of total dental to total health expenditures.

7.4 The managing committee of an association of professionals plans to select a simple random sample of its 15,000 members to gain some insight into the attitude of the membership concerning an association-sponsored life insurance plan now being negotiated with an insurance company. Among the planned questions are the following:

- Is your office located at home? Yes__ No__
- What was your revenue during the last calendar year? __ ($000)
- Are you in favor of the enclosed plan? Yes__ No__

For the purpose of this problem, assume that a simple random sample of five members has been selected. The responses are shown in Table 7.16.

(a) Estimate the total revenue of members with an office at home.

(b) Estimate the number of members with an office at home and in favor of the insurance plan.

(c) Estimate the average revenue of members with an office at home.

(d) Estimate the proportion of members with an office at home and in favor of the insurance plan.

(e) Pretending that the sample size is large, form 90% confidence intervals for the population characteristics in (i) Question (a); (ii) (b); (iii) (c); (iv) (d).

TABLE 7.16   Data for Problem 7.4

Selected Member	Office at Home?	Revenue ($000)	In Favor of Plan?
1	Yes	130	No
2	No	250	No
3	No	320	No
4	Yes	90	Yes
5	No	400	Yes

7.5 The managing director of the city's major newspaper requires some hard facts concerning the appeal of the newspaper's travel section. She commissioned interviews with a simple random sample of subscribers. Among the questions were the following:

- Do you read the travel section regularly? Yes__ No__
- Did you or a member of your family travel abroad last year? Yes__ No__
- How much did you and your family spend on travel last year? __ ($)

The newspaper has 40,000 subscribers. For the purpose of this exercise, assume that a simple random sample of four subscribers was selected, and pretend that the sample size is large. The responses are shown in Table 7.17.

(a) Estimate the proportion and number of subscribers who regularly read the travel section.

(b) Estimate the proportion and number of regular readers of the travel section who traveled abroad.

(c) Estimate the total and average travel expenditure among the newspaper's subscribers.

(d) Estimate the total and average travel expenditure among the regular readers of the travel section.

(e) Calculate 90% confidence intervals for the population characteristics in (i) Question (a); (ii) (b); (iii) (c); (iv) (d).

7.6 Refer to Example 7.3. If the number of elements in the subpopulation (N_d) is known, the obvious estimate of the subpopulation total of a variable Y is $T_d = N_d \bar{Y}_d$, and that of the number of elements in a subpopulation in a given category $T'_d = N_d P_d$.

Pretend you do not know that $\tau_d = 28$ and $\tau'_d = 2$, but you do know that there are four firms in the East side (i.e., $N_d = 4$). Making use of Table 7.8, calculate the expected values of the estimators $T_d = N_d \bar{Y}_d$ and $T' = N_d P_d$, to confirm that

TABLE 7.17 Data for Problem 7.5

Selected Subscriber	Regular Reader?	Travel Expenditure	Travel Abroad?
1	Yes	1,500	No
2	No	400	Yes
3	Yes	800	Yes
4	Yes	300	No

these estimators are unbiased. (Remember that T_d and T_d' are estimated only when $n_d \geq 1$.)

7.7 Refer to Example 7.3. If the number of elements in the subpopulation (N_d) is not known, as is often the case in practice, then a reasonable estimator of $\tau_d = N_d \mu_d$ is

$$T_d = N \frac{n_d}{n} \bar{Y}_d.$$

In effect, T_d applies to the estimated number of elements in the subpopulation (Nn_d/n) the estimated average value of Y in the subpopulation $(\bar{Y}_d)$. A reasonable estimator of the subpopulation total in a category, $\tau_d' = N_d \pi_d$, is

$$T_d' = N \frac{n_d}{n} P_d,$$

which multiplies the estimated number in the subpopulation by the estimate of the subpopulation proportion in the category (P_d).

(a) Show that T_d and T_d' are the same as the estimators $T_2 = N\bar{Y}_2$ and $T_4 = N\bar{Y}_4$ in Example 7.2 and Table 7.5.

(b) You would like to estimate the total number of employees of the firms located in the East side of Ackroyd (τ_d), and the number of East side firms that intend to hire (τ_d').

Pretend you do not know that $\tau_d = 28$ and $\tau_d' = 2$. Making use of Table 7.7, calculate the expected values of the estimators T_d and T_d' based on a simple random sample of size $n = 2$ without replacement. Confirm that T_d and T_d' are unbiased estimators of τ_d and τ_d', respectively, if it is agreed that the estimates will be 0 when $n_d = 0$.

7.8 Consider Example 7.4, but suppose that there are 10 universities in the region with a total of 150,000 students, and that two-stage rather than stratified sampling is used. Specifically, suppose that three of the 10 universities in the region were randomly selected in the first stage; then, in the second stage, simple random samples of students were drawn from each university selected in the first stage. Suppose further that the results of this two-stage sample are as shown in Table 7.9, except that $w_1 = 25{,}400/150{,}000 = 0.169$, $w_2 = 0.045$, and $w_3 = 0.119$.

(a) Calculate the two-stage estimate of the proportion of males among all university students in the region.

(b) Calculate the two-stage estimate of the average income of all university students in the region.

(c) Estimate the average income of male university students in the region on the basis of this two-stage sample.

(d) Estimate the proportion of smokers among male university students in the region.

(e) Compare the estimates in **(c)** and **(d)** with those based on the stratified sample of Example 7.4. Should there be a difference in this case? Explain why.

7.9 Consider Problem 5.21.

(a) Using the alternative estimator of Section 7.4, estimate the proportion and number of residents in the town who visited a physician in the preceding month.

(b) Likewise, estimate the average expenditure on drugs per person and the total expenditure on drugs in the town.

(c) Calculate 90% confidence intervals for the population characteristics in **(a)** and **(b)** using the alternative cluster estimator.

(d) Compare the above calculations with the those of Problem 5.21. Comment.

7.10 A pharmaceutical company commissioned a cluster sample in a certain region for the purpose of estimating the potential size of the market for medicines against indigestion and the characteristics of the potential buyers for such medicines. Of particular interest were the proportion of adults who suffer from indigestion and the average number of hours adults watch television per day on average (adults being persons 20 years old or older). According to the latest census, the region has a population of 1.4 million adults and 0.8 million households.

For the purpose of this problem, suppose that a simple random sample of five households was selected and all adults in the selected households interviewed. The responses are summarized in Table 7.18.

Y_1 is the number of adults in the household, Y_2 the number of adults in the household that suffer from indigestion, and Y_3 the total number of hours adults in the household watch television daily.

(a) Estimate the proportion and number of adults suffering from indigestion in the region.

(b) Estimate the average number of hours adults watch television daily.

TABLE 7.18 Data for Problem 7.10			
Household Number	Y_1	Y_2	Y_3
1	2	1	6.2
2	1	0	2.7
3	2	1	8.2
4	3	1	6.5
5	2	0	4.8

(c) Pretending that the number of sampled households is large, calculate 90% confidence intervals for the population characteristics estimated in (a) and (b).

7.11 Four times a year, a cable television company selects a simple random sample of 1,000 of its 70,000 subscribers to assess their level of satisfaction with the service provided by the company. The selected clients receive a short questionnaire, among the questions of which is the following:

• On a scale from 0 to 10 (0 = very poor, 10 = excellent), indicate how satisfied you are with the quality of the television programs we provide: __

Since the questionnaire is not considerered confidential, it contains a code that allows the company to identify the respondent.

After the latest survey was conducted, it occurred to the designers that the sample should perhaps have been stratified by client address into the four major regions serviced by the company because of different demographics in these regions. The poststratified results of the latest sample are shown in Table 7.19.

The overall average and variance are, respectively, the average ($\bar{Y}$) and variance (S^2) of all 1,000 sample ratings.

(a) Calculate the poststratified estimate of the average level of customer satisfaction.

(b) Calculate a 95% confidence interval for the average level of customer satisfaction based on the poststratified estimate.

(c) Calculate the simple estimate of the average level of customer satisfaction.

(d) Calculate a 95% confidence interval for the average level of customer satisfaction based on the simple estimate.

(e) Should the simple or the poststratified estimate be used in this case?

7.12 In Example 7.6, a simple random sample was stratified after it was selected in order to take advantage of the greater accuracy of a large poststratified sample. An alternative, of course, is to apply the simple estimators.

TABLE 7.19 Data for Problem 7.11

Region	Number of Subscribers	Number Sampled	Average Rating, $\bar{Y}_i$	Variance of Ratings, S_i^2
North	14,000	180	6.2	0.45
East	8,000	120	7.1	0.57
South	37,000	550	4.5	0.51
West	11,000	150	8.3	0.42
Overall	70,000	1,000	5.69	2.58

(a) Using the data in Table 7.12, calculate the simple estimate of the average household expenditure for repairs and improvements in the city. *Hint:* Satisfy yourself that, for any arrangement of n observations into M groups,

$$n\bar{Y} = \sum_{i=1}^{M} n_i \bar{Y}_i. \tag{7.8}$$

(b) It can be shown that, for any arrangement of n observations into M groups,

$$nS^2 = \sum_{i=1}^{M} n_i S_i^2 + \sum_{i=1}^{M} n_i (\bar{Y}_i - \bar{Y})^2. \tag{7.9}$$

Similar relationships between the overall mean or variance and the group means and variances were presented in Section 4.5.

Using the data of Table 7.12, calculate the sample variance S^2 and a 95% confidence interval for the average household expenditure on repairs and improvements in the city.

(c) In this case, should the simple or the poststratified estimates be used?

7.13 Double sampling was used in a survey of households in a certain city. In the first phase, a simple random sample of 500 of the city's 40,000 residences was selected from a commercially compiled list of residences. It was assumed that the list of residences and that of households are identical. The appraised values of the selected residences were then obtained from public municipal assessment rolls. It was thus determined that 145 of the sampled residences (group 1) had an assessed value less than $10,000, 220 (group 2) between $10,000 and $20,000, and 135 (group 3) greater than $20,000. The average assessed value of all sampled residences was $14,650.

In the second phase, a simple random sample of residences was selected from each of the three groups above. The selected households were questioned concerning, among other things, the number of television sets owned and whether or not they used a satellite antenna dish for television reception.

For the purpose of this problem, suppose two residences were selected from each group, with the results shown in Table 7.20.

(a) Calculate the double sample stratified estimate of the average number of TV sets per household in the city.

(b) Calculate the double sample stratified estimate of the proportion of households in the city that use an antenna (... no, this case was not dealt with in Section 7.6).

(c) Calculate the double sample ratio estimate of the average number of TV sets per household in the city.

(d) Calculate the double sample regression estimate of the average number of TV sets per household in the city.

TABLE 7.20 Data for Problem 7.13

Household Number	Assessed Value ($000)	Number of TV Sets	Using Antenna?
1	6.5	0	No
2	7.8	1	No
3	12.4	2	Yes
4	18.5	1	No
5	21.9	2	Yes
6	25.1	3	Yes

(e) Can you think of reasonable double sample ratio and regression estimators of the proportion of households using dish antennas?

7.14 New magazines begin life in a very competitive world with no subscribers and little knowledge of the characteristics of their readers with which to tempt advertisers. However, newstand copies normally contain a card inviting readers to subscribe, often at a substantial discount over the newstand price. New subscribers, in turn, soon receive a confirmation of their subscription and a short questionnaire. "We want to know you better," begins one questionnaire, "so that we can serve you better. Please take a minute to tell us about yourself and what you like or dislike about us. . . ." Not all subscribers, of course, respond to this invitation.

For the purpose of this problem, consider the case of *The Wood Craftsman*, a new magazine addressed to amateur carpenters and woodworkers. After six monthly issues were distributed nationally, the magazine had sold 35,000 copies and received 1,500 subscriptions. Approximately 30% of these subcribers were located in the South, 10% in the North, 40% in the West, and 20% in the East. About 10% of the subscribers responded to the questionnaire. Their responses with respect to two characteristics only are shown in Table 7.21.

TABLE 7.21 Data for Problem 7.14

Region	Number Responding			Average Age	
	Total	Men	Women	Men	Women
South	250	198	52	51	32
West	330	262	68	48	35
North	100	75	25	40	41
East	150	124	26	45	36
	830	659	171		

(a) Is it reasonable to treat the new subscribers as a simple random sample of readers? Explain why.

(b) Is it reasonable to treat those that responded to the questionnaire as a simple random sample of each region's subscribers? Explain why.

(c) Assuming that the verdict concerning **(a)** and **(b)** is affirmative, estimate the proportions of male and female readers, the average age of readers, and the average age of male and female readers of *The Wood Craftsman.*

(d) Assuming the verdict concerning **(a)** and **(b)** is negative, what can be said about the proportions of male and female readers, the average age of readers, and the average age of male and female readers of *The Wood Craftsman.* Explain carefully.

CHAPTER 8

Sampling with Unequal Probabilities

8.1 INTRODUCTION AND SUMMARY

It is evident that the sampling methods described earlier select the population elements in different ways. For example, in stratified sampling the population elements are divided into groups and a sample is selected from each group. Two-stage sampling selects a sample of groups and then a sample of elements from each selected group. Simple sampling, in contrast, does not group the population elements at all.

On further reflection, it becomes evident that changing the selection method changes the probability that a given population element will be included in the sample. We know, for example, that an element has the same chance as any other of being included in a simple random sample, but does not have the same chance when the sample is stratified or two-stage.

This suggests the possibility of characterizing a sampling method not by the physical manner in which the elements are selected but by the method's "inclusion probabilities," the probabilities with which the elements will be included in the sample under that method. There is a further possibility of devising an unbiased estimator for any set of inclusion probabilities, rather than inventing one for each conceivable selection method.

Indeed, as we shall see in this chapter, both possibilities are realities. The inclusion probabilities form a concept that unifies the various sampling methods, and the

"Horvitz–Thompson estimator" is unbiased under any method for which the inclusion probabilities are known. We shall illustrate this use of the Horvitz–Thompson estimator when we examine three-stage sampling and its multistage extensions.

The variance of the Horvitz–Thompson estimator can be expressed as a function of the inclusion probabilities. An examination of this variance will show that in the event the variable of interest is approximately proportional to a known auxiliary variable, an alternative to ratio or regression estimators is to sample with inclusion probabilities proportional to the values of this auxiliary variable. We explain in this chapter why this method, known as sampling with "inclusion probability proportional to size (IPPS)," is desirable and how it may be implemented.

The implementation of sampling with IPPS or arbitrary inclusion probabilities is often not easy when sampling is without replacement. Sampling without replacement is the selection method most frequently used in practice because, among other reasons, it is more accurate than sampling with replacement. If one is willing to forgive the lower accuracy of sampling with replacement, however, sampling with "probability proportional to size (PPS)" is much easier and the results much simpler, as will be shown in the last sections of this chapter. The chapter concludes with an examination of "dollar-unit sampling," a method of sampling with replacement and probability proportional to size that is often applied in sampling for auditing purposes.

8.2 INCLUSION PROBABILITIES

The probability that a given population element will be included in the sample under a given sampling method is called the *inclusion probability* of the element. Table 8.1 shows the inclusion probabilities for the firms in Ackroyd under some of the sampling methods examined earlier.

Firm	**Simple** (Table 3.2)	**Stratified** (Table 4.5)	**Two-Stage** (Table 5.3)	**Modified** (Table 6.7)
A	1/3	1/4	1/4	97/225
B	1/3	1/4	1/4	93/225
C	1/3	1/4	1/4	81/225
D	1/3	1/4	1/2	57/225
E	1/3	1/2	1/2	49/225
F	1/3	1/2	1/4	73/225
Total	2	2	2	450/225=2

TABLE 8.1 Inclusion Probabilities, Ackroyd Population

The source of these probabilities is also indicated in Table 8.1. For example, refer to Table 6.7 and observe that the probability that firm A will be included in the sample is equal to the probability that the sample outcome will be (A,B), (A,C), (A,D), (A,E) or (A,F). This is $(25 + 22 + 16 + 14 + 20)/225$, or $97/225$. All other inclusion probabilities are similarly calculated. In all cases, it will be recalled, the sample size was the same ($n = 2$), and sampling was without replacement.

The inclusion probabilities for many sampling methods can be determined in advance and in general:

- Under simple random sampling, the probability that a given population element will be included in the sample is n/N.
- Under stratified sampling, the inclusion probability of any one of the N_j elements of group j, from which n_j elements are selected without replacement, is n_j/N_j.
- Under two-stage sampling, m of the M groups are selected in the first stage, and n_j of the N_j elements of selected group j are selected in the second stage. The probability that any one element of a given group j will be included in the sample is

$$\frac{m}{M}\frac{n_j}{N_j}.$$

As always, it is assumed that sampling is without replacement. These results can be confirmed using the information in Table 8.1. For example, the probability that any given firm will be included in a simple random sample of size 2 without replacement is 2/6 or 1/3, as shown in Table 8.1.

Figuratively speaking, the inclusion probabilities can be described as the sampling method's "footprint" on the population elements. Each sampling method implies a set of inclusion probabilities.

We have seen that it is possible to manipulate the selection process and the resulting inclusion probabilities if this suits our purpose. Recall, for example, the modified random selection of Section 6.3, under which the first element is selected with probability proportional to its X value, and which permits the ratio estimator to be unbiased.

It would be convenient to have a single unbiased estimator applicable for *any* set of inclusion probabilities, instead of different ones for each sampling method. Such an estimator does indeed exist. It is described, and its properties are examined in the following sections.

8.3 THE HORVITZ–THOMPSON ESTIMATOR (HTE)

The estimator, named after D. G. Horvitz and D. J. Thompson, is defined in Summary 8.1. In this section, we describe its features and illustrate its calculation.

In Section 8.5, we confirm its unbiasedness, examine its variance, and consider some useful implications.

Let us first explain what is meant by "distinct" elements in Summary 8.1. We have been assuming, and shall continue to assume, that the sampling of elements (and of groups in two-stage sampling) is without replacement. Therefore, there is no possibility that a given population element will appear more than once in the sample. In other words, *if sampling is without replacement, all selected elements are distinct, the number of distinct elements in the sample equals the sample size, and n should replace d in Eq. 8.1.*

SUMMARY 8.1 The Horvitz–Thompson estimator

- *Method of sample selection:* Any.
- An *unbiased estimator of the population mean of a variable* Y, μ_y, is the *Horvitz–Thompson estimator (HTE)*, given by

$$\bar{Y}_{ht} = \frac{1}{N} \sum_{i=1}^{d} \frac{Y_i}{p_i}, \tag{8.1}$$

where d is the number of distinct elements in the sample, p_i is the probability that the ith distinct sample element will be included in the sample under a given sampling method (the *inclusion probability* of the element), and Y_i is its Y value.

- An unbiased estimator of the population total of Y is $T_{ht} = N\bar{Y}_{ht}$.
- The variance of $\bar{Y}_{ht}$ is

$$Var(\bar{Y}_{ht}) = \frac{1}{N^2} \sum_{i=1}^{N} \sum_{j>i}^{N} (p_i p_j - p_{ij}) \left[\frac{Y_i}{p_i} - \frac{Y_j}{p_j} \right]^2, \tag{8.2}$$

where p_{ij} denotes the joint inclusion probability of elements i and j.

- An *unbiased estimator of the proportion of elements in the population that fall into a given category*, the Horvitz–Thompson estimator of π, is

$$P_{ht} = \frac{1}{N} \sum_{i=1}^{d} \frac{Y_i'}{p_i}, \tag{8.3}$$

and its variance is

$$Var(P_{ht}) = \frac{1}{N^2} \sum_{i=1}^{N} \sum_{j>i}^{N} (p_i p_j - p_{ij}) \left[\frac{Y_i'}{p_i} - \frac{Y_j'}{p_j} \right]^2, \tag{8.4}$$

where Y_i' (and Y_j') is equal to 1 if the ith (or jth) element belongs to the category, and to 0 if it does not. The number of elements in the population that fall into the category is estimated by $T_{ht}' = N P_{ht}$.

The HTE, however, also applies to sampling with replacement, in which case care has to be taken that only distinct elements are involved in its calculation. In addition, the HTE applies to the case where the number of elements selected is not the same for every sample; this happens, for example, under two-stage sampling, where the number of elements selected in the second stage may vary depending on which group is selected in the first stage.

■ EXAMPLE 8.1 Suppose that a certain sampling method (we need not specify the method or how it is implemented) gives the six firms in the Ackroyd case the following probabilities of being included in a sample of size $n = 2$ without replacement.

Firm	Inclusion Probability, p_i	Number of Employees, Y	Will Hire? (1=Yes, 0=No)
A	1/2	9	1
B	1/4	8	1
C	1/8	6	0
D	1/8	2	0
E	1/2	1	0
F	1/2	5	0
	2		

The HTE of μ (the average number of employees per firm in Ackroyd), is given by

$$\bar{Y}_{ht} = \frac{1}{N} \sum_{i=1}^{2} \frac{Y_i}{p_i} = (\frac{1}{6})[\frac{Y_1}{p_1} + \frac{Y_2}{p_2}].$$

In this expression, Y_1 and Y_2 denote the number of employees of the first and second selected firm, respectively. p_1 is the inclusion probability of the first, and p_2 that of the second selected firm.

Likewise, the HTE of π (the proportion of firms in Ackroyd that intend to hire) is

$$P_{ht} = \frac{1}{N} \sum_{i=1}^{2} \frac{Y_i'}{p_i} = (\frac{1}{6})[\frac{Y_1'}{p_1} + \frac{Y_2'}{p_2}],$$

where Y_1' and Y_2' are 1 or 0 depending on whether or not the firm intends to hire.

For example, if the sample happened to consist of firms B and E, the HT estimate of the average number of employees per firm in Ackroyd would be

$$\bar{Y}_{ht} = (\frac{1}{6})[\frac{8}{(1/4)} + \frac{1}{(1/2)}] = 5.667$$

and the estimate of the proportion of Ackroyd firms intending to hire would be

$$P_{ht} = (\frac{1}{6})[\frac{1}{(1/4)} + \frac{0}{(1/2)}] = \frac{4}{6} = 0.667. \quad \blacksquare$$

If we accept for the moment the claims of this section, we see that it is not necessary to invent an unbiased estimator for every conceivable sampling method. *The HTE serves as an unbiased estimator for any sampling method with known inclusion probabilities.*

But is the HTE an entirely new estimator? Is there any connection between the HTE and the simple, stratified, or two-stage estimators we examined in earlier chapters?

In simple random sampling without replacement, the probability that any given population element will be included in a sample of size n is equal to n/N. Therefore, $p_i = n/N$ for all population elements, and the HTE is

$$
\begin{aligned}
\bar{Y}_{ht} &= \frac{1}{N}[\frac{Y_1}{p_1} + \frac{Y_2}{p_2} + \cdots + \frac{Y_n}{p_n}] \\
&= \frac{1}{N}[\frac{Y_1}{n/N} + \frac{Y_2}{n/N} + \cdots + \frac{Y_n}{n/N}] \\
&= \frac{1}{N}\frac{N}{n}[Y_1 + Y_2 + \cdots + Y_n] \\
&= \frac{1}{n}[Y_1 + Y_2 + \cdots + Y_n] \\
&= \bar{Y}.
\end{aligned}
$$

In other words, the HTE *is* the simple estimator—the ordinary sample average of the Y values.

It should also be clear from the next to last line of the preceding expressions that, if the Y_i are 1 or 0 depending on whether the element is or is not in the category of interest, the sum in brackets is equal to the number of elements, and the HTE is the proportion of elements (P) in the sample that belong to the category of interest.

Under simple random sampling without replacement, therefore, the HT estimators *are* the simple estimators of Chapter 3.

It is a little more complicated, but not difficult to show that if sampling is without replacement and stratified, the HTEs are the stratified estimators ($\bar{Y}_s$ and P_s) discussed in Chapter 4. Likewise, if sampling is without replacement and two-stage, the HTEs are the two-stage estimators ($\bar{Y}_{ts}$ and P_{ts}) of Chapter 5.

The estimators we examined in Chapters 3 to 5, therefore, may be considered special cases of the HTE when the inclusion probabilities are those implied by these three sampling methods.

The HTE, however, is not identical to the ratio estimator under modified random sampling. To see this, suppose for example that the sample consists of firms B and E. The inclusion probabilities of these two firms are 93/225 and 49/225 (see Table 8.1). The HT estimate of μ would be

$$(\frac{1}{6})[\frac{8}{(93/225)} + \frac{1}{(49/225)}] = 3.991.$$

This differs from the ratio estimate, 5.192, for this sample outcome shown in Table 6.6.

8.4 APPLICATION: THREE-STAGE SAMPLING

The HTE can be used with any sampling method provided that the inclusion probabilities are known. We shall illustrate this use of the HTE to derive unbiased estimators of the population characteristics under three-stage sampling, a method frequently used in practice because of its low cost.

Imagine that a region of interest is divided into a number of districts (e.g., census tracks, enumeration areas, city blocks). In each district there are a number of households, and each household consists of a number of household members. In the first stage, a simple random sample of districts will be drawn without replacement. In the second stage, a simple random sample of households will be drawn without replacement from each district selected in the first stage. Finally, in the third stage, a simple random sample of household members will be drawn without replacement from each household selected in the second stage. The ultimate sample, therefore, will consist of a number of individuals. The situation is outlined in Figure 8.1.

The region is composed of L districts. The "typical" district i has M_i households. The typical household j in district i has N_{ij} members. The sampling plan calls for

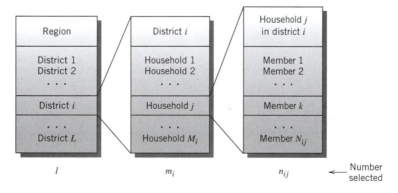

Figure 8.1 Three-stage sample

selecting l of the L districts, m_i of the M_i households in selected district i, and n_{ij} of the N_{ij} members of selected household j in selected district i.

Let us calculate the inclusion probability of a given individual—say, of member k, who belongs to household j, which, in turn, belongs to district i. Clearly, for member k to be included in the ultimate sample: (a) he or she must be included in the sample of n_{ij} members from the N_{ij} in household j; (b) household j must be included in the sample of m_i households from the M_i in district i; and (c) district i must be included in the sample of l districts from the L districts in the region.

Each of these samples is a simple random sample without replacement. Therefore, the probability that district i will be included in the sample of districts is l/L; the probability that household j will be included in the sample of households from district i is m_i/M_i; and the probability that member k will be included in the sample of members from household j is n_{ij}/N_{ij}. It follows that the probability that member k will be included in the ultimate sample is

$$\frac{l}{L} \frac{m_i}{M_i} \frac{n_{ij}}{N_{ij}}.$$

The ultimate sample will consist of a number of individuals drawn from various households and districts. The affiliations, measurements, and inclusion probabilities of the selected individuals will be known after the sample is selected and the individuals interviewed.

Suppose, for instance, that the population consists of 9 persons. Their names and affiliations with households and districts are shown in the first three columns of Table 8.2.

Assume that $l = 1$ district will be selected in the first stage, $m_i = 2$ households will be selected from each district selected in the first stage, and $n_{ij} = 1$ household member will be selected from each household selected in the second stage. The inclusion probabilities are shown in the last column of Table 8.2. For example, the probability that Melinda will be included in the ultimate sample is $1/2 \times 2/3 \times 1/3$ or $1/9$.

Notice, incidentally, that it is not necessary to know in advance the number of households in every district or the number of members of every household. This information is needed only for the *selected* districts and households at each stage. For example, if only the South district is selected in the first stage, it is not necessary to find the number of households in the North district. If only the Jones and Byrd households are selected from the South district, it is not necessary to determine the number of members of the Ryan household.

Ultimately, two individuals will be selected by this procedure and questioned concerning, say, their age and smoking habit.

Suppose Peter Jones and David Byrd happen to be the two individuals in the ultimate sample. Their ages are 25 and 60, respectively; Peter smokes, David does not. To estimate the average age of and the proportion of smokers among the nine

TABLE 8.2   Three-Stage Sampling Illustrated

District	Household	Household Members	l/L	m_i/M_i	n_{ij}/N_{ij}	Inclusion Probability
1. North	1. Brown	1. John	1/2	2/2	1/2	1/4
1. North	1. Brown	2. Mary	1/2	2/2	1/2	1/4
1. North	2. Smith	1. Jane	1/2	2/2	1/1	1/2
2. South	1. Jones	1. Alice	1/2	2/3	1/2	1/6
2. South	1. Jones	2. Peter	1/2	2/3	1/2	1/6
2. South	2. Ryan	1. Jacob	1/2	2/3	1/3	1/9
2. South	2. Ryan	2. Sarah	1/2	2/3	1/3	1/9
2. South	2. Ryan	3. Melinda	1/2	2/3	1/3	1/9
2. South	3. Byrd	1. David	1/2	2/3	1/1	1/3
						2

persons in the region, we need only the inclusion probabilities and the responses of these two selected persons.

The HT estimate of the population average age is

$$\bar{Y}_{ht} = \frac{1}{9}\Big[\frac{25}{1/6} + \frac{60}{1/3}\Big] = 36.67.$$

The HT estimate of the population proportion of smokers is

$$P_{ht} = \frac{1}{9}\Big[\frac{1}{1/6} + \frac{0}{1/3}\Big] = 0.667.$$

This procedure for calculating HT estimates can be extended to more realistic situations, larger sample sizes, and to multistage sampling.

The formula of the HTE of μ under three-stage sampling is:

$$\bar{Y}_{ht} = \frac{1}{N}\frac{L}{l}\sum_{i=1}^{l}\frac{M_i}{m_i}\sum_{j=1}^{m_i}\frac{N_{ij}}{n_{ij}}\sum_{k=1}^{n_{ij}}Y_{ijk}.$$

In the context of the above illustration, Y_{ijk} is the Y value of the kth selected member in the jth selected household in the ith selected district. In general, Y_{ijk} is the Y value of the kth element in the jth selected second-stage group in the ith selected first-stage group. The HTE of π, P_{ht}, is given by the same expression, except that Y_{ijk} is 1 if the kth element is in the category or 0 if it is not.

8.5 THE VARIANCE OF THE HTE AND IMPLICATIONS

The variance of the HTE of μ is given by Eq. 8.2 in Summary 8.1. The *joint inclusion probability* of elements i and j, p_{ij}, is the probability that the (distinct) elements i and j will be included in the sample. p_i is, of course, the ordinary inclusion probability of element i.

The summation notation of Eq. 8.2 is of the form

$$\sum_{i=1}^{N}\sum_{j>i}^{N} a_{ij},$$

which translates as

$$\sum_{j=2}^{N} a_{1j} + \sum_{j=3}^{N} a_{2j} + \cdots + \sum_{j=N-1}^{N} a_{N-1,j},$$

or

$$(a_{12} + a_{13} + \cdots + a_{1N}) + (a_{23} + a_{24} + \cdots + a_{2N}) + \cdots + (a_{N-1,N}).$$

Before we illustrate the manner in which the variance of the HTE is calculated, let us consider the implications of Eq. 8.2.

If we could devise a sampling method such that the inclusion probabilities are proportional to the Y values, that is, if $p_i = cY_i$, where c is a constant, then, for each pair of population elements i and j,

$$\frac{Y_i}{p_i} - \frac{Y_j}{p_j} = \frac{Y_i}{cY_i} - \frac{Y_j}{cY_j} = \frac{1}{c} - \frac{1}{c} = 0$$

and

$$Var(\bar{Y}_{ht}) = 0.$$

In other words, if $p_i = cY_i$, the HTE will always correctly estimate μ.

Any illusion that a perfect estimator was created vanishes, of course, when it is realized that the Y values are not known. If they were known, there would be no point in taking a sample.

Suppose, however, that the variable Y of interest is approximately proportional to a known auxiliary variable X; that is, suppose $Y_i \approx bX_i$ for all population elements. Suppose further that the inclusion probabilities are made proportional to the X values, i.e., $p_i = cX_i$. Then, for all pairs of population elements i and j,

$$\frac{Y_i}{p_i} - \frac{Y_j}{p_j} \approx \frac{bX_i}{p_i} - \frac{bX_j}{p_j} = \frac{bX_i}{cX_i} - \frac{bX_j}{cX_j} = \frac{b}{c} - \frac{b}{c} = 0$$

and $Var(\bar{Y}_{ht}) \approx 0$.

In words, *if $Y_i \approx bX_i$ and the inclusion probabilities are made proportional to the known X values, the variance of the HTE of μ is approximately zero.* Just how close is the variance to zero depends, of course, on the degree to which bX approximates Y.

This observation forms the basis for a sampling method that gives each element inclusion probability proportional to the value of an auxiliary variable. The method is examined in the next section.

In the remainder of this section, we use the Ackroyd data to confirm the unbiasedness of the HTE and the theoretical expression of its variance.

■ EXAMPLE 8.2 To keep the calculations simple and short, let us assume that there are only three firms in Ackroyd, as shown in Table 8.3.

Suppose that under a given method of selecting a sample of size $n = 2$ (we need not specify the method or the manner in which it is implemented), the possible sample outcomes and their probabilities are as shown in the first two columns of Table 8.4 (the remaining columns will be explained shortly).

TABLE 8.3 Half-of-Ackroyd Population and Its Characteristics

Firm	Employees, Y	Hire? (1=Yes, 0=No)
A	9	1
B	8	1
C	6	0
	23	2

Number of firms, $N = 3$
Average number of employees, $\mu = 23/3 \approx 7.6667$
Proportion of firms intending to hire, $\pi = 2/3 \approx 0.6667$

TABLE 8.4 Sample Outcomes and HTEs

Outcome	Probability	HTE of μ, $\bar{Y}_{ht}$	HTE of π, P_{ht}
A,B	0.5	7.5595	0.8928
A,C	0.3	7.7500	0.4167
B,C	0.2	7.8095	0.4762
	1.0		

The inclusion probabilities, p_i, are:

Firm	i	p_i
A	1	0.8
B	2	0.7
C	3	0.5
		2

and the joint inclusion probabilities, p_{ij}:

Pair	i, j	p_{ij}
A,B	1,2	$p_{12} = 0.5$
A,C	1,3	$p_{13} = 0.3$
B,C	2,3	$p_{23} = 0.2$

For example, the probability that firm A will be included in the sample is $0.5 + 0.3$ or 0.8; the probability that firms A and B will be included in the sample is 0.5; and so on. (In this simple example, the joint inclusion probabilities are the probabilities of the sample outcomes, but this is not always the case.)

The variance of the HTE of μ, calculated according to Eq. 8.2, is:

$$Var(\bar{Y}_{ht}) = (\frac{1}{3^2})\Big[(p_1 p_2 - p_{12})(\frac{Y_1}{p_1} - \frac{Y_2}{p_2})^2 + (p_1 p_3 - p_{13})(\frac{Y_1}{p_1} - \frac{Y_3}{p_3})^2$$

$$+ (p_2 p_3 - p_{23})(\frac{Y_2}{p_2} - \frac{Y_3}{p_3})^2\Big]$$

$$= (\frac{1}{9})\Big[(0.8 \times 0.7 - 0.5)(\frac{9}{0.8} - \frac{8}{0.7})^2$$

$$+ (0.8 \times 0.5 - 0.3)(\frac{9}{0.8} - \frac{6}{0.5})^2$$

$$+ (0.7 \times 0.5 - 0.2)(\frac{8}{0.7} - \frac{6}{0.5})^2\Big]$$

$$= 0.0119.$$

The HT estimates of μ for each possible sample outcome are shown in the third column of Table 8.4. For example, if the sample consists of firms A and B, the HT estimate of μ is

$$\bar{Y}_{ht} = (\frac{1}{3})\Big[\frac{9}{0.8} + \frac{8}{0.7}\Big] = 7.5595.$$

It is now straightforward to verify by direct calculation that the HTE of μ is unbiased and that its variance is indeed as calculated using Eq. 8.2:

$$E(\bar{Y}_{ht}) = (7.5595)(0.5) + \cdots + (7.8095)(0.2) = 7.6667$$
$$Var(\bar{Y}_{ht}) = (7.5595 - 7.6667)^2(0.5) + \cdots + (7.8095 - 7.6667)^2(0.2)$$
$$= 0.0119.$$

If the sample consists of firms A and B, the HT estimate of π is

$$P_{ht} = (\frac{1}{3})[\frac{1}{0.8} + \frac{1}{0.7}] = 0.8928.$$

It can be easily verified that the HT estimates of π, P_{ht}, for each other sample outcome are as shown in the last column of Table 8.4, that P_{ht} is indeed unbiased, and that its variance (whether calculated directly or using Eq. 8.4) is 0.0516.

As may be expected, the simple estimators are not unbiased when the inclusion probabilities are not the same for all elements. Table 8.5 shows the simple estimates, $\bar{Y}$ and P, for all the sample outcomes of this example.

It is easy to verify that $E(\bar{Y}) = 7.9 \neq 7.667$ and $E(P) = 0.75 \neq 0.667$. Therefore, $\bar{Y}$ and P are not unbiased.   ◨

8.6 INCLUSION PROBABILITIES PROPORTIONAL TO SIZE (IPPS)

In the last section we noted that if the variable of interest is approximately proportional to a known auxiliary variable, and if the inclusion probabilities can be made proportional to the values of this auxiliary variable, the variance of the HT estimator of the population mean of the variable of interest will be small. Just how small depends, of course, on the quality of the approximation. In this section we consider how to select a random sample so that the probability that a given element is included in the sample is proportional to its X value. This method is known as sampling with IPPS, short for *Inclusion Probability Proportional to Size*.

For the remainder of this chapter, we shall assume that the auxiliary variable X takes *positive* values only. As always, p_i is the probability that population element i will be included in the sample.

TABLE 8.5 Simple Estimators, Example 8.2			
Outcome	Probability	$\bar{Y}$	P
A,B	0.5	8.5	1.0
A,C	0.3	7.5	0.5
B,C	0.2	7.0	0.5
	1.0		

We want $p_i = cX_i$, where c is a constant. We are not free, however, to use any c. If sampling is without replacement, and the total sample size is the same (say, n) for every possible sample under the sampling method, it can be shown that the sum of the inclusion probabilities for all elements in the population must equal the sample size, that is,

$$\sum_{j=1}^{N} p_j = n.$$

This can be confirmed by referring to Table 8.1. In our illustrations based on the Ackroyd case, the sample size under simple, stratified, and modified random sampling was equal to 2. As can be observed in Table 8.1, the sum of the inclusion probabilities is indeed equal to 2 for all these methods.

A two-stage sample will yield n observations in total, no matter which groups are selected in the first stage, if the number of elements drawn from each of the m groups selected in the first stage is equal to n/m. That was the case in the Ackroyd illustration (recall that two elements were drawn from the one group selected in the first stage); it is for this reason that the sum of the inclusion probabilities under this method is also equal to 2 in this case.

The constant c, therefore, must satisfy $\sum_{j=1}^{N} p_j = \sum_{j=1}^{N} cX_j = n$, or $c = n/\sum_{j=1}^{N} X_j$. We conclude that when sampling with IPPS the inclusion probabilities are

$$p_i = \frac{nX_i}{\sum_{j=1}^{N} X_j}. \tag{8.5}$$

Obviously, the inclusion probabilities cannot be negative or exceed 1. Under the assumption that all X_i are positive, all $p_i > 0$, as required. In addition, in order for all $p_i \leq 1$, it is necessary that

$$nX_i \leq \sum_{j=1}^{N} X_j \tag{8.6}$$

for each population element i. This last requirement *must* be satisfied in order to implement sampling with IPPS.

If, then, we can devise a scheme for selecting a sample with IPPS (we describe such a scheme in the next section), the population average value of variable Y will be estimated by:

$$\bar{Y}_{ht} = \frac{1}{N}\Big[\frac{Y_1}{p_1} + \frac{Y_2}{p_2} + \cdots + \frac{Y_n}{p_n}\Big],$$

where $p_i = nX_i/\sum X$. To simplify the notation, we have written $\sum_{j=1}^{N} X_j$ as $\sum X$, and will do so for the rest of this chapter.

■ EXAMPLE 8.3 In Example 6.1, we assumed a simple linear relationship between a hospital's purchases of a given pharmaceutical product (Y) and the size of the hospital as measured by the number of its beds (X). In Section 6.2, we described three methods for estimating the parameter b of the relationship $Y \approx bX$ and the average and total purchases of Product Y by all hospitals. Another method for utilizing this relationship is to sample with IPPS and use the HTE.

The first three columns of Table 8.6 outline the population of hospitals and the hospital purchases of Product Y.

The number of beds of each hospital is known, as is the total number of beds of all hospitals (186,030).

Suppose that a random sample of $n = 3$ hospitals will be selected with IPPS, using the number of beds as the measure of the size of a hospital. The desired inclusion probabilities, $p_i = 3X_i/186,030$, are shown in the last column of Table 8.6.

Suppose further that a sample has somehow been selected with IPPS, and that it happened to consist of hospitals 1, 2, and 1,158. The following table shows the information that will be available after the selected hospitals are questioned concerning their purchases of Product Y.

Sampled Hospital	Number of Beds, X_i	Purchases of Product Y, Y_i ($000)
1	675	500
2	450	350
1,158	1,500	1,100

The HT estimate of the average purchases of Product Y per hospital is

$$\bar{Y}_{ht} = (\frac{1}{1,158})[\frac{500}{0.0109} + \frac{350}{0.0073} + \frac{1,100}{0.0242}] = 120.269 \ (\$000).$$

The estimate of the total hospital purchases of Product Y is $(1,158)(120.269)$, or 139,271 ($000). ■

TABLE 8.6 Population of Hospitals, Example 8.3

Hospital No., i	Number of Beds, X_i	Purchases of Y, Y_i	$p_i = nX_i/\sum X$
1	675	?	0.0109
2	450	?	0.0073
⋮	⋮	⋮	⋮
1,158	1,500	?	0.0242
Total	186,030	?	3

8.7 SAMPLING WITH IPPS

But *how* does one select a sample of size n with IPPS? A little reflection will show that the answer is not obvious if the sample is to be without replacement. The following procedure is due to M. R. Sampford.

To select a random sample of size n with IPPS,

(1) Select the first element with probability $f_i = X_i / \sum X$.
(2) Select each of the remaining $n - 1$ elements with probability g_i proportional to $f_i/(1 - nf_i)$.
(3) Examine the resulting sample:

    (3a) If *all* the sampled elements are distinct, the selected elements form the desired sample

    (3b) Otherwise, reject the entire selection and repeat the procedure from the start.

Example 8.4 illustrates the mechanics of this selection procedure.

■ EXAMPLE 8.4 Consider the population of hospitals described in Example 8.3. Suppose a sample of $n = 100$ hospitals will be selected with IPPS. We begin by calculating the probabilities f_i and g_i as shown in cols. (3), (5), and (6) of Table 8.7.

Next, we assign four-digit numbers to hospitals, as shown in cols. (4) and (7) of Table 8.7, using the cumulative probabilities as guides.

Finally, we generate a sequence of 100 four-digit random numbers. The first random number should be compared to the numbers in col. (4), the remaining ones to the numbers in col. (7). For example, if the sequence of random numbers begins with 9,958, 0020, 0015, ... , the tentative sample list would begin with hospital 1,158, 2, 1, ... If all the hospitals in the list are distinct, the sample should consist of these hospitals. If not, the entire tentative list should be rejected and the selection

TABLE 8.7 Assigned Numbers, Example 8.4, Partial Listing

Hospital No., i (1)	Number of Beds, X_i (2)	$f_i = X_i / \sum X$ (3)	Assigned Numbers (4)	$f_i/(1 - 100f_i)$ (5)	$g_i = $ (5)/3.1052 (6)	Assigned Numbers (7)
1	675	0.0036	0001–0036	0.0056	0.0018	0001–0018
2	450	0.0024	0037–0060	0.0031	0.0010	0019–0028
⋮	⋮	⋮	⋮	⋮	⋮	⋮
1,158	1,500	0.0081	9920–0000	0.0426	0.0137	9864–0000
	$\sum X = 186{,}030$	1.0000		3.1052	1.0000	

process begin anew. It is not, of course, necessary to generate a complete tentative list of 100 hospitals before examining it for duplications—as soon as a hospital is selected a second time, the entire partial tentative list is rejected, and the process begins from scratch. ◪

It is not at all difficult to confirm with the help of a simple example that this selection procedure does give each element IPPS.

◪ EXAMPLE 8.5 Consider the "Half-of-Ackroyd" population, and suppose it is desired to select a random sample of size $n = 2$ with inclusion probability proportional to the firms' annual sales, as given in Table 6.4. The relevant data are as follows:

Firm	X_i	Y_i	$p_i = 2X_i/34$
A	13	9	0.7647
B	12	8	0.7059
C	9	6	0.5294
	$\sum X = 34$		2

We want the inclusion probabilities, $p_i = nX_i / \sum X$, to be as shown in the last column.

The selection procedure for this case is:

(1) Select the first firm with probability $f_i = X_i/34$,

(2) Select the second firm with probability g_i proportional to $f_i/(1 - 2f_i)$,

(3) If the two selected firms are distinct, the sample is accepted; otherwise, it is rejected.

The preliminary calculations are shown in Table 8.8.

The probability tree in Figure 8.2 shows the probabilities of the nine possible outcomes under this selection procedure.

In this list of outcomes, order does matter; for example (A,B) means first A, then B. If the outcomes are (A,A), (B,B), or (C,C), the sample is rejected; the probability

TABLE 8.8 Sample Selection, Example 8.2

Firm (1)	X_i (2)	$f_i = X_i/34$ (3)=(2)/34	$f_i/(1 - 2f_i)$ (4)	g_i (5)=(4)/3.3878
A	13	0.3824	1.6258	0.4799
B	12	0.3529	1.1995	0.3541
C	9	0.2647	0.5625	0.1660
	34	1.0000	3.3878	1.0000

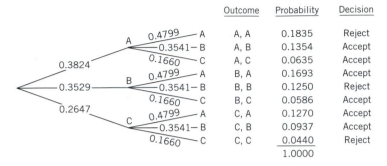

			Outcome	Probability	Decision
			A, A	0.1835	Reject
			A, B	0.1354	Accept
			A, C	0.0635	Accept
			B, A	0.1693	Accept
			B, B	0.1250	Reject
			B, C	0.0586	Accept
			C, A	0.1270	Accept
			C, B	0.0937	Accept
			C, C	0.0440	Reject
				1.0000	

Figure 8.2   Probability tree, IPPS procedure

of this occurring is equal to 0.1835+0.1250+0.0440 or 0.3525. The probability that the sample will be accepted is $1 - 0.3525$ or 0.6475. The probability that an acceptable sample will consist of firms A and B, in this order, is 0.1354/0.6475 or 0.2091. Put differently, out of every 10,000 samples selected according to this procedure, 6,475 are expected to be acceptable; of these, 1,354 are expected to consist of firms A and B, in this order; thus, the proportion of acceptable samples consisting of A and B, in this order, is expected to be 1,354/6,475 or 0.2091. The following table shows the probabilities of all valid sample outcomes under this procedure:

Outcome	Probability
A,B	0.2091
A,C	0.0981
B,A	0.2615
B,C	0.0905
C,A	0.1961
C,B	0.1447
	1.0000

The probability that the sample will include A, therefore, is $0.2091 + 0.0981 + 0.2615 + 0.1961$, or 0.7648. The probability that it will include B is 0.7058, and that it will include C is 0.5294. The slight differences with the desired probabilities are due to rounding errors. We conclude that the selection procedure does indeed yield the desired inclusion probabilities.   ◼

8.8 SOME OBSERVATIONS ON THE HTE

- Different sampling methods may result in the same inclusion probabilities. In other words, the inclusion probabilities do not identify uniquely a sampling method.

- The joint inclusion probabilities, p_{ij}, and the variance of the HTE are often hard to compute. It is often difficult, therefore, to compare sampling with unequal probabilities to other sampling methods.
- The HTE is not the only unbiased estimator. It follows that the HTE may not be the best unbiased estimator, that is, the unbiased estimator with the smallest possible variance. However, the HTE is best for estimating the population mean of a variable when that variable is proportionally related to a known auxiliary variable and the sample is selected with IPPS.

8.9 INTERVAL ESTIMATES BASED ON THE HTE

Approximate large-sample confidence intervals based on the HTE and any given inclusion probabilities are given in Summary 8.2.

◪ EXAMPLE 8.6   Three of the 1,158 hospitals in Table 8.6 were selected without replacement and IPPS. The selected hospitals were questioned regarding their purchases of Product Y and whether or not they purchase another product, Z. Their responses are shown in Table 8.9.

The HT estimates of the average purchases of Product Y per hospital (μ) and of the proportion of hospitals which purchase Product Z (π) are

$$\bar{Y}_{ht} = \frac{1}{1,158}\left[\frac{500}{0.0109} + \frac{350}{0.0073} + \frac{1,100}{0.0242}\right] = 120.269,$$

and

$$P_{ht} = \frac{1}{1,158}\left[\frac{1}{0.0109} + \frac{0}{0.0073} + \frac{1}{0.0242}\right] = 0.1149.$$

In order to construct a confidence interval for μ, we need

$$
\begin{aligned}
S_{ht}^2 &= \frac{1}{d-1}\sum_{i=1}^{d}\left(\frac{dY_i}{Np_i} - \bar{Y}_{ht}\right)^2 \\
&= \frac{1}{3-1}\left[\left(\frac{3 \times 500}{1,158 \times 0.0109} - 120.269\right)^2 \right.\\
&\quad + \left(\frac{3 \times 350}{1,158 \times 0.0073} - 120.269\right)^2 \\
&\quad \left. + \left(\frac{3 \times 1,100}{1,158 \times 0.0242} - 120.269\right)^2\right] \\
&= 11.9436,
\end{aligned}
$$

SUMMARY 8.2   Confidence intervals based on the HTE

- For large n and $N - n$, an approximate $100(1 - \alpha)\%$ confidence interval for μ, the population mean of a variable Y, is

$$\bar{Y}_{ht} \pm Z_{\alpha/2}\sqrt{\widehat{Var(\bar{Y}_{ht})}}, \tag{8.7}$$

where $\widehat{Var(\bar{Y}_{ht})}$ is an estimator of $Var(\bar{Y}_{ht})$ given by

$$\widehat{Var(\bar{Y}_{ht})} = \frac{1}{n}\frac{N - n}{N - 1}S_{ht}^2,$$

and

$$S_{ht}^2 = \frac{1}{d - 1}\sum_{i=1}^{d}\left(\frac{dY_i}{Np_i} - \bar{Y}_{ht}\right)^2, \tag{8.8}$$

where d is the number of distinct elements in the sample and p_i the inclusion probability of the ith element.
- For large n and $N - n$, an approximate $100(1 - \alpha)\%$ confidence interval for π, the proportion of elements in the population that belong to a given category, is

$$P_{ht} \pm Z_{\alpha/2}\sqrt{\widehat{Var(P_{ht})}}, \tag{8.9}$$

where $\widehat{Var(P_{ht})}$ is an estimator of $Var(P_{ht})$ given by

$$\widehat{Var(P_{ht})} = \frac{1}{n}\frac{N - n}{N - 1}S_{ht}'^2,$$

and

$$S_{ht}'^2 = \frac{1}{d - 1}\sum_{i=1}^{d}\left(\frac{dY_i'}{Np_i} - P_{ht}\right)^2. \tag{8.10}$$

In the last expression, Y_i' is 1 if the sample element falls into the category, or 0 if it does not.
- The corresponding confidence intervals for the population totals, $\tau = N\mu$ and $\tau' = N\pi$, are formed by multiplying by N the limits of Eqs. 8.7 and 8.9, respectively.

---

and

$$\widehat{Var(\bar{Y}_{ht})} = \frac{1}{n}\frac{N - n}{N - 1}S_{ht}^2 = \left(\frac{1}{3}\right)\left(\frac{1,158 - 3}{1,158 - 1}\right)(11.9436) = 3.9743.$$

If this were a large sample, an approximate 95% confidence interval for μ would be

$$(120.269) \pm (1.96)\sqrt{3.9743},$$

or from 116.362 to 124.176 ($000).

TABLE 8.9 Sample Results, Example 8.6

Sampled Hospital	Number of Beds, X_i	p_i	Purchases of Product Y, Y_i ($000)	Purchasing Product Z? (1=Yes, 0=No)
1	675	0.0109	500	1
2	450	0.0073	350	0
1,158	1,500	0.0242	1,100	1

A confidence interval for π requires

$$S_{ht}'^2 = \frac{1}{d-1} \sum_{i=1}^{d} \left(\frac{dY_i'}{Np_i} - P_{ht}\right)^2$$

$$= \frac{1}{3-1} \left[\left(\frac{3 \times 1}{1,158 \times 0.0109} - 0.1149\right)^2 \right.$$

$$+ \left(\frac{3 \times 0}{1,158 \times 0.0073} - 0.1149\right)^2$$

$$\left. + \left(\frac{3 \times 1}{1,158 \times 0.0242} - 0.1149\right)^2 \right]$$

$$= 0.0141,$$

and

$$\widehat{Var(P_{ht})} = \frac{1}{n} \frac{N-n}{N-1} S_{ht}'^2 = \left(\frac{1}{3}\right)\left(\frac{1,158-3}{1,158-1}\right)(0.0141) = 0.00469.$$

If this were a large sample, an approximate 95% confidence interval for π would be

$$(0.1149) \pm (1.96)\sqrt{0.00469},$$

or from -0.02 to 24.91 (%). We know very well that a proportion cannot be negative, so the stated interval estimate would be from 0 to 24.91%. In this case, of course, Eqs. 8.7 and 8.9 may not be applied as the sample size is too small. However, inconsistent confidence limits may occasionally be produced even by properly applied formulas because the latter are not specifically designed to observe logical constraints on the estimates. ◼

8.10 SAMPLING WITH REPLACEMENT AND PPS

We observed earlier that when sampling is without replacement, it is often difficult to devise a selection procedure ensuring that the inclusion probabilities equal the desired ones. In addition, the joint inclusion probabilities and the variance of the

HTE may be quite complicated. Other factors being equal, sampling with replacement tends not to be as accurate as sampling without replacement, but on occasion the ease with which the former method can be applied may compensate for the loss of accuracy, especially when the samples are large. If one is willing to consider sampling with replacement to implement selection with unequal probabilities, the selection procedure, the estimator, and its variance are all much simpler.

Suppose that n elements are to be drawn one at a time and with replacement from a population of N elements. Let q_i be the probability that a given element i will be selected in any one draw, where q_i remains constant for all draws ($q_i = 1/N$ for simple random sampling with replacement). An unbiased estimator under this sampling method is the Hansen–Hurwitz estimator (HHE) defined in Summary 8.3. The estimator is named after M. H. Hansen and W. N. Hurwitz.

The selection probabilities q_i can be any positive numbers adding up to 1. In the special case where they are made proportional to the values of a known positive auxiliary variable ($q_i = X_i / \sum X$), the selection method is referred to as sampling with *Probability Proportional to Size (PPS)*.

SUMMARY 8.3   The Hansen–Hurwitz estimator (HHE)

- *Method of sample selection:* Random sampling with replacement, unequal probabilities.
- *An unbiased estimator of the population mean of a variable Y, μ, is the Hansen–Hurwitz estimator (HHE):*

$$\bar{Y}_{hh} = \frac{1}{nN} \sum_{i=1}^{n} \frac{Y_i}{q_i}, \qquad (8.11)$$

  where q_i is the selection probability of element i in any given draw.
- In the special case of sampling with PPS, $q_i = X_i / \sum_{j=1}^{N} X_j$.
- An unbiased estimator of the population total of Y is $T_{hh} = N\bar{Y}_{hh}$.
- The variance of $\bar{Y}_{hh}$ is

$$Var(\bar{Y}_{hh}) = \frac{1}{n} \sum_{i=1}^{N} \left[\frac{Y_i}{Nq_i} - \mu\right]^2 q_i = \frac{1}{n}\left[\frac{1}{N^2} \sum_{i=1}^{N} \frac{Y_i^2}{q_i} - \mu^2\right]. \qquad (8.12)$$

- The same expressions give an unbiased estimator, P_{hh}, of the proportion of elements in the population that belong to a given category (π), and its variance, except that μ in Eq. 8.12 is replaced by π, and Y_i in Eqs. 8.11 and 8.12 is equal to 1 if the ith element belongs to the category, or to 0 if it does not.
- An unbiased estimator of the number of elements in the population that belong to the category is $T'_{hh} = NP_{hh}$.

■ EXAMPLE 8.7   Consider again the population of hospitals last discussed in Example 8.6. The problem is to estimate the average and total purchases of a given pharmaceutical product by hospitals. The number of beds is the known auxiliary variable. The sample will be of size n, with replacement and PPS.

We begin by assigning six-digit numbers to hospitals as shown in Table 8.10.

In the third column of Table 8.10, 675 six-digit numbers are assigned to hospital 1, 450 to hospital 2, and so on, down to 1,500 for hospital 1,158.

Next, n six-digit random numbers are generated, ignoring 000000 and any numbers greater than 186030. If the list of these numbers happens to begin with, say, 001052, 185953, 000600, 000987, ..., then the list of selected hospitals would begin with nos. 2, 1,158, 1, 2, ... Note that hospital 2 will appear (at least) twice in the sample. Clearly, the procedure gives each hospital a probability of selection in any draw proportional to the number of its beds.

Suppose that this procedure was applied to select a random sample with replacement of $n = 4$ hospitals. Suppose further that the selected hospitals are nos. 2, 1,158, 1, and 2. The following information will be known after the selected hospitals are interviewed.

Sampled Hospital	Number of Beds, X_i	q_i	Purchases of Product Y, Y_i ($000)	Purchasing Product Z? (1=Yes, 0=No)
2	450	0.0024	350	0
1,158	1,500	0.0081	1,100	1
1	675	0.0036	500	1
2	450	0.0024	350	0

The HH estimate of the average purchases of Product Y per hospital is

$$\bar{Y}_{hh} = \frac{1}{(4)(1,158)}\left[\frac{350}{0.0024} + \frac{1,100}{0.0081} + \frac{500}{0.0036} + \frac{350}{0.0024}\right] = 122.271 \ (\$000),$$

TABLE 8.10   Sampling with PPS, Example 8.7

Hospital No., i	Number of Beds, X_i	Assigned Numbers	$q_i = X_i / \sum X$
1	675	000001–000675	0.0036
2	450	000676–001125	0.0024
⋮	⋮	⋮	⋮
1,158	1,500	184531–186030	0.0081
	186,030		1.0000

and the estimated total hospital purchases of Product Y is $(1,158)(122.271)$ or $141,590$ ($000).

The HH estimate of the proportion of hospitals which purchase Product Z is

$$P_{hh} = \frac{1}{(4)(1,158)} \left[\frac{0}{0.0024} + \frac{1}{0.0081} + \frac{1}{0.0036} + \frac{0}{0.0024} \right] = 0.0866,$$

and the estimate of the number of hospitals that purchase Product Z is $(1,158)$ (0.0866) or about 100. ◼

The variance of the HHE of μ becomes zero if the selection probabilities are made proportional to the Y values: $q_i = Y_i / \sum_{i=1}^{N} Y_i = Y_i / N\mu$. To see this, substitute the last expression for q_i in Eq. 8.12 and note that

$$Var(\bar{Y}_{hh}) = \frac{1}{n} \sum_{i=1}^{N} \left(\frac{Y_i N\mu}{N Y_i} - \mu \right)^2 = \frac{1}{n} \sum_{i=1}^{N} (\mu - \mu)^2 = 0.$$

It is not possible, of course, to implement such selection probabilities because the Y values are not known. However, as with the HTE and sampling without replacement, the variance of the HHE of μ becomes approximately zero if the Y values are approximately proportional to the values of a known auxiliary variable X, and the selection probabilities are made proportional to these X values.

To see this, observe that if $Y_i \approx bX_i$, then $\sum_{i=1}^{N} Y_i \approx b \sum_{i=1}^{N} X_i$. Dividing this last expression by N, we find $\mu_y \approx b\mu_x$. If $q_i = X_i / \sum X = X_i / N\mu_x$, we have

$$\frac{Y_i}{N q_i} - \mu_y \approx \frac{bX_i}{N(X_i / N\mu_x)} - b\mu_x = b\mu_x - b\mu_x = 0.$$

As usual, we fall back on the Ackroyd case to confirm the unbiasedness of the HH estimator and its variance.

◼ **EXAMPLE 8.8** Consider once again the half-of-Ackroyd population, and suppose it is desired to select a random sample of size $n = 2$ with replacement and probability proportional to the firms' annual sales. The relevant data are as follows:

Firm	Sales, X_i	$q_i = X_i/34$	Employees, Y_i	Hire? Y_i'
A	13	13/34=0.3824	9	1
B	12	12/34=0.3529	8	1
C	9	9/34=0.2647	6	0
	34	1.0000	23	2

We have $N = 3$, $\mu = 23/3 = 7.6667$, and $\pi = 2/3 = 0.6667$. The possible sample outcomes and their probabilities are shown in Figure 8.3.

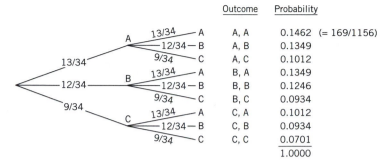

	Outcome	Probability
	A, A	0.1462 (= 169/1156)
	A, B	0.1349
	A, C	0.1012
	B, A	0.1349
	B, B	0.1246
	B, C	0.0934
	C, A	0.1012
	C, B	0.0934
	C, C	0.0701
		1.0000

Figure 8.3 Probability tree, PPS illustration

The HH estimates of μ and π for each sample outcome are shown in Table 8.11.

For example, if the outcome is (A,A), the HH estimate of μ is

$$\bar{Y}_{hh} = \frac{1}{(2)(3)}\left[\frac{9}{13/34} + \frac{9}{13/34}\right] = 7.8461,$$

and the HH estimate of π is

$$P_{hh} = \frac{1}{(2)(3)}\left[\frac{1}{13/34} + \frac{1}{13/34}\right] = 0.8718.$$

(The final results of all calculations in this example are shown to four decimal places, but the intermediate calculations were carried out to greater precision.)

TABLE 8.11 Sample Outcomes and HH Estimates, Example 8.8

Outcome	Probability	HHE of μ, $\bar{Y}_{hh}$	HHE of π, P_{hh}
A,A	0.1462	7.8461	0.8718
A,B	0.1349	7.7008	0.9081
A,C	0.1012	7.7008	0.4359
B,A	0.1349	7.7008	0.9081
B,B	0.1246	7.5555	0.9444
B,C	0.0934	7.5555	0.4722
C,A	0.1012	7.7008	0.4359
C,B	0.0934	7.5551	0.4722
C,C	0.0701	7.5551	0.0000
	1.0000		

It is straightforward to calculate the expected average values and variances of these estimates:

$$E(\bar{Y}_{hh}) = (7.8461)(0.1462) + \cdots + (7.5551)(0.0701) = 7.6667,$$
$$E(P_{hh}) = (0.8718)(0.1462) + \cdots + (0.0000)(0.0701) = 0.6667,$$
$$Var(\bar{Y}_{hh}) = (7.8461 - 7.6667)^2(0.1462) + \cdots + (7.555 - 7.6667)^2(0.0701)$$
$$= 0.00997,$$
$$Var(P_{hh}) = (0.8718 - 0.6667)^2(0.1462) + \cdots + (0.0000 - 0.6667)^2(0.0701)$$
$$= 0.08048.$$

Note that the estimators are indeed unbiased, and that the direct calculation of their variances agrees with Eq. 8.12:

$$Var(\bar{Y}_{hh}) = \frac{1}{2}\Big[\frac{9^2}{(3^2)(13/34)} + \frac{8^2}{(3^2)(12/34)} + \frac{6^2}{(3^2)(9/34)} - (\frac{23}{3})^2\Big] = 0.00997,$$

$$Var(P_{hh}) = \frac{1}{2}\Big[\frac{1^2}{(3^2)(13/34)} + \frac{1^2}{(3^2)(12/34)} + \frac{0^2}{(3^2)(9/34)} - (\frac{2}{3})^2\Big] = 0.08048. \blacksquare$$

8.11 INTERVAL ESTIMATES BASED ON THE HHE

If the sample is large, with replacement and unequal but constant probabilities of selection in every draw, approximate confidence intervals are given in Summary 8.4.

◪ EXAMPLE 8.9 We repeat below the results of Example 8.7 concerning the sample of four hospitals, selected with replacement and with PPS:

Sampled Hospital	Number of Beds, X_i	q_i	Purchases of Product Y, Y_i ($000)	Purchasing Product Z? (1=Yes, 0=No)
2	450	0.0024	350	0
1,158	1,500	0.0081	1,100	1
1	675	0.0036	500	1
2	450	0.0024	350	0

The HH estimates of the average purchases of Product Y per hospital (μ) and of the proportion of hospitals which purchase Product Z (π) were calculated earlier as 122.271 and 0.0866, respectively.

A sample of only four hospitals is much too small—certainly smaller than the recommended rule-of-thumb size of 50. We pretend it is large enough in order to illustrate the calculation of the confidence intervals.

SUMMARY 8.4   Confidence intervals based on the HHE

- For large n, an approximate $100(1 - \alpha)\%$ confidence interval for μ, the population mean of a variable Y, is

$$\bar{Y}_{hh} \pm Z_{\alpha/2}\sqrt{\widehat{Var(\bar{Y}_{hh})}} \qquad (8.13)$$

where $\widehat{Var(\bar{Y}_{hh})}$ is an estimator of $Var(\bar{Y}_{hh})$ given by

$$\widehat{Var(\bar{Y}_{hh})} = \frac{1}{n(n-1)} \sum_{i=1}^{n} \left(\frac{Y_i}{Nq_i} - \bar{Y}_{hh}\right)^2. \qquad (8.14)$$

- For large n, an approximate $100(1 - \alpha)\%$ confidence interval for π, the proportion of elements in the population that belong to a given category, is

$$P_{hh} \pm Z_{\alpha/2}\sqrt{\widehat{Var(P_{hh})}} \qquad (8.15)$$

where $\widehat{Var(P_{hh})}$ is an estimator of $Var(P_{hh})$ given by

$$\widehat{Var(P_{hh})} = \frac{1}{n(n-1)} \sum_{i=1}^{n} \left(\frac{Y_i'}{Nq_i} - P_{hh}\right)^2. \qquad (8.16)$$

In the last expression, Y_i' is 1 if the sample element falls into the category, or 0 if it does not.
- The corresponding confidence intervals for the population totals, $\tau = N\mu$ and $\tau' = N\pi$, are formed by multiplying the limits of 8.13 and 8.15, respectively, by N.

For a confidence interval for μ, we need

$$\begin{aligned}
\widehat{Var(\bar{Y}_{hh})} &= \frac{1}{n(n-1)} \sum_{i=1}^{n} \left(\frac{Y_i}{Nq_i} - \bar{Y}_{hh}\right)^2 \\
&= \frac{1}{(4)(4-1)}\Big[\big(\frac{350}{1{,}158 \times 0.0024} - 122.271\big)^2 \\
&\quad + \big(\frac{1{,}100}{1{,}158 \times 0.0081} - 122.271\big)^2 \\
&\quad + \big(\frac{500}{1{,}158 \times 0.0036} - 122.271\big)^2 \\
&\quad + \big(\frac{350}{1{,}158 \times 0.0024} - 122.271\big)^2\Big] \\
&= 4.7729.
\end{aligned}$$

If this were a large sample, an approximate 95% confidence interval for μ would be

$$(122.271) \pm (1.96)\sqrt{4.7729},$$

or from 117.989 to 126.553 ($000).

For a confidence interval for π, we require

$$\widehat{Var(P_{hh})} = \frac{1}{n(n-1)} \sum_{i=1}^{n} \left(\frac{Y_i'}{Nq_i} - P_{hh}\right)^2$$

$$= \frac{1}{(4)(4-1)} \left[\left(\frac{0}{1,158 \times 0.0024} - 0.0866\right)^2\right.$$

$$+ \left(\frac{1}{1,158 \times 0.0081} - 0.0866\right)^2$$

$$+ \left(\frac{1}{1,158 \times 0.0036} - 0.0866\right)^2$$

$$\left. + \left(\frac{0}{1,158 \times 0.0024} - 0.0866\right)^2\right]$$

$$= 0.00324.$$

If this were a large sample, an approximate 95% confidence interval for π would be

$$(0.0866) \pm (1.96)\sqrt{0.00324},$$

or from -0.02 to 19.82%. The earlier comments concerning negative estimates of proportions apply here as well. ◼

8.12 APPLICATION: DOLLAR-UNIT SAMPLING

Consider an auditor examining the accounts receivable of a firm. With each account there is associated a "book" value, the balance of the account at a given point in time. For example, the firm's records may show that Able Enterprises owes $328.15 to the firm on December 31st as a result of a recent sale charged to the account and not yet paid. The book value of the account may or may not be "true" because of error or fraud. One task of the auditor is to determine the true value of all accounts receivable. When the number of accounts is large and other transactions or items of the firm's balance sheet must also be audited, an examination of all accounts receivable may be impossible—a sample must be used.

A random sample of accounts receivable may be selected using one of the familiar methods (for example, simple, stratified, two-stage, etc.), but a method popular and unique in auditing is *dollar-unit sampling (DUS)*. Under this method, the population elements are the individual dollars of the account balances. A

sample consists of a number of these dollars. A simple example will help clarify this method.

Suppose the firm has three accounts receivable with the following book balances as of December 31st (the last column will be explained shortly):

Account	Book Value ($)	Assigned Numbers
A	10	01–10
B	20	11–30
C	5	31–35
	35	

Under DUS, the population elements are the 35 dollars making up the total balance "... as if there were [35] dollar bills spread out on a table ...".[1] A simple random sample of these dollar units is selected with replacement. The selected dollar "... acts like a hook and pulls down the entire [account] with it ...".[2] The pulled down accounts are audited and their true balances are determined.

To implement DUS in this example, one can imagine marking 10 otherwise identical chips with the letter A, 20 chips with B, and 5 with C. These 35 chips are put into a bowl and thoroughly mixed. One chip is then drawn at random. If it is marked 'A,' account A is audited; if marked 'B,' account B is audited, and so on. The chip is now reinserted into the bowl, the 35 chips are again thoroughly mixed, and another chip is drawn and similarly interpreted. The procedure is repeated n times to yield n dollar units and the corresponding accounts. Obviously, the same result is achieved by using two-digit random numbers and the assignment shown in the last column of the preceding table.

Suppose three two-digit random numbers were generated, and, in consequence, two accounts (A and C) were selected and audited. The audited ("true") values of these accounts were 9 and 7, respectively. The book value of A was 10, and that of C was 5. In DUS, the true value of each *dollar* of account A is deemed to be 9/10 or $0.90, and that of account B 7/5 or $1.40. These calculations are repeated in the following table:

Dollar Selected	Account	Book Value of Dollar	True Value of Dollar
09	A	1	9/10=0.90
34	C	1	7/ 5=1.40
05	A	1	9/10=0.90
			3.20

---

[1]D. A. Leslie, A. D. Teitlebaum, and R. J. Anderson, *Dollar-Unit Sampling: A Practical Guide for Auditors* (Toronto: Copp Clark Pitman, 1979), p. 86.

[2]*ibid.*, p. 84.

The three dollar units are treated as a simple random sample from a population of 35 such units. Thus, the estimated average true value per population *dollar*, is 3.20/3, and the estimated total true value of population dollars, that is, the total true value of all accounts receivable is (35)(3.20/3) or $37.33.

It should be clear that a dollar-unit sample of size n pulls down a simple random sample of n accounts with replacement. In other words, *under DUS, the probability of selecting an account is proportional to the book value of the account.* It may not be clear that *the DUS estimate of the total true value of accounts receivable is the HH estimate based on this sample of accounts.*

To see this, let us look at the sample as one with replacement and PPS of three *accounts* from a population of three *accounts receivable*:

	Population			**Sample**	
Account	Book Value, X_j	$q_j = X_j/\sum X$	Account	True Value, Y_i	q_i
A	10	10/35	A	9	10/35
B	20	20/35	C	7	5/35
C	5	5/35	A	9	10/35
	35				

The HH estimate of the population average true value *per account* is

$$\bar{Y}_{hh} = \frac{1}{(3)(3)}\left[\frac{9}{10/35} + \frac{7}{5/35} + \frac{9}{10/35}\right]$$

$$= \frac{35}{(3)(3)}\left[\frac{9}{10} + \frac{7}{5} + \frac{9}{10}\right]$$

$$= \frac{35}{(3)(3)}(3.20),$$

and the estimate of the population total true value of accounts receivable is

$$T_{hh} = N\bar{Y}_{hh} = (3)\frac{35}{(3)(3)}(3.20) = (35)\frac{3.20}{3} = 37.33,$$

which is identical to the DUS estimate.

This is not a numerical accident. Let X_j be the book value of account j. The population size of dollar units is $N' = \sum_{j=1}^{N} X_j$; we shall write this last sum as $\sum X$. Let Y_i' be the true value of the ith selected dollar unit. If Y_i and X_i are the true and book values of the corresponding "pulled down" account, then $Y_i' = Y_i/X_i$. The DUS estimate of the population total true value of accounts receivable is

$$N'\frac{1}{n}\sum_{i=1}^{n} Y_i' = \frac{\sum X}{n}\sum_{i=1}^{n}\frac{Y_i}{X_i}.$$

If we look at the sample as one of accounts selected with replacement and PPS, the HH estimate of the total true value of all accounts is

$$T_{hh} = N \bar{Y}_{hh} = \frac{1}{n} \sum_{i=1}^{n} \frac{Y_i}{q_i} = \frac{1}{n} \sum_{i=1}^{n} \frac{Y_i}{X_i / \sum X} = \frac{\sum X}{n} \sum_{i=1}^{n} \frac{Y_i}{X_i},$$

and is identical to the DUS estimate.

Dollar-unit sampling—or, more generally, *monetary-unit sampling (MUS)*— also provides a method for constructing confidence intervals for the average and total true value of an accounting population that is not described in this book. This method and numerous others investigated in the accounting literature were prompted by dissatisfaction with the poor performance of the "classical" method described in Summary 8.4 when this is applied to populations having few accounts in error and a skewed distribution of error amount for accounts that have an error.[3]

8.13 TO SUM UP

- The probability that a given population element will be included in the sample is called the inclusion probability of the element. The inclusion probabilities depend on the sampling method. Each sampling method implies a set of inclusion probabilities.
- The Horvitz–Thompson estimator (HTE) serves as an unbiased estimator for any sampling method with known inclusion probabilities.
- The HTE *is* the simple, stratified, or two-stage estimator under simple, stratified, or two-stage sampling, respectively.
- If the variable of interest is approximately proportional to a known variable, that is, $Y \approx bX$, and the inclusion probabilities can be made proportional to the X values, then the variance of the HTE is approximately zero. In this case, sampling with inclusion probabilities proportional to the X values (IPPS) is an alternative to using simple random sampling with ratio or regression estimators or stratified sampling for the purpose of exploiting the available auxiliary information.
- If sampling is with replacement, the Hansen–Hurwitz estimator (HHE) is an unbiased estimator for any sampling method under which the probability of selection of any element in each draw is known.
- If the variable of interest is approximately proportional to a known auxiliary variable, then sampling with replacement and selection probabilities proportional

---

[3]For additional information on MUS see, for example, D. R. Gwilliam, *A Survey of Auditing Research* (Upper Saddle River, NJ: Prentice/Hall International, 1987), Ch. 16; H. Arkin, *Sampling Methods for the Auditor: An Advanced Treatment* (New York: McGraw-Hill, 1982); D. M. Guy and D. R. Carmichael, *Audit Sampling: An Introduction to Statistical Sampling*, 2nd ed. (New York: Wiley, 1986).

to the size of this variable (PPS) makes the variance of the HHE approximately zero.
- Dollar-unit sampling (DUS) is in effect sampling with PPS, and the DUS estimator is the HHE of the true total value of the variable of interest.

PROBLEMS

8.1 *Computing exercise:* Using the program SCALC, verify the numerical results presented in this chapter. Note any deviations, and determine if these are the result of rounding at intermediate stages.

8.2 Table 4.7 shows a stratified random sample of two firms drawn from the Ackroyd population. The groups were formed according to the firms' nature of operations as shown in Table 4.6.

(a) Determine the inclusion probabilities of the firms in Ackroyd under this sample.

(b) Verify that these inclusion probabilities agree with the general results for stratified samples given in Section 8.2.

8.3 Table 5.5 shows a two-stage random sample of two firms drawn from the Ackroyd population stratified according to the firms' nature of operations.

(a) Determine the inclusion probabilities of the firms in Ackroyd under this sample.

(b) Verify that these inclusion probabilities agree with the general results for two-stage samples given in Section 8.2.

8.4 Refer to Example 8.1.

(a) Calculate the HT estimates of μ and π in the event the sample consists of firms A and C.

(b) Calculate the HT estimates of μ and π in the event the sample consists of firms A and E.

8.5 (a) Suppose the elements of a population are classified into $M = 2$ groups. Show that the HT estimators of μ and π are the stratified estimators.

(b) Same as **(a)**, except that the population consists of $M = 3$ groups.

(c) Same as **(a)**, except the elements are classified into an arbitrary number (M) of groups.

8.6 Show that the HT estimators of μ and π *are* the two-stage estimators, $\bar{Y}_{ts}$ and P_{ts}, in the following situations.

(a) One group is selected in the first stage $(m = 1)$; then in the second stage n elements are drawn from the selected group.

(b) Two groups are selected first $(m = 2)$; then n_1 and n_2 elements are drawn from the two selected groups.

(c) In the first stage, m groups are selected; in the second stage, n_i elements are drawn from each selected group i.

8.7 (a) Refer to Section 8.4 and suppose that Mary Brown and Melinda Ryan are the two persons selected by the three-stage sample. Mary is 14 and smokes, Melinda is 40 and does not smoke. Estimate the average age of and the proportion of smokers among the nine persons in the region.

(b) Same as **(a)**, except that the sample consists of Peter Jones and Melinda Ryan.

8.8 Refer to Example 8.2.

(a) Verify that the HT estimates of π for each sample outcome are as shown in Table 8.4.

(b) Confirm that P_{ht} is an unbiased estimator of π.

(c) Confirm that $Var(P_{ht}) = 0.0516$, in agreement with Eq. 8.4.

8.9 Consider the "Other-half-of-Ackroyd" population described in Table 8.12.

Suppose that under a certain method of selecting a random sample of two firms without replacement, the possible sample outcomes and their probabilities are as follows:

Outcome	Probability
D,E	0.1
D,F	0.6
E,F	0.3

(a) Calculate the population average number of employees (μ) and the population proportion of firms intending to hire (π).

(b) Calculate the HT estimates $\bar{Y}_{ht}$ and P_{ht} for each possible sample outcome.

(c) Calculate the expected values of $\bar{Y}_{ht}$ and P_{ht}. Are they unbiased estimators of μ and π in this case?

(d) Calculate the variances of $\bar{Y}_{ht}$ and P_{ht}. Confirm that they agree with Eqs. 8.2 and 8.4.

TABLE 8.12 "Other-Half-of-Ackroyd" Population

Firm	Sales, X	Employees, Y	Will Hire?
D	3	2	No
E	1	1	No
F	7	5	No

8.10 The three largest of the 1,158 hospitals of the population outlined in Table 8.6 have 2,100, 1,600, and 1,750 beds. In view of this fact, is it possible to take a sample of 100 hospitals without replacement and IPPS as described in Example 8.4?

8.11 The values of a known auxiliary variable X for the five elements of a population are as follows:

Element No.	X
1	10
2	200
3	40
4	100
5	300

In the manner of Example 8.4, select a random sample of two elements without replacement and IPPS. Show the assigned numbers, the random numbers used, and the selected elements.

8.12 (a) Consider the "Other-half-of-Ackroyd" population described in Table 8.12, and suppose that a random sample of two firms will be selected without replacement and inclusion probability proportional to the firms' annual sales. In the manner of Example 8.5, confirm that the selection procedure of Section 8.7 does give each firm IPPS.

(b) Same as **(a)**, except that last year's sales of Firm F is 3, not 7.

8.13 Refer to Example 8.6. A random sample of four hospitals was selected without replacement from the population of 1,158 hospitals. The sampled hospitals, their inclusion probabilities, and their purchases of products Y and Z are shown in Table 8.13.

(a) Calculate the HT estimates of the population average purchases of Product Y per hospital (μ) and of the proportion of hospitals which purchase Product Z (π).

(b) Estimate the corresponding population totals $\tau = N\mu$ and $\tau' = N\pi$.

(c) Calculate S_{ht}^2 and $S_{ht}'^2$. Calculate $\widehat{Var}(\bar{Y}_{ht})$ and $\widehat{Var}(P_{ht})$.

TABLE 8.13   Sample Results, Problem 8.13

Sampled Hospital	p_i	Purchases of Product Y, Y_i ($000)	Purchasing Product Z? (1=Yes, 0=No)
2	0.010	350	0
567	0.025	1,150	1
801	0.030	1,300	0
1,158	0.035	1,100	1

(d) Pretending that n and $N - n$ are large, calculate 90% confidence intervals for μ, π, τ, and τ'.

8.14 Same as Problem 8.13, except that the sample results are as shown in Table 8.14.

8.15 Table 8.15 shows the five elements of a population and their values of a known auxiliary variable X.

(a) Using random numbers generated from a table or by computer, select a random sample of three elements with replacement and PPS. Show the assigned numbers, the random numbers used, and the selected elements.

(b) Suppose that the sample in **(a)** consists of elements 2, 4, and 5, with values of three variables (Y_1 to Y_3) and two categories (C_1 and C_2) of interest as follows:

Element	Y_1	Y_2	Y_3	C_1	C_2
2	−8	− 5	16	Yes	No
4	−6	−15	31	Yes	Yes
5	−4	−23	46	No	No

Calculate the HH estimates of the population means of the variables and of the population proportions in the categories. Calculate the HH estimates of the population totals.

TABLE 8.14   Sample Results, Problem 8.14

Sampled Hospital	p_i	Purchases of Product Y, Y_i ($000)	Purchasing Product Z? (1=Yes, 0=No)
155	0.040	135	1
321	0.015	170	1
801	0.020	1,300	0
967	0.030	505	1

TABLE 8.15   Popula-tion, Problem 8.15

Element No.	X
1	20.7
2	5.2
3	9.6
4	10.3
5	14.2

8.16 Same as Problem 8.15, except that the population consists of eight elements with the values of an auxiliary variable X shown in Table 8.16.

8.17 Suppose a random sample of size 2 will be selected with replacement and PPS from the "Other-half-of-Ackroyd" population shown in Table 8.12.

(a) Calculate the population average number of employees, μ, and the population proportion of firms intending to hire, π.

(b) With the help of a probability tree, list the possible sample outcomes, their probabilities, and the values of the HT estimates $\bar{Y}_{hh}$ and P_{hh} of μ and π.

(c) Calculate the expected values of $\bar{Y}_{hh}$ and P_{hh}. Are they unbiased estimators of μ and π in this case?

(d) Calculate the variances of $\bar{Y}_{ht}$ and P_{ht}. Confirm that they agree with Eq. 8.12.

8.18 A random sample of three elements was drawn with replacement and unequal but constant selection probabilities from a population of size 100. Table 8.17 shows the selected elements, their selection probabilities in each draw (q_i), and the values of the variable (Y) and category (C) of interest.

(a) Calculate the HH estimates of the population mean of Y (μ) and proportion in C (π).

(b) Calculate $\widehat{Var}(\bar{Y}_{hh})$ and $\widehat{Var}(P_{hh})$ according to Eqs. 8.14 and 8.16.

(c) Pretending that the sample size is large, calculate 90% confidence intervals for $\mu, \pi, \tau = N\mu$ and $\tau' = N\pi$.

TABLE 8.16   Population, Problem 8.16

Element No.	X	Element No.	X
1	9.07	5	24.46
2	12.82	6	4.64
3	3.12	7	18.03
4	15.55	8	10.15

TABLE 8.17   Sample Results, Problem 8.18

Element, i	q_i	Y_i	In C?
1	0.015	40	Yes
2	0.006	50	No
3	0.010	30	Yes

8.19 Same as Problem 8.18, except that the sample results are as shown in Table 8.18.

8.20 A firm has four accounts receivable with the following current book balances.

Account	Book Value, $
A	200
B	400
C	100
D	300

(a) Using random numbers generated from a table or by computer, select a dollar-unit sample (DUS) of size 3. Show the assigned numbers, the random numbers used, and the accounts selected.

(b) Suppose the selected dollars "belong to" accounts B, D, and B. The audited ("true") value of account B is $350, and that of D $320. Calculate the DUS estimates of the population average and total true value of all accounts receivable.

(c) Consider the sample as one of three accounts, selected with replacement and PPS. Calculate the HH estimates of the population average and total true value of all accounts receivable.

8.21 During a certain period of time, a consulting firm issued four invoices for services rendered for the following amounts:

Invoice	Amount ($)
A	240
B	50
C	30
D	180

(a) Using random numbers generated from a table or by computer, select a dollar-unit sample (DUS) of three invoices. Show the assigned numbers, the random numbers used, and the accounts selected.

TABLE 8.18 Sample results, Problem 8.19

Element, i	q_i	Y_i	In C?
1	0.008	35	No
2	0.013	45	Yes
3	0.015	60	No

(b) Suppose the selected dollars "belong to" invoices A, C, and D. The audited ("true") values of these invoices are $250, $30, and $160, respectively. Calculate the DUS estimates of the population average and total true amount of all issued invoices.

(c) Consider the sample as one of three invoices, selected with replacement and PPS. Calculate the HH estimates of the population average and total true amount of all issued invoices.

8.22 Under cluster sampling, it will be recalled, a simple random sample of m groups is selected, and all their elements are measured or questioned. Suppose that the selection of groups is with replacement and probability proportional to group size, that is, $q_i = N_i/N$.

(a) Show that the HHE of the population total of a variable Y is

$$T_{hh} = \frac{N}{m}\sum_{i=1}^{m}\bar{Y}_i = \frac{N}{m}\sum_{i=1}^{m}\mu_i.$$

(b) Show that the HHE of the population mean of Y is the ordinary average of the sample means of the selected groups.

(c) Show that the HHE of the number of population elements falling into a given category is

$$T'_{hh} = \frac{N}{m}\sum_{i=1}^{m}P_i = \frac{N}{m}\sum_{i=1}^{m}\pi_i.$$

(d) Show that the HHE of the population proportion in the category is the ordinary average of the sample proportions in the selected groups.

8.23 Suppose that the selection of groups in the first of a two-stage sample is without replacement but with probability proportional to their size, rather than with equal probability as was assumed in Chapter 5. Suppose group i has N_i elements, so that the desired probability that it will be included in the first-stage sample of m of the M groups is cN_i. Suppose also that n_i of the N_i elements of group i will be selected at random, without replacement, and with equal probability if group i is selected in the first stage.

(a) Show that c must be equal to m/N.

(b) Show that the inclusion probability of any element of group i (that is, the probability that this element will be included in the ultimate sample) is mn_i/N.

(c) Show that the HT estimator of the population mean (μ) of a variable Y is

$$\bar{Y}_{ht} = \frac{1}{m}\sum_{i=1}^{m}\bar{Y}_i,$$

where $\bar{Y}_i$ is the sample mean of Y in selected group i.

(d) Show that the HT estimator of the population proportion (π) in a certain category is

$$\bar{Y}_{ht} = \frac{1}{m} \sum_{i=1}^{m} P_i,$$

where P_i is the sample proportion in the category in selected group i.

(e) Show that when only one group is selected in the first stage, the HT estimator of μ is simply the sample mean of Y in the one selected group.

(f) Show that when only one group is selected in the first stage, the HT estimator of π is simply the sample mean of Y in the one selected group.

(g) Consider the Ackroyd population of six firms, stratified according to geographical location as shown in Tables 4.3 or 5.2. Suppose that a two-stage sample will be taken: in the first stage, one of the two groups (East, West) will be selected with probability proportional to the number of firms in the group; in the second stage, two firms will be selected at random and without replacement from the group selected in the first stage. Determine the probability distribution of the HT estimator of the average number of employees per firm in Ackroyd, and show that it is indeed unbiased. Determine the probability distribution of the HT estimator of the proportion of firms in Ackroyd that intend to hire, and show that it is indeed unbiased.

8.24 A sample of adults (persons 18 years old or older) was to be selected from the 5,250 adults known to reside in a certain geographical area. Among the questions the selected adults were to be asked were:

Q1: How much did you spend on entertainment last month? $\$$___

Q2: Do you smoke? ___ Yes ___ No

The area was divided into three regions, and each region was subdivided into districts. The number of districts in each region is shown in the first two columns of the following table.

Region	Number of Districts	Selected Districts
R1	50	D1
R2	20	D2, D3
R3	30	D4, D5
	100	

It was decided to select a simple random sample without replacement of districts from each region: one from R1, two from R2, and two from R3. The selected

districts are identified as D1 through D5 in the last column of the preceding table.

The following table shows the number of households in each selected district, and the labels of the households selected by simple random sampling without replacement from each selected district.

District	Number of Households	Selected Households
D1	30	H1, H2, H3
D2	25	H4, H5
D3	15	H6
D4	40	H7, H8, H9
D5	10	H0
	120	

For example, two households (labeled H4 and H5) were selected at random and without replacement from D2.

All the adults in the selected households were interviewed. The following table shows the number of adults in each selected household and their identification labels:

Household	Number of Adults	Selected Adults
H1	1	A01
H2	2	A02, A03
H3	3	A04, A05, A06
H4	1	A07
H5	3	A08, A09, A10
H6	2	A11, A12
H7	1	A13
H8	4	A14, A15, A16, A17
H9	1	A18
H0	2	A19, A20
	20	

The ultimate sample, therefore, was a sample of 20 adults. The selected adults were interviewed, and their responses to the questionnaire were as follows:

Adult	Q1	Q2 (1=Yes, 0=No)
A01	100	1
A02	60	0
A03	30	1
A04	50	0
A05	40	0
A06	0	0
A07	75	0
A08	10	1
A09	0	1
A10	0	0
A11	20	1
A12	90	1
A13	30	0
A14	40	0
A15	60	1
A16	0	0
A17	15	0
A18	350	0
A19	20	1
A20	30	1

(a) In the manner of Section 8.4, calculate the inclusion probabilities of all adults in the ultimate sample.

(b) Calculate the HT estimates of the average expenditure on entertainment per adult and of the total expenditure on entertainment by adults in the area.

(c) Calculate the HT estimates of the proportion and number of adults that smoke in the area.

CHAPTER 9

Sampling from a Process

9.1 INTRODUCTION AND SUMMARY

The prediction approach to sampling provides a framework different from the randomization approach described in earlier chapters. Rather than treat the population as a collection of elements with unknown but fixed values of the variables and categories of interest, the prediction approach assumes that these values are themselves determined by some sort of process.

The starting point of the prediction approach is that any variable of interest is determined by other variables and attributes. The variable of interest can be thought of as determined in part by known variables and in part by a random process with certain long run properties. The relationship between the variable of interest and the known variables can be estimated with the help of the sample observations, but the sample itself need not be random. This relationship, in turn, can be used to predict the values of the variable of interest for all nonsampled elements, hence also the population total or mean of this variable.

The prediction approach, as will be explained in this chapter, rests on different foundations from those of the randomization approach and may on occasion prescribe different solutions. Thus, for example, under certain circumstances, the best sample under the prediction approach is not random but purposive—for example, it consists of the elements having the largest values of an auxiliary variable. The

prediction approach thus seems to give legitimacy to other, possibly less costly, selection methods than random sampling.

The variable of interest can, of course, be related to known variables in many different ways, and there may be many types of random processes affecting the remainder. In what follows, we begin with the simplest possible relationships and random processes, and gradually work up to more complicated ones. The chapter concludes with a review of the foundations of the two approaches and a comparison of their strengths and weaknesses.

9.2 THE PREDICTION APPROACH

We have been describing a population as a collection of elements, to each of which are "attached" values of the variables and categories of the attributes of interest. Until now, these attachments were assumed to be unknown but given. We did not inquire into the *process* that may have generated these values and attributes.

Consider once again, for example, the problem of estimating the purchases of pharmaceutical products by hospitals. Table 9.1 outlines this population of hospitals and their purchases of a certain pharmaceutical product, Y.

The question marks represent the purchases (in $) of Product Y by each of the 1,158 hospitals in the population. These purchases are unknown but have been assumed to be given numbers. If all hospitals are interviewed and all respond truthfully, it has been assumed, all the question marks will be removed and the total purchases of Product Y will be determined by simple addition.

It can be argued, however, that a hospital's purchases of a pharmaceutical product do not materialize in a vacuum but are *determined* by a number of factors: the size of the hospital, the number of its patients, its type (e.g., general, specialized), its location, the characteristics of the patient population it serves, and others. It may be reasonable, therefore, to claim that the purchases of the product, Y, by a hospital are a function of factors $X_1, X_2, X_3, \ldots$:

$$Y = f(X_1, X_2, X_3, \ldots). \tag{9.1}$$

Y is the *dependent variable*, and $X_1, X_2, X_3, \ldots$ the *explanatory variables* of this relationship.[1]

In principle, $f(\ )$ describes a deterministic relationship; that is, one in which for each set of values of the explanatory variables, there is one and only one value of the dependent variable. In practice, the number of explanatory variables may be large, many may be unknown, and the exact form of the function $f(\ )$ may not be known. Often, however, Y can be *approximated* well by a function of k principal

---

[1]The terminology and assumptions of this chapter are those of regression analysis.

TABLE 9.1	Population of Hospitals	
Hospital No., i	**Number of Beds, X_i**	**Purchases of Product Y, Y_i**
1	675	?
2	450	?
⋮	⋮	⋮
1,158	1,500	?
Total	186,030	?

known explanatory variables $X_1, X_2, \ldots, X_k$. For example, the approximating relationship could be linear:

$$Y \approx \alpha + \beta_1 X_1 + \beta_2 X_2 + \cdots + \beta_k X_k,$$

where the βs are *parameters* to be estimated.

We can write, therefore,

$$Y = \alpha + \beta_1 X_1 + \beta_2 X_2 + \cdots + \beta_k X_k + U. \tag{9.2}$$

U can be thought of as the *deviation* between the actual value of Y and what is explained by the principal variables:

$$U = Y - (\alpha + \beta_1 X_1 + \beta_2 X_2 + \cdots + \beta_k X_k).$$

U represents the combined effect of the omitted explanatory variables and of the unknown functional form, $f(\ )$. Equation 9.2 is a *model* of the true relationship (9.1).

Obviously, the linear model (9.2) is not the only possible model, but it is simple and will serve us well for a start.

A sample will reveal the Y values of the selected elements and will provide the information needed to estimate the parameters of Model 9.2 (in a little while we shall see how this may be done). The estimated Y value of a population element i that is not in the sample is

$$\hat{Y}_i = a + b_1 X_{1i} + b_2 X_{2i} + \cdots + b_k X_{ki},$$

where b_j is the estimate of the parameter β_j, and X_{ji} the value of explanatory variable j for population element i.

The population total of variable Y, $\tau = \sum_{i=1}^{N} Y_i$, is, as always, a population characteristic of interest.

An estimator, T, of the population total of Y is the sum of two quantities: (a) the sum of the actual Y values of the sample elements (which will be known after the sample is taken), and (b) the sum of the estimated Y values of the elements not

included in the sample. Rearranging the population elements so that the sampled elements are listed first,

$$T = \underbrace{Y_1 + Y_2 + \cdots + Y_n}_{\text{sample}} + \underbrace{(\hat{Y}_{n+1} + \hat{Y}_{n+2} + \cdots + \hat{Y}_N)}_{\text{remainder}}$$

$$= \sum_{i=1}^{n} Y_i + \sum_{i=n+1}^{N} \hat{Y}_i. \tag{9.3}$$

The estimator of the population average value of Y is, of course, $\bar{Y} = T/N$.

The *prediction error*, $\tilde{E}$, is the difference between estimated and true population totals,

$$\tilde{E} = T - \tau = \left(\sum_{i=1}^{n} Y_i + \sum_{i=n+1}^{N} \hat{Y}_i \right) - \left(\sum_{i=1}^{n} Y_i + \sum_{i=n+1}^{N} Y_i \right)$$

$$= \sum_{i=n+1}^{N} \hat{Y}_i - \sum_{i=n+1}^{N} Y_i \tag{9.4}$$

$$= \sum_{i=n+1}^{N} (\hat{Y}_i - Y_i).$$

In words, the prediction error is, reasonably enough, the sum of the differences between the estimated and actual Y values of the elements that were not sampled.

This error, intuition may tell us, should depend on the methods by which the sampled elements are selected and the model's parameters estimated. It should also depend on the model's ability to approximate the dependent variable, that is, on the deviations U. These deviations are, of course, unknown. However, if it can be assumed that they are *generated by a process* with certain properties (by a "mechanism," if you like, with certain known features), then it is possible to determine the behavior of the prediction error in the long run, and to choose methods for selecting the sample and for estimating the parameters of the model that minimize the prediction error in the long run.

This is, in essence, the *prediction approach* to sampling, developed by R. M. Royall and his associates, and which we shall begin to examine in the following section. The mechanism generating the deviations is the critical element of this approach, having the same function as the various random selection mechanisms (the "mix and shuffle" variants) of the previous chapters. However, the two approaches rest on entirely different foundations and may produce conflicting recommendations: for example, the prediction approach may suggest that the sample elements be selected in a purposive, nonrandom fashion.

In the following sections we describe the prediction approach in the context of the simplest possible models, but the approach itself is quite general.

9.3 A SIMPLE LINEAR MODEL WITH NO INTERCEPT

Suppose that there is a single known principal explanatory variable, X, and that the assumed relationship between the variable of interest, Y, and X for each population element i is

$$Y_i = \beta X_i + U_i. \tag{9.5}$$

Equation 9.5 is the simplest form of Eq. 9.2. The X values of all N population elements are assumed known.

For example, Y_i could be the purchases of Product Y by hospital i, and X_i the size of the hospital, as measured by the number of its beds. The number of beds for each hospital is assumed to be known (cf. Table 9.1).

As stated earlier, the deviations U_i represent the combined effect of the omitted explanatory variables and of the true form of the relationship between Y and all explanatory variables. They are unknown. In its simplest version, the prediction approach assumes that these deviations are generated *as if* by random draws with replacement from a population of U values having mean 0 and a certain variance, σ^2.

To grasp the mechanism that is supposed to generate the deviations, think of Eq. 9.5 as describing the relationship between the purchases of Product Y by a given hospital and the number of its beds. For the sake of illustration, suppose that the best description of this relationship is provided by $\beta = 0.1$, that is, $Y_i = 0.1X_i + U_i$.

Imagine now that 10 identical tags are placed in a box. Three of these tags bear the number -10, four the number 0, and three the number $+10$. The mean of the numbers in the box, it can be confirmed, is 0, and their variance is $\sigma^2 = 60$.

Hospital 1 has $X_1 = 675$ beds (see Table 9.1). This hospital's purchases of Product Y, it is assumed, are determined in part by the number of its beds, and in part by other factors, the combined effect of which behaves like the outcome of a random draw from the box: if the tag drawn happens to bear the number -10, then $Y_1 = (0.1)(675) + (-10) = 57.5$; if the tag bears the value 0, $Y_1 = (0.1)(675) + 0 = 67.5$; and if it bears the number $+10$, $Y_1 = (0.1)(675) + 10 = 77.5$. The possible values of Y_1, therefore, are 57.5, 67.5, and 77.5. Under the assumptions of the model, the probabilities of these values are 0.3, 0.4, and 0.3, respectively. The outcome of the first draw (U_1) does not depend on (is independent of) X_1.

The second hospital has 450 beds. Its purchases of Product Y will be $(0.1)(450)$ plus the number borne by the tag in another random draw from the box. (The draws are with replacement, so the box always contains the same 10 tags.) Clearly, Y_2 could be 35, 45, or 55, with probabilities 0.3, 0.4, and 0.3, respectively. The outcome of the second draw from the box (U_2) is unrelated to the first (U_1) and to X_1 or X_2. Furthermore, Y_2 is unrelated to Y_1; for example, the probability that Y_2 will take the value 55 does not depend on the value of Y_1.

The purchases of each of the remaining 1,156 hospitals in the population are assumed to be determined in like fashion, *as if* by the same mechanism. Of course,

β could be any number, the box could contain any number of tags, and the tags could bear any numbers, as long as their average value is zero.

Under these assumptions, the population total, $\tau = Y_1 + Y_2 + \cdots + Y_N$, is not a constant as was assumed in earlier chapters, but a variable that cannot be predicted in advance with certainty.

Note the emphasis on the words *as if*. It is not claimed that the deviations are in fact outcomes of draws of tags from a box, but rather that they behave as outcomes of such draws are expected to behave: they cannot be predicted with certainty, do not depend on one another or the X values, and their long-run average value is zero. These assumptions define the model of this section, which we shall call *Model A*, and summarize as follows:

$$\text{Model A:} \begin{cases} Y_i = \beta X_i + U_i, \\ U_i \text{ independent, with } E(U_i) = 0 \text{ and } Var(U_i) = \sigma^2. \end{cases}$$

It should be noted that the symbol σ^2 has a different meaning in this chapter; it does not represent the population variance of the Y values, as was the case in all preceding chapters.

A sample of n elements will reveal the Y values of the selected elements. This sample need not be random: *any n of the N population elements may be selected*. If the assumptions of this section are satisfied, a scatter diagram of the n pairs of Y and X values will probably have the appearance of Figure 9.1.

There will probably be a tendency for the points to lie along a straight line through the origin (0,0) and within a band of uniform width about that line. We say "probably" because the appearance of the scatter diagram will vary from sample to sample.

Let b be any estimator of β based on the n observations (Y_i, X_i). For example, b could be one of the three estimators described in Section 6.2: the average of ratios $b = (1/n) \sum (Y_i / X_i)$, the ratio of averages $b = \bar{Y}/\bar{X}$, the ordinary least squares estimator $b = (\sum X_i Y_i)/(\sum X_i^2)$, or any other estimator. The estimated Y value of an element i not in the sample is

$$\hat{Y}_i = bX_i,$$

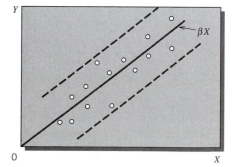

Figure 9.1 Scatter diagram, Model A

and the estimate of the population total of the Y values is calculated according to Eq. 9.3:

$$T = \sum_{i=1}^{n} Y_i + \sum_{i=n+1}^{N} \hat{Y}_i = \sum_{i=1}^{n} Y_i + b \sum_{i=n+1}^{N} X_i.$$

The notation used in the last line is a little awkward and assumes that the population elements are arranged so that those sampled are listed first. We shall use instead the simpler notation

$$T = \sum_{i=1}^{n} Y_i + b \sum_{j \in R} X_j.$$

The index i thus refers to the sampled elements. The shorthand "$j \in R$" should be read as "for all elements j in the remainder (the rest of the population)." Thus, $\sum_{j \in R} X_j$ is the sum of the X values of the nonsampled elements.

The prediction error, $\tilde{E}$, is the difference between T and the population total, τ,

$$\tilde{E} = T - \tau = \left(\sum_{i=1}^{n} Y_i + b \sum_{j \in R} X_j\right) - \left(\sum_{i=1}^{n} Y_i + \sum_{j \in R} Y_j\right) \tag{9.6}$$

$$= b \sum_{j \in R} X_j - \sum_{j \in R} Y_j.$$

The long-run behavior of this error depends on the choice of b, hence also of T. Two reasonable requirements for this choice are: (1) that the expected value of $\tilde{E}$ be zero, and (2) that the variance of $\tilde{E}$ about zero be as small as possible. In other words, it is reasonable to search for a method of estimating β and τ that will tend to produce neither positive nor negative errors, and that will tend to minimize the variability of these errors about zero.

Unfortunately, there is no known solution to this problem. There is, however, a simple solution if we restrict ourselves to estimators T that can be written as linear functions of the sample Y values:

$$T = \sum_{i=1}^{n} d_i Y_i,$$

where d_i are constants or functions of the known X values. Such estimators are said to be *linear estimators* of τ.

A linear estimator T with zero expected error, that is, such that $E(\tilde{E}) = E(T - \tau) = 0$, is called a *linear unbiased estimator* of τ. That linear unbiased estimator which has the smallest error variance, $Var(\tilde{E})$, among all linear unbiased estimators is the *best linear unbiased estimator (BLUE)* of τ. The BLUE of τ under Model A and its error variance are given in Summary 9.1.

These results hold for *any* sample of n elements, no matter how it is selected. Do, however, examine carefully Eq. 9.9. The population size (N), the sample size (n),

---

SUMMARY 9.1 BLUE of τ and error variance, Model A

- The best linear unbiased estimator, T^*, of the population total of a variable Y under Model A is

$$T^* = \sum_{i=1}^{n} Y_i + b \sum_{j \in R} X_j, \qquad (9.7)$$

where b is the least squares estimator of β:

$$b = \frac{\sum_{i=1}^{n} X_i Y_i}{\sum_{i=1}^{n} X_i^2}. \qquad (9.8)$$

The error variance associated with this estimator is

$$Var(\tilde{E}^*) = Var(T^* - \tau) = \sigma^2 \left[\frac{(\sum_{j \in R} X_j)^2}{(\sum_{i=1}^{n} X_i^2)} + N - n \right]. \qquad (9.9)$$

- The best estimator of the population mean of variable Y is T^*/N.

---

and the variance of the deviations (σ^2) do not depend on the manner in which the sample is chosen. The ratio within the square brackets of Eq. 9.9 does. The numerator of this ratio is the square of the sum of the known X values in the remainder; the denominator is the sum of the squared X values in the sample. It is clear that the numerator is minimized, the denominator maximized, hence also $Var(\tilde{E}^*)$ minimized when the sample consists of the n elements having the largest X values. In other words, *the best sample for Model A is not a random sample, but one purposely selected so as to consist of the elements having the largest X values.*

◪ EXAMPLE 9.1 Let us return to the problem of estimating the total hospital purchases of pharmaceutical Product Y. If it is believed that the assumptions of Model A are satisfied, the best sample of size n should consist of the n largest hospitals in terms of number of beds. The best estimate of the total hospital purchases of Product Y is given by Eq. 9.7 with b the least squares estimate, Eq. 9.8.

Suppose, for example, that the three largest hospitals have identification numbers 234, 567, and 801. If only three hospitals can be sampled, these hospitals will be questioned concerning their purchases of Product Y. Assume that their responses are as shown in the third column of Table 9.2.

The least squares estimate is calculated from the sums in cols. (4) and (5):

$$b = \frac{\sum X_i Y_i}{\sum X_i^2} = \frac{7,475,000}{10,032,500} = 0.7451.$$

TABLE 9.2 Sample of Hospitals and Calculations, Example 9.1				
Sampled Hospital (1)	Number of Beds, X_i (2)	Purchases of Product Y, Y_i ($000) (3)	$X_i Y_i$ (4)	X_i^2 (5)
234	2,100	1,600	3,360,000	4,410,000
567	1,600	1,150	1,840,000	2,560,000
801	1,750	1,300	2,275,000	3,062,500
	5,450	4,050	7,475,000	10,032,500

Recall that the total number of beds in all 1,158 hospitals is 186,030 (see Table 9.1). Therefore, the total number of beds in the nonsampled hospitals is $\sum_{j \in R} X_j = 186,030 - 5,450 = 180,580$. Therefore, the best estimate of the total hospital purchases of Product Y is

$$T^* = \sum_{i=1}^{n} Y_i + b \sum_{j \in R} X_j = (4,050) + (0.7451)(180,580) = 138,600.$$

The best estimate of the average purchases of Product Y per hospital is $T^*/N = 138,600/1,158 = 119.689$ ($000).

The sample of the largest hospitals will be best for estimating the total purchases of *any* pharmaceutical product for which the assumptions of Model A are satisfied. "Best," of course, means best in terms of the criteria of this section. ◼

It should be clearly understood that Eq. 9.8 gives the best b and Eq. 9.7 the best T based on *any* sample of n elements, no matter how selected. When the sample consists of the elements with the largest X values, Eqs. 9.8 and 9.7 give the very best T, that is, the estimator of the population total of a variable with the smallest error variance among all linear unbiased estimators and sample selections.

In the rest of this section, we demonstrate that T^* is a linear estimator of τ. Indeed, we show more generally that if b is a linear estimator of β, then T utilizing this b is a linear estimator of τ, and that all three estimators of β described earlier (the average of ratios, the ratio of averages, and the least squares estimator) are linear. Therefore, the class of linear estimators is not artificial but includes among its members these and other reasonable estimators. Readers who do not care to question these assertions may skip the material in the rest of this section without fear.

We say that b is a linear estimator of β if it can be written as a linear function of the sample Y values:

$$b = \sum_{i=1}^{n} e_i Y_i,$$

where e_i are constants or functions of the known X values. If this is the case, then

$$T = \sum_{i=1}^{n} Y_i + b \sum_{j \in R} X_j$$

$$= \sum_{i=1}^{n} Y_i + (\sum_{i=1}^{n} e_i Y_i)(\sum_{j \in R} X_j)$$

$$= \sum_{i=1}^{n} \underbrace{[1 + (\sum_{j \in R} X_j) e_i]}_{d_i} Y_i$$

$$= \sum_{i=1}^{n} d_i Y_i.$$

Therefore, if b is a linear estimator of β, T is a linear estimator of τ.

All three estimators of β mentioned earlier are linear. To see this, note that the average of the ratios can be written as

$$b = \frac{1}{n} \sum_{i=1}^{n} \frac{Y_i}{X_i} = \sum_{i=1}^{n} (\frac{1}{n X_i}) Y_i = \sum_{i=1}^{n} e_i Y_i,$$

where $e_i = 1/n X_i$. Likewise, the ratio of averages is

$$b = \frac{\bar{Y}}{\bar{X}} = \frac{\sum_{i=1}^{n} Y_i}{\sum_{i=1}^{n} X_i} = \sum_{i=1}^{n} (\frac{1}{\sum_{i=1}^{n} X_i}) Y_i = \sum_{i=1}^{n} e_i Y_i,$$

where $e_i = 1/\sum_{i=1}^{n} X_i$. Finally, the least squares estimator,

$$b = \frac{\sum_{i=1}^{n} X_i Y_i}{\sum_{i=1}^{n} X_i^2} = \sum_{i=1}^{n} (\frac{X_i}{\sum_{i=1}^{n} X_i^2}) Y_i = \sum_{i=1}^{n} e_i Y_i,$$

where $e_i = X_i/\sum_{i=1}^{n} X_i^2$.

It follows that any estimator T using one of these b's is linear. In particular, T^* using the least squares estimator b is linear, as claimed earlier.

We see that the class of linear estimators T is not artificial, but includes among its members estimators which common sense suggests are reasonable. The fact that T^* is best among these members, therefore, acquires some additional practical significance.

9.4 BEST LINEAR UNBIASED ESTIMATOR CONFIRMED

To provide some insight into the results just presented cookbook-fashion, let us fall back, once more, on the Ackroyd illustration. There are three firms, it will be recalled, in Half-of-Ackroyd. These firms are A, B, and C, and their sales (X) during last year were as follows:

Firm	Sales ($000), X
A	13
B	12
C	9

The variable of interest is the current number of employees, Y, and the characteristic of interest the total number employed by the three firms.

Let us suppose that a firm's number of employees is approximately related to the firm's sales, as follows:

$$Y_i = (0.5)X_i + U_i.$$

Suppose further that the U's are independent of one another and the X's, with the following common probability distribution:

U	**Probability**
-1	0.3
0	0.4
$+1$	0.3
	1.0

We can think of each U_i as the outcome of a random draw with replacement from a population of U values, 30% of which have the value -1, 40% have the value 0, and 30% the value $+1$. The mean U value is 0, and the variance of U values is $\sigma^2 = 0.6$. In short, the assumptions of Model A apply.

The possible values of Y_1 (the number of employees of firm A) are $(0.5)(13) - 1 = 5.5$, $(0.5)(13) + 0 = 6.5$, and $(0.5)(13) + 1 = 7.5$; the probabilities of these values are 0.3, 0.4, and 0.3, respectively. Likewise, the possible values of Y_2 are $(0.5)(12) + U_2$, that is, 5, 6, or 7, with the same probabilities; and the possible values of Y_3 are 3.5, 4.5, and 5.5, again with probabilities 0.3, 0.4, and 0.3, respectively. Y_1, Y_2, and Y_3 are independent of one another and the X values by virtue of the assumed generating process.

There are thus three possible values for each of the three Y's, therefore $3 \times 3 \times 3 = 27$ possible combinations of values. Figure 9.2 shows the top of the probability tree describing these process outcomes and their probabilities.

If the assumed model is correct, the number of employees of the three firms in the town could be *any one* of these 27 combinations. Which combination happens to materialize is a matter of chance. The total number of employees in the town is the sum of these Y values, and is also random. No particular significance need be attached to any one sum value; by no means, according to the prediction approach, is the population total "given," as assumed in earlier chapters.

Now, suppose it has been decided to select a sample of two firms and that these two firms are to be A and C. These two firms are not selected randomly but arbitrarily. A and C will be interviewed, the number of their employees determined, and the total number of employees of the three firms in Half-of-Ackroyd estimated.

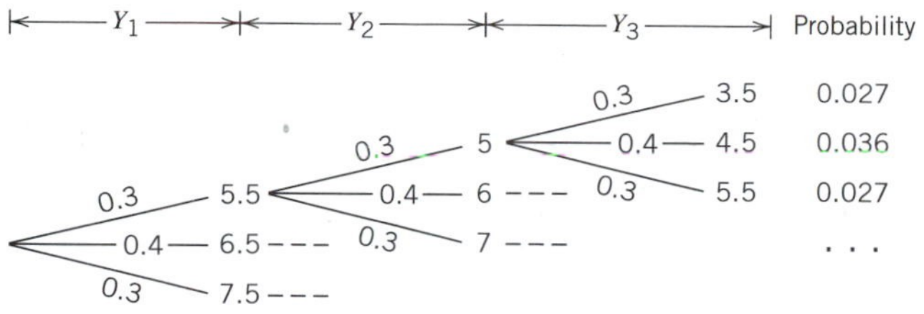

Figure 9.2 Partial probability tree, Half-of-Ackroyd case

In making these estimates, we pretend not to know β or the common distribution of U.

The estimate of the total number of employees will be

$$T = \sum_{i=1}^{n} Y_i + b \sum_{j \in R} X_j = (Y_1 + Y_3) + bX_2 = (Y_1 + Y_3) + 12b,$$

where b is the estimate of β. The prediction error will be

$$\tilde{E} = T - \tau = [(Y_1 + Y_3) + 12b] - [Y_1 + Y_2 + Y_3] = 12b - Y_2.$$

If we use the least squares estimator of β, then

$$b = b_1 = \frac{\sum_{i=1}^{n} X_i Y_i}{\sum_{i=1}^{n} X_i^2} = \frac{X_1 Y_1 + X_3 Y_3}{X_1^2 + X_3^2} = \frac{13Y_1 + 9Y_3}{13^2 + 9^2} = \frac{13Y_1 + 9Y_3}{250}.$$

If we use the average of ratios, then

$$b = b_2 = \frac{1}{n} \sum_{i=1}^{n} \frac{Y_i}{X_i} = \frac{1}{2}(\frac{Y_1}{X_1} + \frac{Y_3}{X_3}) = \frac{1}{2}(\frac{Y_1}{13} + \frac{Y_3}{9}).$$

Lastly, if we use the ratio of averages (or sums),

$$b = b_3 = \frac{\sum_{i=1}^{n} Y_i}{\sum_{i=1}^{n} X_i} = \frac{Y_1 + Y_3}{X_1 + X_3} = \frac{Y_1 + Y_3}{13 + 9} = \frac{Y_1 + Y_3}{22}.$$

We would like to confirm that T using b_1 is the best linear unbiased estimator of the total number of employees.

Consider the first of the 27 possible combinations of Y values shown in Figure 9.2: $Y_1 = 5.5$, $Y_2 = 5.0$, and $Y_3 = 3.5$. This combination will occur with probability 0.027. Given this set of Y values, the three estimates of β are $b_1 = [(13)(5.5) + (9)(3.5)]/250 = 0.4120$, $b_2 = (0.5)[(5.5/13) + (3.5/9)] = 0.4060$, and $b_3 = (5.5 + 3.5)/22 = 0.4091$.

The estimate of the total number of employees is $T = (Y_1 + Y_3) + 12b = 9 + 12b$, and the prediction error $\tilde{E} = 12b - Y_2 = 12b - 5$. If we use b_1, then $T = T_1 = 9 + (12)(0.4120) = 13.9440$ and $\tilde{E} = \tilde{E}_1 = (12)(0.4120) - 5 = -0.0560$.

If we use b_2, $T_2 = 13.8718$ and $\tilde{E}_2 = -0.1282$. Finally, if we use b_3, $T_3 = 13.9091$ and $\tilde{E}_3 = -0.0909$.

Table 9.3 shows the estimates and prediction errors for this and the other 26 possible process outcomes.

It is straightforward (and, with the help of a computer program, easy) to show that

$$E(\tilde{E}_1) = E(\tilde{E}_2) = E(\tilde{E}_3) = 0.0; \text{ and}$$
$$Var(\tilde{E}_1) = 0.9456, \quad Var(\tilde{E}_2) = 0.9945, \quad Var(\tilde{E}_3) = 0.9570.$$

For example,

$$E(\tilde{E}_1) = (-0.0560)(0.027) + (0.3760)(0.036) + \cdots + (0.0560)(0.027) = 0,$$
$$Var(\tilde{E}_1) = (-0.0560)^2(0.027) + (0.3760)^2(0.036) + \cdots$$
$$+ (0.0560)^2(0.027) - (0)^2 = 0.9456.$$

We see that all three estimators of the population total (T_1, T_2 and T_3) are unbiased, in the sense that their expected error—the expected difference from the population total, τ—is zero. In the long run and on average, therefore, none of these estimators tends to overestimate or underestimate the population total. Further, and as claimed in the previous section, of the three estimators the best is T_1, that is, the one using the least squares estimator b_1 of β; this is so because T_1 has the smallest error variance, 0.9456, among the estimators examined. Finally, we see that the error variance of T_1, as calculated directly, agrees with Eq. 9.9:

$$Var(\tilde{E}_1) = \sigma^2 \Big[\frac{(\sum_{j \in R} X_j)^2}{(\sum_{i=1}^{n} X_i^2)} + N - n \Big] = (0.6) \Big[\frac{12^2}{13^2 + 9^2} + 3 - 2 \Big] = 0.9456.$$

The decision to select A and C for the sample was arbitrary. Similar results will be obtained with samples consisting of A and B, or B and C. In fact, as the reader is asked to show in Problems 9.12 and 9.13, the best sample should consist of firms A and B, as these two firms have the two largest annual sales.

The illustration of this section should have made clear the main difference between the prediction approach and the conventional or *randomization approach* described in earlier chapters.

Under the randomization approach, the Y and X values of the population elements are assumed to be unknown but given numbers. So are the total and mean of these numbers. The variability of the estimators is due to the random selection of the sample elements: different samples select different elements, and the estimators based on the selected elements will vary from sample to sample.

Under the prediction approach, the Y values of the population elements as well as the population total or mean of the variable of interest are random. So are the estimators of these population characteristics, which will take different values in different realizations of the generating process even when the same elements are selected for the sample. The source of randomness is the process generating the deviations.

TABLE 9.3   Process Outcomes, Estimates, and Prediction Errors

Y_1	Y_2	Y_3	Probability	b_1	T_1	$\tilde{E}_1$	b_2	T_2	$\tilde{E}_2$	b_3	T_3	$\tilde{E}_3$
5.5	5.0	3.5	0.027	0.4120	13.9440	−0.0560	0.4060	13.8718	−0.1282	0.4091	13.9091	−0.0909
5.5	5.0	4.5	0.036	0.4480	15.3760	0.3760	0.4615	15.5385	0.5385	0.4545	15.4545	0.4545
5.5	5.0	5.5	0.027	0.4840	16.8080	0.8080	0.5171	17.2051	1.2051	0.5000	17.0000	1.0000
5.5	6.0	3.5	0.036	0.4120	13.9440	−1.0560	0.4060	13.8718	−1.1282	0.4091	13.9091	−1.0909
5.5	6.0	4.5	0.048	0.4480	15.3760	−0.6240	0.4615	15.5385	−0.4615	0.4545	15.4545	−0.5455
5.5	6.0	5.5	0.036	0.4840	16.8080	−0.1920	0.5171	17.2051	0.2051	0.5000	17.0000	0.0000
5.5	7.0	3.5	0.027	0.4120	13.9440	−2.0560	0.4060	13.8718	−2.1282	0.4091	13.9091	−2.0909
5.5	7.0	4.5	0.036	0.4480	15.3760	−1.6240	0.4615	15.5385	−1.4615	0.4545	15.4545	−1.5455
5.5	7.0	5.5	0.027	0.4840	16.8080	−1.1920	0.5171	17.2051	−0.7949	0.5000	17.0000	−1.0000
6.5	5.0	3.5	0.036	0.4640	15.5680	0.5680	0.4444	15.3333	0.3333	0.4545	15.4545	0.4545
6.5	5.0	4.5	0.048	0.5000	17.0000	1.0000	0.5000	17.0000	1.0000	0.5000	17.0000	1.0000
6.5	5.0	5.5	0.036	0.5360	18.4320	1.4320	0.5556	18.6667	1.6667	0.5455	18.5455	1.5455
6.5	6.0	3.5	0.048	0.4640	15.5680	−0.4320	0.4444	15.3333	−0.6667	0.4545	15.4545	−0.5455
6.5	6.0	4.5	0.064	0.5000	17.0000	0.0000	0.5000	17.0000	0.0000	0.5000	17.0000	0.0000
6.5	6.0	5.5	0.048	0.5360	18.4320	0.4320	0.5556	18.6667	0.6667	0.5455	18.5455	0.5455
6.5	7.0	3.5	0.036	0.4640	15.5680	−1.4320	0.4444	15.3333	−1.6667	0.4545	15.4545	−1.5455
6.5	7.0	4.5	0.048	0.5000	17.0000	−1.0000	0.5000	17.0000	−1.0000	0.5000	17.0000	−1.0000
6.5	7.0	5.5	0.036	0.5360	18.4320	−0.5680	0.5556	18.6667	−0.3333	0.5455	18.5455	−0.4545
7.5	5.0	3.5	0.027	0.5160	17.1920	1.1920	0.4829	16.7949	0.7949	0.5000	17.0000	1.0000
7.5	5.0	4.5	0.036	0.5520	18.6240	1.6240	0.5385	18.4615	1.4615	0.5455	18.5455	1.5455
7.5	5.0	5.5	0.027	0.5880	20.0560	2.0560	0.5940	20.1282	2.1282	0.5909	20.0909	2.0909
7.5	6.0	3.5	0.036	0.5160	17.1920	0.1920	0.4829	16.7949	−0.2051	0.5000	17.0000	0.0000
7.5	6.0	4.5	0.048	0.5520	18.6240	0.6240	0.5385	18.4615	0.4615	0.5455	18.5455	0.5455
7.5	6.0	5.5	0.036	0.5880	20.0560	1.0560	0.5940	20.1282	1.1282	0.5909	20.0909	1.0909
7.5	7.0	3.5	0.027	0.5160	17.1920	−0.8080	0.4829	16.7949	−1.2051	0.5000	17.0000	−1.0000
7.5	7.0	4.5	0.036	0.5520	18.6240	−0.3760	0.5385	18.4615	−0.5385	0.5455	18.5455	−0.4545
7.5	7.0	5.5	0.027	0.5880	20.0560	0.0560	0.5940	20.1282	0.1282	0.5909	20.0909	0.0909

9.5 SOME VARIANTS OF THE SIMPLE LINEAR MODEL WITH NO INTERCEPT

An important feature of the prediction approach is the nature of the "mechanism" that is assumed to generate the deviations U_i, in this case, of the model

$$Y_i = \beta X_i + U_i.$$

In the previous sections, it was assumed that the U_i behave *like* outcomes of random draws with replacement from a common population of U values having mean zero and a certain variance.

A sample of n elements will reveal the Y values of the selected elements, and the scatter diagram of the n pairs (Y_i, X_i) will provide a rough check of these

assumptions. A scatter diagram that has the pattern shown in panel (A) of Figure 9.3 is consistent with these assumptions; scatter diagrams with the patterns shown in panels (B) and (C) are not.

In all three panels, there appears to be a tendency for the Y values to lie along a straight line through the origin. It appears that the deviations about the line tend to cancel each other out, which is consistent with the assumption that $E(U_i) = 0$.

In panels (B) and (C), however, the scatter of the deviations from that line is not uniform as in panel (A), but appears to depend on X. In panel (B) the scatter appears to increase with X, and in (C) it appears to increase with X^2.

In cases (B) and (C), it may be reasonable to assume that the deviations U_i are generated *as if* by random draws with replacement from *separate* populations of U values, one for each possible X value. The mean U value in all these populations is 0, but the variances increase proportionately with X in the case of panel (B), or proportionately with X^2 in the case of panel (C).

Model B, therefore, can be defined as follows:

$$\text{Model B:} \quad \begin{cases} Y_i = \beta X_i + U_i, \\ U_i \text{ independent, with } E(U_i) = 0, \text{ and } Var(U_i) = \sigma^2 X_i, \end{cases}$$

and *Model C* as

$$\text{Model C:} \quad \begin{cases} Y_i = \beta X_i + U_i, \\ U_i \text{ independent, with } E(U_i) = 0, \text{ and } Var(U_i) = \sigma^2 X_i^2. \end{cases}$$

Model A of Section 9.3 and Models B and C of this section are variants of the same simple model, differing only with respect to the assumed variance of the deviations. The goal is the same as in the last section, namely, to find the best linear unbiased estimator of τ. The best estimator b, the BLUE T^* of τ, and the error variance of this BLU estimator, $Var(\tilde{E}^*) = Var(T^* - \tau)$, are shown in Summary 9.2.

Several conclusions may be drawn from an inspection of Summary 9.2.

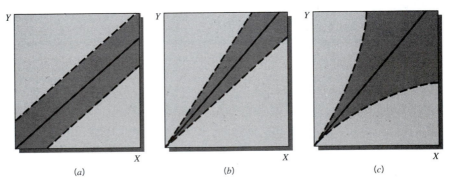

Figure 9.3 Scatter diagrams for variants of simple model

SUMMARY 9.2   Simple linear models with no intercept

	Model A	Model B	Model C
Sample selection method:	Any	Any	Any
Relationship:	$Y_i = \beta X_i + U_i$	$Y_i = \beta X_i + U_i$	$Y_i = \beta X_i + U_i$
U_i are:	independent	independent	independent
$E(U_i) =$	0	0	0
$Var(U_i) =$	σ^2	$\sigma^2 X_i$	$\sigma^2 X_i^2$
Best b, $b^* =$	$\dfrac{\sum_{i=1}^{n} X_i Y_i}{\sum_{i=1}^{n} X_i^2}$	$\dfrac{\sum_{i=1}^{n} Y_i}{\sum_{i=1}^{n} X_i} = \dfrac{\bar{Y}}{\bar{X}}$	$\dfrac{1}{n}\sum_{i=1}^{n} \dfrac{Y_i}{X_i}$
BLUE, $T^* =$	$\sum_{i=1}^{n} Y_i + b^*(\sum_{j\in R} X_j)$	$\sum_{i=1}^{n} Y_i + b^*(\sum_{j\in R} X_j)$	$\sum_{i=1}^{n} Y_i + b^*(\sum_{j\in R} X_j)$
$Var(\tilde{E}^*) =$	$\sigma^2\left[\dfrac{(\sum_{j\in R} X_j)^2}{(\sum_{i=1}^{n} X_i^2)} + N - n\right]$	$\sigma^2\left[\dfrac{N^2}{n}(1 - \dfrac{n}{N})\dfrac{\bar{X}_R}{\bar{X}}\mu_x\right]$	$\sigma^2\left[\dfrac{1}{n}(\sum_{j\in R} X_j)^2 + (\sum_{j\in R} X_j^2)\right]$

Note: $\bar{X} = \frac{1}{n}\sum_{i=1}^{n} X_i$; $\bar{X}_R = \frac{1}{N-n}\sum_{j\in R} X_j$; $\mu_x = \frac{1}{N}\sum_{i=1}^{N} X_i = \bar{X} + \bar{X}_R$.

- The best estimator of β under Model B is the *ratio estimator* examined at length in Chapter 6, while the best estimator under Model C is *the average of ratios* Y_i / X_i. The *least squares estimator* is best for Model A. We see that each of the three reasonable estimators of the parameter β of the simple model is best in turn under different assumptions concerning the variance of the deviations U_i.
- For all three variants of the simple model, *the best sample is not random, but one consisting of the n elements having the largest X values.* This conclusion follows the observation that $Var(\tilde{E}^*)$ in all three variants is minimized if the sample consists of the *n* elements having the largest *X* values (and the remainder of the $N - n$ elements having the smallest *X* values).

9.6 INTERVAL ESTIMATES, SIMPLE LINEAR MODELS WITH NO INTERCEPT

When the sample size, n, and the population remainder, $N - n$, are large, an approximate $100(1 - \alpha)\%$ confidence interval for the population total of a variable Y is

$$T^* \pm Z_{\alpha/2}\sqrt{\widehat{Var(\tilde{E}^*)}}, \qquad (9.10)$$

where $\widehat{Var(\tilde{E}^*)}$ is an estimator of $Var(\tilde{E}^*)$, obtained by replacing σ^2 in the expression for $Var(\tilde{E}^*)$ with

$$S^2 = \frac{1}{n-1} \sum_{i=1}^{n} \frac{(Y_i - bX_i)^2}{v_i},$$

where $v_i = 1$ for Model A, $v_i = X_i$ for Model B, and $v_i = X_i^2$ for Model C. The expressions of $Var(\tilde{E}^*)$ for models A, B, and C are listed in Summary 9.2. Note that S^2 has a different meaning in this chapter; it does not represent the sample variance of Y as was the case in all preceding chapters.

A confidence interval for the population mean of the variable is calculated by dividing the limits of (9.10) by N.

■ EXAMPLE 9.1
(Continued)

We shall illustrate the calculations using the data of Table 9.2, and assuming that Model A applies. These calculations are strictly for the purpose of illustration, because the sample size is not large enough to justify the application of the formulas.

Earlier in Section 9.3, we had determined that $b^* = 0.7451$, $T^* = 138,600$, $\sum_{i=1}^{n} X_i^2 = 10,032,500$, and $\sum_{j \in R} X_j = 180,580$. The estimated values of Y are given by

$$\hat{Y}_i = 0.7451 X_i.$$

Table 9.4 shows the steps leading to the calculation of S^2 as

$$S^2 = \frac{1}{3-1}(3,038.22) = 1,519.11.$$

Next, we find

$$\widehat{Var(\tilde{E}^*)} = S^2 \Big[\frac{(\sum_{j \in R} X_j)^2}{(\sum_{i=1}^{n} X_i^2)} + N - n\Big]$$

$$= (1,519.11)\Big[\frac{(180,580)^2}{10,032,500} + 1,158 - 3\Big]$$

$$= 6,692,211.$$

If this were a large sample, a, say, 95% confidence interval for the total hospital purchases of Product Y would be

$$138,600 \pm (1.96)\sqrt{6,692,211},$$

or from 133,530 to 143,670 ($000). A 95% confidence interval for the average purchases of Product Y per hospital would be from 133,530/1,158 to 143,670/1,158, or from 115.31 to 124.07 ($000).

These results could have been obtained with the help of a computer program for regression having a no-intercept option (that is, having the ability to force the intercept to equal zero). Figure 9.4 shows the output of one such program. The square root of S^2 appears under the label S=. ◾

TABLE 9.4   Estimated Values and Residuals, Example 9.1

Sampled Hospital	Number of Beds, X_i	Purchases of Product Y, Y_i ($000)	$\hat{Y}_i$	$Y_i - \hat{Y}_i$	$(Y_i - \hat{Y}_i)^2$
234	2,100	1,600	1,564.71	35.29	1,245.38
567	1,600	1,150	1,192.16	−42.16	1,777.46
801	1,750	1,300	1,303.92	−3.92	15.37
	5,450	4,050			3,038.22

```
Dependent Variable: Y


                          STANDARD
VARIABLE    ESTIMATE        ERROR     T-RATIO    P-VALUE
      X      0.74508      0.01231      60.55      0.000

S = 38.98
```

Figure 9.4   Computer output, Example 9.1

9.7 A SIMPLE LINEAR MODEL WITH INTERCEPT

Next to Model A in complexity is *Model D*:

$$\text{Model D:} \quad \begin{cases} Y_i = \alpha + \beta X_i + U_i, \\ U_i \text{ independent, with } E(U_i) = 0, \text{ and } Var(U_i) = \sigma^2. \end{cases}$$

Model D differs from A only with respect to the constant term, α; Model A is a special case of D with $\alpha = 0$. If Model D applies, a scatter diagram of the sample (Y_i, X_i) values should still show a tendency for the points to lie within a band of constant width along a straight line, but one now not necessarily through the origin. A scatter diagram consistent with Model D is shown in Figure 9.5.

As in Model A, the U_i may be thought of as outcomes of random draws with replacement from a common population of U values having zero mean, and some variance σ^2.

The BLUE of τ and its error variance are shown in Summary 9.3.

The results in Summary 9.3 apply to any sample, whether randomly or otherwise selected. The second expression for T^* in Eq. 9.11 is obtained from the first after substituting $a = \bar{Y} - b\bar{X}$ in the manner of Section 6.5.

Before we illustrate the calculations, let us examine carefully Eq. 9.13. The first term within the square brackets depends on the sample and population sizes, but not on which elements are selected in the sample. The second term is always nonnegative, and zero if $\bar{X} = \mu_x$, in which case $Var(\tilde{E}^*)$ attains its minimum value. In other words, *the best sample for Model D is one in which the n elements*

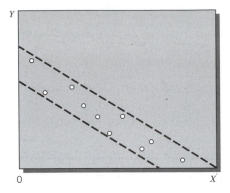

Figure 9.5 Scatter diagram consistent with Model D

SUMMARY 9.3 BLUE of population total and error variance, Model D

- The best linear unbiased estimator of the population total of a variable Y under Model D is

$$T^* = \sum_{i=1}^{n} Y_i + \sum_{j \in R}(a + bX_j) = N[\bar{Y} + b(\mu_x - \bar{X})], \qquad (9.11)$$

where a and b are the least squares estimators:

$$b = \frac{\sum_{i=1}^{n} X_i Y_i - n\bar{X}\bar{Y}}{\sum_{i=1}^{n} X_i^2 - n\bar{X}^2}, \qquad (9.12)$$

$$a = \bar{Y} - b\bar{X}.$$

The error variance associated with this estimator is

$$Var(\tilde{E}^*) = \sigma^2 N^2\Big[\big(\frac{1}{n} - \frac{1}{N}\big) + \frac{(\bar{X} - \mu_x)^2}{\sum_{i=1}^{n}(X_i - \bar{X})^2}\Big]. \qquad (9.13)$$

In the above expressions, $\bar{X}$ and $\bar{Y}$ are the averages of the sample X and Y values, respectively, and μ_x is the population average of the known X values.

- The best estimator of the population mean of the variable is T^*/N.

---

are selected so as to make the sample average equal to the population average of the explanatory variable X. Such a sample is said to be *balanced* with respect to X. The best sample for Model D, therefore, does *not* consist of the elements with the largest X values, as was the case for Models A, B, and C.

If the sample is balanced, then, from Eq. 9.11, $T^* = N\bar{Y}$, that is, the best estimator of the population total is the simple estimator.

It may not be possible, of course, to achieve exact balance, but it is always possible to select a sample so that $\bar{X}$ is approximately equal to μ_x. In this case, it would be advisable to spread out the X values as much as possible, as this would tend to further reduce the last term of Eq. 9.13 and $Var(\tilde{E}^*)$.

■ EXAMPLE 9.2 Suppose it is believed that the hospital purchases of Product Z follow more closely the assumptions of Model D rather than those of Model A. Product Z, for example, could be used for sterilizing hospital equipment and its usage could be expected to have a substantial fixed component. The number of beds is the explanatory variable X. Recall that the average number of beds of all 1,158 hospitals is 186,030/1,158, or 160.65.

Suppose that a sample of three hospitals is desired. The hospitals with identification numbers 155, 321, and 967 have 100, 150, and 230 beds, respectively. If the sample were to consist of these hospitals, therefore, it would be approximately balanced with respect to the number of beds, since $\bar{X} = 160$ is quite close to the population average 160.65. The purchases of Product Z of the three selected hospitals shown in col. (3) of Table 9.5 will be known after the hospitals are questioned. We calculate $\bar{X} = 480/3 = 160$, $\bar{Y} = 505/3 = 168.33$,

$$b = \frac{\sum_{i=1}^{n} X_i Y_i - n\bar{X}\bar{Y}}{\sum_{i=1}^{n} X_i^2 - n\bar{X}^2} = \frac{85,000 - 3(160)(168.33)}{85,400 - 3(160)^2} = 0.488,$$

and

$$a = \bar{Y} - b\bar{X} = 168.33 - (0.488)(160) = 90.2.$$

The estimate of the purchases of Product Z by a hospital not in the sample is given by

$$\hat{Y}_i = 90.2 + 0.488 X_i.$$

TABLE 9.5 Sample of Hospitals and Calculations

Sampled Hospital (1)	Number of Beds, X_i (2)	Purchases of Product Z, Y_i ($000) (3)	$X_i Y_i$ (4)	X_i^2 (5)
155	100	135	13,500	10,000
321	150	170	25,500	22,500
967	230	200	46,000	52,900
	480	505	85,000	85,400

The estimate of the total hospital purchases of Product Z is

$$
\begin{aligned}
T^* &= N[\bar{Y} + b(\mu_x - \bar{X})] \\
&= (1{,}158)[168.33 + (0.488)(160.65 - 160)] \\
&= (1{,}158)(168.647) \\
&= 195{,}293 \ (\$000).
\end{aligned}
$$

The estimate of the average purchases of Product Z per hospital is $T^*/N = 195{,}293/1{,}158 = 168.647$ ($000).

The least squares estimates can also be calculated with the help of a computer program for regression. The results of regressing Y against X using the data of Table 9.5 are shown in Figure 9.6. ◪

For large n and $N - n$, an approximate $100(1 - \alpha)\%$ confidence interval for the population total of a variable Y is given by

$$
T^* \pm Z_{\alpha/2}\sqrt{\widehat{Var(\tilde{E}^*)}}, \tag{9.14}
$$

where $\widehat{Var(\tilde{E}^*)}$ is an estimate of $Var(\tilde{E}^*)$, obtained by replacing in Eq. 9.13 σ^2 with

$$
S^2 = \frac{1}{n - 2}\sum_{i=1}^{n}(Y_i - \hat{Y}_i)^2.
$$

The square root of S^2 is usually printed by computer programs under the label S=, as shown in Figure 9.6.

The corresponding confidence interval for the population mean of the variable is obtained by dividing the limits of Eq. 9.14 by N.

```
Dependent Variable: Y

                        STANDARD
VARIABLE    ESTIMATE      ERROR     T-RATIO    P-VALUE
INTERCEP       90.19      14.72        6.13      0.103
       X     0.48837     0.08727        5.60      0.113

S = 8.093    R-SQUARED = 0.969
```

Figure 9.6 Computer output, Example 9.2

■ EXAMPLE 9.2
(Continued)

To calculate a confidence interval for the total hospital purchases of Product Z, we observe first in the computer output that $S = 8.093$. We then calculate

$$\sum_{i=1}^{n}(X_i - \bar{X})^2 = (100 - 160)^2 + \cdots + (230 - 160)^2 = 8,600,$$

and

$$\widehat{Var(\tilde{E}^*)} = S^2 N^2\Big[\Big(\frac{1}{n} - \frac{1}{N}\Big) + \frac{(\bar{X} - \mu_x)^2}{\sum_{i=1}^{n}(X_i - \bar{X})^2}\Big]$$

$$= (8.093)^2(1,158)^2\Big[\Big(\frac{1}{3} - \frac{1}{1158}\Big) + \frac{(160 - 160.65)^2}{8600}\Big]$$

$$= 29,204,686.$$

If this were a large sample, an approximate 95% confidence interval for the total hospital purchases of Product Z would be

$$195,293 \pm (1.96)\sqrt{29,204,686},$$

which is the interval from 184,701 to 205,885 ($000). An approximate 95% confidence interval for the average purchases of Product Z per hospital is from 184,701/1,158 to 205,885/1,158, or from 159.500 to 177.794 ($000). ■

9.8 A LINEAR MODEL WITH MORE THAN ONE EXPLANATORY VARIABLE

The model of the last section can be extended to accommodate any number k of explanatory variables. We shall call this *Model E:*

Model E: $\begin{cases} Y_i = \alpha + \beta_1 X_{1i} + \beta_2 X_{2i} + \cdots + \beta_k X_{ki} + U_i, \\ U_i \text{ independent, with } E(U_i) = 0, \text{ and } Var(U_i) = \sigma^2. \end{cases}$

For example, the purchases of a pharmaceutical product by a hospital can be assumed to be influenced, in addition to the number of beds, by the number of its patients, the type of the hospital, its location, and other factors.

The BLUE of the population total under Model E is given in Summary 9.4.

We do not write here the formulas for the least squares estimators and the error variance because these cannot be expressed simply without vector and matrix notation. In any event, the least squares estimates are almost always calculated by computer programs for regression, as illustrated in the following example.

■ EXAMPLE 9.3

Suppose that the purpose of the hospital survey is to estimate the purchases of pharmaceutical products in a given month. Two explanatory variables will be used:

SUMMARY 9.4 BLUE and error variance, Model E

- The best linear unbiased estimator of the population total of a variable Y under Model E is

$$T^* = N[\bar{Y} + \sum_{j=1}^{k} b_j(\mu_j - \bar{X}_j)], \tag{9.15}$$

where b_j is the least squares estimator of β_j, μ_j is the population mean, and $\bar{X}_j$ is the sample mean of explanatory variable X_j.
- The best estimator of the population mean of variable Y is T^*/N.

---

the number of beds at the beginning of the month and the number of patients who were hospitalized during the month. It is assumed that the values of both explanatory variables will be known for all 1,158 hospitals when the sample is taken at the end of the month. The population is outlined in Table 9.6.

In the notation of this section, we have $\mu_1 = 186{,}030/1{,}158 = 160.65$, and $\mu_2 = 205{,}915/1{,}158 = 177.82$.

Suppose that nine hospitals are arbitrarily selected and questioned concerning their purchases of Product Y with the results shown in Table 9.7.

Assuming that Model E applies, a computer program is used to regress Y against X_1 and X_2. The results are shown in Figure 9.7.

We read $a = 27.09$, $b_1 = 0.4012$, and $b_2 = 0.3945$. The estimated purchases of Product Y by a hospital i which is not included in the sample are given by

$$\hat{Y}_i = 27.09 + 0.4012X_{1i} + 0.3945X_{2i},$$

where X_{1i} and X_{2i} are the number of beds and the number of patients of this hospital.

TABLE 9.6 Population of Hospitals, Example 9.3

Hospital No., i	Number of Beds, X_{1i}	Number of Patients, X_{2i}	Purchases of Product Y, Y_i
1	675	800	?
2	450	375	?
⋮	⋮	⋮	⋮
1,158	1,500	1,650	?
Total	186,030	205,915	?

TABLE 9.7	Sample of Hospitals, Example 9.3		
Selected Hospital	Number of Beds, X_{1i}	Number of Patients, X_{2i}	Purchases of Product Y, Y_i ($000)
1	675	600	500
2	450	400	350
155	100	120	135
234	2,100	1,750	1,600
321	150	160	170
567	1,600	1,250	1,150
801	1,750	1,500	1,300
967	230	190	200
1,158	1,500	1,200	1,100
Average	950.56	796.67	722.78

```
Dependent Variable: Y


                                 STANDARD
     VARIABLE      ESTIMATE         ERROR     T-RATIO     P-VALUE
     INTERCEPT        27.09         16.54        1.64       0.152
            X1       0.4012        0.1997        2.01       0.091
            X2       0.3945        0.2470        1.60       0.161


     S = 27.27      R-SQUARED = 0.998
```

Figure 9.7 Computer output, Example 9.3

The BLU estimate of the total hospital purchases of Product Y is

$$T^* = N\left[\bar{Y} + \sum_{j=1}^{2} b_j(\mu_j - \bar{X}_j)\right]$$
$$= (1,158)\left[722.78 + (0.4012)(160.65 - 950.56)\right.$$
$$\left. + (0.3945)(177.82 - 796.67)\right]$$
$$= (1,158)(161.732)$$
$$= 187,286\ (\$000).$$

The best estimate of the average purchases of Product Y per hospital is T^*/N or 161.732 ($000). ◪

A sample is said to be *balanced* if the sample and population means of all explanatory variables are equal (that is, all $\bar{X}_j = \mu_j$). If this is the case, it can be

seen from Eq. 9.15 that the best estimator of the population total is the simple estimator $N\bar{Y}$.

Among the factors that influence a variable of interest are often attributes. As is well-known from applied regression, an attribute can be handled by means of "dummy" or "indicator" variables. For each attribute, a set of dummy variables is introduced, one dummy variable for each category of the attribute. Each dummy variable takes the value 1 if the observation belongs to the category that the dummy variable represents or the value 0 if it does not. From each such set of dummy variables, one is normally left out of the model to ensure that the least squares estimates are unique. The regression coefficients of the remaining dummy variables measure the effect of the corresponding category on the dependent variable over that of the base category—the category corresponding to the dummy variable that was left out.

◢ EXAMPLE 9.4 Suppose that in addition to the number of beds and the number of patients, a hospital's purchases of a pharmaceutical product depend also on the type of the hospital. Two hospital types are distinguished: general and specialized. The types of the nine selected hospitals are listed together with other sample information in Table 9.8.

With respect to type, a hospital belongs to one of two categories: general (G) or specialized (S), as shown in col. (4). Two dummy variables are introduced in cols. (5) and (6): X_3 represents general hospitals and takes the value 1 if the hospital is general or the value 0 if it is not; X_4 represents specialized hospitals and takes

TABLE 9.8   Sample of Hospitals, Example 9.4

Selected Hospital (1)	Number of Beds, X_{1i} (2)	Number of Patients, X_{2i} (3)	Type (4)	Dummy Variables X_{3i} (5)	Dummy Variables X_{4i} (6)	Purchases of Product Y, Y_i (7)
1	675	600	S	0	1	500
2	450	400	S	0	1	350
155	100	120	S	0	1	135
234	2,100	1,750	G	1	0	1,600
321	150	160	S	0	1	170
567	1,600	1,250	G	1	0	1,150
801	1,750	1,500	G	1	0	1,300
967	230	190	S	0	1	200
1,158	1,500	1,200	G	1	0	1,100
Average	950.56	796.67		0.4444	0.5556	722.78

the value 1 if the hospital is specialized or the value 0 if it is not. Assume that Model E applies with $k = 4$.

One of the two dummy variables should be left out of the model to ensure unique least squares estimates. Either one will do—the choice is arbitrary. Let us suppose X_4 is left out, and that Y is regressed against X_1, X_2, and X_3 with the results shown in Figure 9.8.

The estimate of the purchases of Product Y by a hospital i not included in the sample is

$$\hat{Y}_i = 30.30 + 0.3253X_{1i} + 0.4636X_{2i} + 31.46X_{3i}.$$

The number of beds and the number of patients being equal, it is estimated that a general hospital ($X_3 = 1$) purchases 31.46 ($000) more of Product Y than does a specialized hospital ($X_3 = 0$).

The average of the sample X_3 values, $\bar{X}_3 = 0.4444$, is the proportion of general hospitals in the sample. The population mean of X_3 is the proportion of general hospitals in the population. Assume this is $\mu_3 = 0.55$.

In view of the sample results, therefore, the best estimate of the total hospital purchases of Product Y is

$$
\begin{aligned}
T^* &= N\Big[\bar{Y} + \sum_{j=1}^{3} b_j(\mu_j - \bar{X}_j)\Big] \\
&= (1{,}158)\big[722.78 + (0.3253)(160.65 - 950.56) \\
&\quad + (0.4636)(177.82 - 796.67) + (31.46)(0.55 - 0.4444)\big] \\
&= (1{,}158)(182.246) \\
&= 211{,}041 \ (\$000).
\end{aligned}
$$

The best estimate of the average purchases of Product Y per hospital is T^*/N or 182.246 ($000). ◾

```
Dependent Variable: Y

                          STANDARD
VARIABLE      ESTIMATE      ERROR      T-RATIO     P-VALUE
INTERCEPT       30.30       19.56        1.55       0.182
       X1      0.3253      0.2875        1.13       0.309
       X2      0.4636      0.3177        1.46       0.204
       X3       31.46       78.92        0.40       0.707

S = 29.41          R-SQUARED = 0.998
```

Figure 9.8 Computer output, Example 9.4

9.9 THE PREDICTION APPROACH IN GENERAL

Model E can be further extended. The best linear unbiased estimator, T^*, its error variance, $Var(\tilde{E}^*)$, and approximate confidence intervals for the population total or mean of a variable can also be determined when:

(a) the U_i are independent with $E(U_i) = 0$ but arbitrary variances (for example, $Var(U_i) = \sigma^2 X_{1i}^2$, $Var(U_i) = \sigma^2 e_i$ where e_i are given numbers, etc.); and/or

(b) the U_i are not independent but correlated with given arbitrary correlation coefficients.[2]

9.10 SIMPLE DIAGNOSTIC TOOLS

The best methods for selecting the sample and for estimating the population total or mean of a variable depend on the underlying model. For example, if Model A holds, it is best to select the elements with the largest X values and to use Eqs. 9.8 and 9.7 to estimate the population total. In contrast, if the slightly more general Model D applies, the best methods are quite different: it is best to select a sample balanced in X and to estimate the population total using Eqs. 9.12 and 9.11.

Two questions may therefore be raised at this point. First, how does one determine which is the underlying model, and, second, what are the consequences of using methods that are best under a given model when that is not the model?

The scatter diagram is a simple, although by no means infallible, diagnostic tool. In the case where, for example, Model E is considered, k scatter diagrams may be inspected, each plotting Y against one explanatory variable X_j. If these diagrams indicate that the relationship between the dependent and each explanatory variable is approximately linear, then one of the assumptions of Model E—linearity—would appear to be satisfied.

If the second assumption of Model E—constant variance—is satisfied, the same scatter diagrams should indicate no tendency for the scatter to vary with the explanatory variable.

The third assumption of Model E—independence—is more difficult to check as the deviations U_i are not observable and independence may be violated in many different ways. If the deviations are indeed like random selections with replacement from a population of U values, then there should be no noticeable pattern in the residuals:

$$\hat{U}_i = Y_i - \hat{Y}_i = Y_i - (a + b_1 X_{1i} + \cdots + b_k X_{ki}).$$

---

[2]For these extensions, the reader familiar with vector and matrix algebra may consult, for example, Richard M. Royall, "The Prediction Approach to Sampling Theory," in P. R. Krishnaiah and C. R. Rao (eds.), *Handbook of Statistics, Vol. 6* (Amsterdam: Elsevier Science Publishers B.V. 1988), pp. 399–413.

A plot of the residuals in the order of the observations i may reveal one form of violation of independence. Independence is consistent with Figure 9.9a, but not with the exaggerated patterns of Figures 9.9b, 9.9c, or 9.9d.

It is clear that the preceding are rough checks of a model's assumptions, and it is possible that the model will not be identified correctly. The second question, therefore, concerning the consequences of incorrect identification acquires importance.

Obviously, this is a large question for there are many possible models and the consequences depend on which model is used and which one holds. We shall assume that Model E is used and briefly describe some consequences of three violations of its assumptions.

One violation occurs if the true model is E but the variances of the deviations are not constant and/or the deviations are not independent. It can be shown that T^* calculated according to Summary 9.4 remains unbiased but is no longer best. That is, the expected error of this estimator is still zero, but its error variance is not the least among linear unbiased estimators of the population total. In addition,

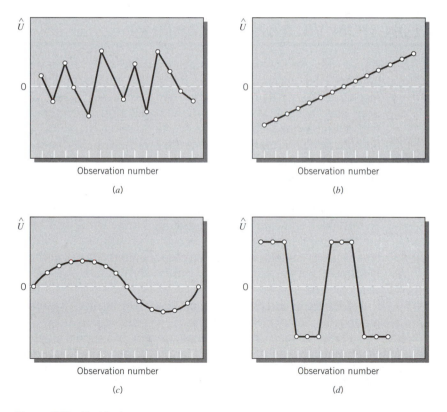

Figure 9.9 Residual patterns

confidence intervals such as those of Model D can be shown not to include the population mean with the nominal probability. Problems 9.15 and 9.16 confirm in part these conclusions.

Two other violations arise when the assumed linear relationship either fails to include relevant explanatory variables or includes irrelevant ones. Let us explain.

If hospital purchases of a given product depend on the number of beds and the number of patients of the hospital but the model used does not include the number of patients, we say that a relevant variable has been omitted. Conversely, if the model used includes, say, the age of the hospital manager's niece (or another variable having nothing to do with the purchases of the product), we say that an irrelevant variable has been included.

As can be intuitively surmised, the latter case is the less serious of the two. If Model E includes irrelevant variables, the estimator T^* of Summary 9.4 remains unbiased, although it is no longer best. However, if Model E fails to include a relevant variable, the estimator is not even unbiased. Problems 9.18 and 9.19 confirm these conclusions.

9.11 PREDICTION VS. RANDOMIZATION

As we observed at the beginning of this text, the main problem in sampling is how to infer characteristics of the whole (the population) on the basis of an inspection of a part (the sample). Any estimates based on the sample observations may be close to or far from the population characteristics—we cannot know with certainty how far because the population characteristics are rarely known. Sample selection methods and estimators have been evaluated according to their expected behavior (specifically, with respect to bias and variability) in a large number of repetitions of the "process" (the "mechanism") that generates the sample observations.

Under the randomization approach elaborated in Chapters 3 to 8, the values of all variables and categories are assumed to be fixed, and the sample observations are those of a random sample of elements selected by means of marked chips or tags in a hat, computer-generated random numbers, or another random mechanism. It follows that the sample observations are the same whenever the same elements are selected.

Under the prediction approach described in this chapter, the mechanism generating the sample observations is assumed rather than observed. The variable of interest is seen as determined in part by known explanatory variables; the other, unexplained part (the deviation) is viewed as generated by a random process with certain long-run properties. Consequently, the same elements may take on different values in different realizations of the process. A model describes the assumed form of the relationship between the variable of interest and the known

explanatory variables and the assumed properties of the process generating the deviations. The sample elements, which need not be selected at random, provide the information needed to predict the values of the nonsampled elements and hence also of such characteristics as the population total or mean of an unknown variable.

There are, of course, many possible models, and there may be little or no information to help decide which one applies. The consequences of assuming that one model applies when it does not range from mild to very serious.

This can be regarded as the main weakness of the prediction approach. By contrast, if the sample is selected as prescribed by the randomization approach, the desirable long-run properties of the estimators are assured; no other assumptions need be satisfied.

There is at least one area, however, where the two approaches strengthen the results of one another. We noted that the best sample under Model D is a sample balanced in X, in which case the best estimator of the population total of the variable of interest Y is the simple estimator $N\bar{Y}$. We also remarked that it may be difficult to achieve exact balance, approximate balance being all that can be hoped for. A simple random sample of elements (especially a large one) is likely to be just that—a balanced sample, that is, one in which $\bar{X} \approx \mu_x$. Therefore, the simple estimator $N\bar{Y}$ based on a simple random sample is likely to be approximately optimal under Model D.

The simple estimator is best also under the more general Model E, provided that the sample is balanced with respect to all explanatory variables. Again, a simple random sample is likely to be approximately balanced. In other words, selecting a simple random sample and estimating the population total as $N\bar{Y}$ is a desirable procedure for two different reasons: the estimator is unbiased by virtue of the randomization employed, and best linear unbiased in the event the assumptions of Model E are satisfied.

PROBLEMS

9.1 *Computing exercise:* Using the program SCALC, verify the numerical results presented in this chapter. Note any deviations, and determine if these are the result of rounding at intermediate stages.

9.2 The true relationship between a variable Y and two determining variables X_1 and X_2 is

$$Y = (10)X_1^{1.5}X_2^{0.001}.$$

(a) Calculate the values of Y corresponding to the following values of X_1 and X_2:

X_1	X_2
200	225
210	256
220	360
230	220
240	353
250	485
260	231
270	442
280	403
290	386
300	234

(b) In a scatter diagram, plot the pairs of (Y, X_1) values. Do you think Y can be well approximated using X_1 only? Why?

(c) Assuming X_1 is a known variable, approximate Y as a linear function of X_1 using the data in (a).

9.3 Same as Problem 9.2, except that the relationship between Y on the one hand and X_1 and X_2 on the other is

$$Y = (20)(1.015)^{X_1}(0.99)^{X_2}.$$

9.4 It is thought that the purchases by hospitals of pharmaceutical Product Z follow the assumptions of Model A. The purchases of this product by the three largest hospitals are shown in Table 9.9.

In the manner of Example 9.1, calculate the BLU estimates of the total and average hospital purchases of this product.

9.5 Same as Problem 9.4, except that the product is V, the purchases of which by the sampled hospitals are shown in Table 9.10.

9.6 A special case of Model A is that in which all $X_i = 1$, that is,

$$Y_i = \beta + U_i,$$

where the U_i are independent, with $E(U_i) = 0$, and $Var(U_i) = \sigma^2$.

TABLE 9.9 Data for Problem 9.4

Sampled Hospital	Number of Beds, X_i	Purchases of Product Z, Y_i ($000)
234	2,100	310
567	1,600	250
801	1,750	260

TABLE 9.10   Data for Problem 9.5

Sampled Hospital	Number of Beds, X_i	Purchases of Product V, Y_i ($000)
234	2,100	1,000
567	1,600	900
801	1,750	800

(a) What is the probable pattern of the sample (Y_i, X_i) values?

(b) Apply Eq. 9.8 to show that

$$b = \frac{1}{n} \sum_{i=1}^{n} Y_i = \bar{Y},$$

that is, b is the average of the Y values in the sample.

(c) Show that the BLU estimator of τ is $T^* = N\bar{Y}$. Compare this with the estimator of the population total under simple random sampling.

(d) Apply Eq. 9.9 to show that

$$Var(\tilde{E}^*) = \frac{N(N-n)\sigma^2}{n}.$$

(e) In this case, does it matter how a sample of size n is selected?

9.7 (a) Using the data in Table 9.2, and assuming Model B applies, calculate the BLU estimates of the total and average hospital purchases of Product Y.

(b) Again assuming Model B holds, calculate 90% confidence intervals for the total and average hospital purchases of Product Y. Pretend that the sample size is large.

9.8 Same as Problem 9.7, except assume that Model C holds. For the calculations in **(b)** assume $\sum_{j \in R} X_j^2 = 29 \times 10^6$.

9.9 It is believed that the hospital purchases of pharmaceutical Product V follow the assumptions of Model D. A sample of three hospitals, balanced with respect to the number of beds, yielded the data shown in Table 9.11.

(a) Plot the three pairs of (Y_i, X_i) values in a scatter diagram. Is the plot consistent with the assumptions of Model D?

(b) Calculate the BLU estimates of the total and average hospital purchases of Product V.

(c) Pretending that the sample size is large, calculate 90% confidence intervals for the population total and average hospital purchases of Product V.

TABLE 9.11	Data for Problem 9.9	
Sampled Hospital	Number of Beds, X_i	Purchases of Product V, Y_i ($000)
155	100	240
321	150	280
967	230	300

9.10 Same as Problem 9.9, except that the sample data are as shown in Table 9.12.

9.11 Refer to Example 9.3, and assume that the sample of nine hospitals also yielded information concerning the hospital purchases of Product Z, as shown in Table 9.13.

(a) Assuming that the purchases of Product Z (Y) are approximately linearly related to the number of beds (X_1) and the number of patients (X_2) according to

TABLE 9.12	Data for Problem 9.10	
Sampled Hospital	Number of Beds, X_i	Purchases of Product V, Y_i ($000)
155	100	160
321	150	120
967	230	80

TABLE 9.13	Data for Problem 9.11			
Selected Hospital	Number of Beds, X_{1i}	Number of Patients, X_{2i}	Type	Purchases of Product Z, Y_i
1	675	600	S	220
2	450	400	S	195
155	100	120	S	135
234	2,100	1,750	G	310
321	150	160	S	170
567	1,600	1,250	G	250
801	1,750	1,500	G	260
967	230	190	S	200
1,158	1,500	1,200	G	300

Model E, calculate the BLU estimates of the total and average hospital purchases of Product Z.

(b) Same as **(a)**, except use the type of hospital as an additional explanatory variable.

9.12 Consider the "Half-of-Ackroyd" case described in Section 9.4, but suppose that the sample will consist of firms B and C.

(a) Show that $T = (Y_2 + Y_3) + 13b$, and $\tilde{E} = 13b - Y_1$.

(b) Show that

$$b_1 = \frac{12Y_2 + 9Y_3}{225},$$

$$b_2 = \frac{1}{2}(\frac{Y_2}{12} + \frac{Y_3}{9}),$$

$$b_3 = \frac{Y_2 + Y_3}{21}.$$

(c) Show that $E(\tilde{E}_1) = E(b_1 X_1 - Y_1) = 0$ and $Var(\tilde{E}_1) = 1.0507$. (*Hint:* The calculations for **(c)** to **(e)** are long and prone to error. Consider using a computer program.)

(d) Show that $E(\tilde{E}_2) = E(b_2 X_1 - Y_1) = 0$ and $Var(\tilde{E}_2) = 1.0890$.

(e) Show that $E(\tilde{E}_3) = E(b_3 X_1 - Y_1) = 0$ and $Var(\tilde{E}_3) = 1.0599$.

(f) Interpret the results obtained in light of the theory presented in Section 9.3.

9.13 Consider the "Half-of-Ackroyd" case described in Section 9.4, but suppose that the sample will consist of firms A and B.

(a) Show that $T = (Y_1 + Y_2) + 9b$, and $\tilde{E} = 9b - Y_3$.

(b) Show that

$$b_1 = \frac{13Y_1 + 12Y_2}{313},$$

$$b_2 = \frac{1}{2}(\frac{Y_1}{13} + \frac{Y_2}{12}),$$

$$b_3 = \frac{Y_1 + Y_2}{25}.$$

(c) Calculate $E(\tilde{E}_1) = E(b_1 X_1 - Y_1)$ and $Var(\tilde{E}_1)$.

(d) Calculate $E(\tilde{E}_2) = E(b_2 X_1 - Y_1)$ and $Var(\tilde{E}_2)$.

(e) Calculate $E(\tilde{E}_3) = E(b_3 X_1 - Y_1)$ and $Var(\tilde{E}_3)$. (*Hint:* Consider using a computer program for the calculations in **(c)** through **(e)**.)

(f) Interpret the results obtained in light of the theory presented in Section 9.3, and the results of Problem 9.12.

9.14 Consider the "Half-of-Ackroyd" case described in Section 9.4 and Table 9.3.

(a) Show that $E(Y_1 + Y_2 + Y_3) = \beta(X_1 + X_2 + X_3) = 17$. Interpret this process characteristic.

(b) Show that T_1, T_2, and T_3 are all unbiased estimators of the above process characteristic.

(c) Show that $Var(T_1) = 2.8128$, $Var(T_2) = 2.9483$, and $Var(T_3) = 2.8661$. (*Hint:* Consider using a computer program for the calculations in **(a)**, **(b)**, and **(c)**.)

(d) Why are these variances not equal to the variances of $\tilde{E}_1$, $\tilde{E}_2$, and $\tilde{E}_3$, respectively? Elaborate on the implications of these comparisons.

9.15 Consider the "Half-of-Ackroyd" population described in Section 9.4, but assume that Model B applies. Specifically, suppose that $Y_i = (0.5)X_i + U_i$, where the U_i are independent but with probability distributions that depend on X_i, as follows:

$X_1 = 13$		$X_2 = 12$		$X_3 = 9$	
U_1	**Prob.**	U_2	**Prob.**	U_3	**Prob.**
−1	0.325	−1	0.3	−1	0.225
0	0.350	0	0.4	0	0.550
+1	0.325	+1	0.3	+1	0.225
	1.000		1.0		1.000

This means, for example, that the possible values of Y_1 are 5.5, 6.5, and 7.5, with probabilities 0.325, 0.350, and 0.325, respectively. As in Section 9.4, the sample consists of Firms A and C. (*Hint:* Use a computer program for the following calculations.)

(a) Calculate the means and variances of U_1, U_2, and U_3. Show that $E(U_i) = 0$ and $Var(U_i) = (0.05)X_i$, hence that Model B applies.

(b) In the manner of Table 9.3, show the possible combinations of values of Y_1, Y_2, and Y_3, and the corresponding probabilities. (*Hint:* The probability that $Y_1 = 5.5$, $Y_2 = 5.0$, and $Y_3 = 3.5$ is 0.0219375.)

(c) Show that for each combination of Y_1, Y_2, and Y_3, the values of b_1 (least squares), b_2 (average of ratios), and b_3 (ratio of averages) as well as the corresponding T's and $\tilde{E}$'s are the same as in Table 9.3. In other words, that the change in the underlying model in this case affects only the probabilities calculated in **(b)**.

(d) Calculate the expected value and variance of $\tilde{E}_1$, $\tilde{E}_2$, and $\tilde{E}_3$.

(e) In view of **(d)**, which is the BLUE of τ under Model B?

(f) Again in view of the results in **(d)**, what are the consequences of erroneously assuming that the true model is A or C, when in fact it is B?

9.16 Consider the "Half-of-Ackroyd" population described in Section 9.4, but assume that Model C applies. Specifically, suppose that $Y_i = (0.5)X_i + U_i$, where the U_i are independent but with probability distributions that depend on X_i, as follows:

$X_1 = 13$		$X_2 = 12$		$X_3 = 9$	
U_1	**Prob.**	U_2	**Prob.**	U_3	**Prob.**
-1	169/600	-1	144/600	-1	81/600
0	262/600	0	312/600	0	438/600
$+1$	169/600	$+1$	144/600	$+1$	81/600
	600/600		600/600		600/600

This means, for example, that the possible values of Y_2 are 5, 6, and 7, with probabilities 0.240, 0.520, and 0.240, respectively. As in Section 9.4, the sample consists of firms A and C. (*Hint:* Use a computer program for the following calculations.)

(a) Calculate the means and variances of U_1, U_2, and U_3. Show that $E(U_i) = 0$ and $Var(U_i) = (1/300)X_i^2$, hence that Model C applies.

(b) In the manner of Table 9.3, show the possible combinations of values of Y_1, Y_2, and Y_3, and the corresponding probabilities. (*Hint:* The probability that $Y_1 = 5.5$, $Y_2 = 5.0$, and $Y_3 = 3.5$ is 0.009126.)

(c) Show that for each combination of Y_1, Y_2, and Y_3, the values of b_1 (least squares), b_2 (average of ratios), and b_3 (ratio of averages) as well as the corresponding T's and $\tilde{E}$'s are the same as in Table 9.3—in other words, that the change in the underlying model in this case affects only the probabilities calculated in **(b)**.

(d) Calculate the expected value and variance of $\tilde{E}_1$, $\tilde{E}_2$, and $\tilde{E}_3$.

(e) In view of **(d)**, which is the BLUE of τ under Model C?

(f) Again in view of the results in **(d)**, what are the consequences of erroneously assuming that the true model is A or B, when in fact it is C?

9.17 Consider the "Half-of-Ackroyd" population. Assume that Model D applies with $Y_i = 10 + 0.5X_i + U_i$, where U_i are independent of one another, each U_i taking the values -1, 0, or $+1$ with probabilities 0.3, 0.4, and 0.3, respectively. The sample consists of firms A and C. (*Hint:* A computer program will make the following calculations easier and more accurate.)

(a) Show that $b = (Y_1 - Y_3)/4$, $T^* = (1.5)(Y_1 + Y_3) + b$, and $\tilde{E}^* = (0.5)(Y_1 + Y_3) + b - Y_2$.

(b) In the manner of Table 9.3, show the possible combinations of values of Y_1, Y_2 and Y_3, their probabilities, and the corresponding values of b, T^*, and $\tilde{E}^*$.

(c) Calculate the expected error, $E(\tilde{E}^*)$. Is this T^* unbiased?

(d) Calculate the error variance, $Var(\tilde{E}^*)$, and show that the result agrees with Eq. 9.13.

9.18 Consider the "Half-of-Ackroyd" population. Assume that A is the true model with $Y_i = 0.5X_i + U_i$, where the U_i are independent of one another, each U_i taking the values -1, 0, or $+1$ with probabilities 0.3, 0.4, and 0.3, respectively. However, Model D is believed to hold and Eqs. 9.12 and 9.11 are used to estimate the total number of employees in the town. As in Section 9.4, the sample consists of firms A and C. (*Hint:* A computer program will make the following calculations easier and more accurate.)

(a) Show that Eqs. 9.12 and 9.11 give $b = (Y_1 - Y_3)/4$, $T^* = (1.5)(Y_1 + Y_3) + b$, and $\tilde{E}^* = (0.5)(Y_1 + Y_3) + b - Y_2$.

(b) In the manner of Table 9.3, show the possible combinations of values of Y_1, Y_2, and Y_3, their probabilities, and the corresponding values of b, T^*, and $\tilde{E}^*$.

(c) Calculate the expected error, $E(\tilde{E}^*)$, and error variance, $Var(\tilde{E}^*)$.

(d) The presumed relationship can be written as $Y = \alpha X_0 + \beta X + U$, where X_0 is always equal to 1. In relation to the true model, therefore, the presumed model can be said to include an irrelevant variable (see Section 9.10). What are the consequences of this type of error?

9.19 Consider the "Half-of-Ackroyd" population. Assume that Model D is the true model and $Y_i = 10 + 0.5X_i + U_i$, where U_i are independent of one another, each U_i taking the values -1, 0, or $+1$ with probabilities 0.3, 0.4, and 0.3, respectively. However, Model A is believed to apply and Eqs. 9.8 and 9.7 are used to estimate the total number of employees in the town. As in Section 9.4, the sample consists of firms A and C. (*Hint:* A computer program should make the following calculations easier and more accurate.)

(a) In the manner of Table 9.3, show the possible combinations of values of Y_1, Y_2, and Y_3, their probabilities, and the corresponding values of b, T^*, and $\tilde{E}^*$ calculated according to Eqs. 9.8 and 9.7.

(b) Calculate the expected error, $E(\tilde{E}^*)$, and error variance, $Var(\tilde{E}^*)$.

(c) The true relationship can be written as $Y = \alpha X_0 + \beta X + U$, where X_0 is always equal to 1. In relation to the true model, therefore, the presumed model fails to include a relevant variable (see Section 9.10). What are the consequences of this type of error?

9.20 During the Peloponnesian War (431 to 404 B.C.), the Plataeans, besieged by the Peloponnesians and their allies, devised a plan by which some of them eventually escaped. A critical part of the plan involved the scaling of the wall that surrounded the city. The Plataeans constructed ladders to reach the top of the

encircling wall "... and they did this by calculating the height of the wall from the number of layers of bricks at a point which was facing in their direction and had not been plastered. The layers were counted by a lot of people at the same time, and, though some were likely to get the figure wrong, the majority would get it right, especially as they counted the layers frequently and were not so far away from the wall that they could not see it well enough for their purpose. Thus, guessing what the thickness of a single brick was, they calculated how long their ladders would have to be." (Thucydides, *The Peloponnesian War* (Baltimore: Penguin Classics, 1967), p.172.)

Dull as it may be, interpret this passage in the light of the theory presented in this chapter.

9.21 State/provincial and federal grants to municipalities (their share of income taxes) are often based on the total value of all properties in the municipality as a percentage of the total value of all the properties in the region. For example, if the total value of all properties in Municipality A is 10 and the total value of all properties in the region is 100, then A's share of the grants is 10%. Property values, of course, change over time, and these shares must be reappraised periodically. The estimation of a municipality's total property value is an important and occasionally contentious exercise.

The value of a real property is usually defined as the price at which it can be expected to sell in a free market, and the total value of all properties in a municipality as the sum of these estimates. Relatively few properties, however, change hands in a period of time such as a month, quarter, or year. Assuming free-market conditions prevail, the values of the properties that were sold are their selling prices; the values of the properties that were not sold, however, must be estimated on the basis of the information that the sold properties provide.

(a) In Problem 6.20, the following statement appeared. "Many factors prompt an owner to sell and a buyer to buy. These include financial need or desire to invest, relocation, change of income, standard of living, family size, and a host of others. It is reasonable to assume that one property is just as likely to be offered for sale and sold as another. It is also reasonable to assume that whether or not a particular property is offered for sale and sold does not depend on whether or not any other property is offered for sale and sold. In other words, it is reasonable to treat the properties that change ownership in a period of time as a simple random sample with replacement from the population of all the properties in the region, even though no act of selection is involved." Comment on this statement.

(b) For the purpose of this problem, assume that a municipality has 1,000 real properties. Five of these properties were sold last month. The selling price and lot size of these properties are shown in Table 9.14.

TABLE 9.14	Data for Problem 9.21	
Property	Lot Size (sq. ft.)	Selling Price ($000)
1	1,256	166
2	2,500	195
3	6,400	320
4	3,300	249
5	7,900	650

Bear in mind that every real estate property (even a condominium apartment) has claim to some land. The land under private ownership in the municipality amounts to about 6.5 million square feet.

Estimate the total value of all properties in the municipality. State carefully any assumptions you are forced to make, and explain in detail your choice of estimation method.

(c) Calculate a 95% confidence interval for the total value of all properties in the municipality. If you find that you need a quantity that should be known but is not provided, use a contrived but reasonable figure.

(d) Explain any reservations you may have about the method described above for estimating the total value of all properties in a municipality.

9.22 One difficulty in implementing the approach described in Problem 6.23 for estimating the commercial viability of a new product stems from the fact that substantial promotional effort is usually needed for a new product to have a chance of being successful. It is not sufficient, in other words, to display the product alongside similar products at selected retail outlets. Obviously, if these retail outlets are a simple random sample of the retail outlets in the country, the promotional expenditure necessary to reach all their customers would be very high.

For this reason, a new product is often evaluated in one "test market"—for example, a selected city or county. Advertising is restricted to the newspapers and television and radio stations serving this market. A number of retail outlets within the test market are selected to carry the new product and to display it on their shelves along with similar other products. Whatever other promotional material is needed (e.g., flyers, coupons, etc.) can be tailored to the selected outlets and test market. The period of time over which the sales of the new product are monitored is long enough to allow the full impact of the promotional campaign to manifest itself.

The test market is often selected so as to resemble as much as possible, on average, the country at large with respect to key demographic and economic characteristics (for example, income, proportion unemployed, age, proportion of women in the labor force, etc.).

The price of the product and the advertising budget are set at levels comparable to those that would apply if the product were sold nationally. The advertising campaign is launched, the new product sales observed, and the national sales of the product estimated by multiplying the sales in the test region by the number of such regions in the country.

Is there any merit to this practice? Could any one region (not necessarily one resembling the country on average) serve equally well as a test market? Under what conditions does the above practice make sense in the light of the material presented in this chapter? Which problems does this practice not solve? In your opinion, how should new products be evaluated?

9.23 Burns and Howell (B&H) is a consulting company for executive recruitment and compensation. B&H provides its clients information concerning "normal" compensation levels of chief executive officers (CEOs) of large companies in the country, and updates this information at frequent intervals. B&H also compiles an index showing trends in CEO compensation over time.

B&H concentrates on the largest companies in terms of total assets. A list of these companies is readily available, as is information concerning their total assets and sales. However, published information regarding the compensation of their CEOs is not always complete or accurate, calling for considerable research on the part of B&H. For this reason, the number of companies forming part of the B&H sample is relatively small. This sample is not, of course, random.

For the purpose of this problem, assume that B&H concentrates on the 100 largest companies, and that its sample is of size 5. Table 9.15 shows the annual CEO compensation, the total assets, and the total sales of the five selected firms.

If, in responding to the following questions, you find you need additional information that should be known but is not given, carry out your calculations as far as possible without it and make reasonable assumptions to complete them.

(a) If this were a simple *random* sample of 5 companies from the population of the 100 largest companies, what would be your estimate of the average annual CEO compensation among the 100 largest companies?

TABLE 9.15 Data for Problem 9.23

Company	Annual CEO Compensation ($000)	Total Assets ($billion)	Total Sales ($billion)
1	700	900	1,500
2	600	800	1,200
3	650	700	1,300
4	675	600	1,400
5	600	500	1,250

(b) Given that the sample is not random, under which assumptions can an estimate with desirable properties be made of the average annual CEO compensation among the 100 largest companies? What exactly are these properties?

(c) In view of the sample results, do you think these assumptions are satisfied? Explain why.

(d) If your answer to **(c)** is affirmative, estimate the average annual CEO compensation among the 100 largest companies.

(e) Using the data in Table 9.15, is it possible to estimate the average annual CEO compensation among the *1,000* largest companies? If so, under what conditions?

CHAPTER 10

Nonsampling Errors

10.1 INTRODUCTION AND SUMMARY

We have been assuming that the population from which the sample is selected *is* the population of interest; that all selected elements *can* be measured, questioned, or interviewed; and that these measurements *do* reveal the "true" value of any variable or category of any attribute of interest. We noted earlier that these assumptions are frequently violated in practice.

Consider as an illustration a market research firm interviewing selected shoppers at malls in order to estimate how many would buy a new product demonstrated on site. In the first place, not all shoppers do their shopping in malls, so the population of all shoppers and that of shoppers in malls are not identical. Further, some of the selected shoppers could refuse to be questioned because they are busy, not interested, or unwilling to participate in the demonstrations. Still further, their responses could be different from their real intentions; for example, some could say they would buy the product when in fact they have no such intention. These "untrue" responses could be due to a misunderstanding of the product, its price and features, to a lack of attention at the demonstration, and many other factors—including a malicious desire to confuse the questioner. For all these reasons, the resulting estimates of, say, the proportion and number of individuals who would buy the product could be misleading even when the sample is technically well designed and the estimators used have desirable long-term properties.

In this chapter, we examine some of the problems created by ill-targeted populations (also described as "coverage" or "sampling frame" errors), and by nonresponse and measurement errors. It will become apparent that there are no magic formulas to correct inferences about the characteristics of the population of interest when the sample is selected from a different population, when a substantial number of selected people cannot or will not be questioned, or when the questions are poorly worded and lend themselves to different interpretations. The aim of every investigation should be to *avoid* these problems as much as possible through careful planning of the study and of the questions.

We shall also examine telephone surveys (a case study contrasting two sampling frames for approximating a target population of residential telephone numbers), randomized response (a technique for reducing nonresponse and measurement error), and radio and television audience measurement (a case study of the compromise that must frequently be made in practice among cost, choice of list from which to select the sample, response rate, and accurate measurement).

We begin this chapter, however, with yet another, easily avoidable error, arising when an entire population is treated as a sample.

10.2 ENTIRE POPULATION AVAILABLE

Before implementing methods applicable to samples, it is useful to inquire whether there is a call for such methods in the first place. Not infrequently, all the elements of the population of interest are accessible at reasonable cost, and the population characteristics can be determined with certainty. Although it is rare to observe samples taken when the entire population can be easily investigated, it is not at all rare to observe an entire population treated as a sample from an unspecified whole.

For instance, in an internal study of compensation equity in a large organization, a list of all employees was compiled, and each employee's name, position, gender, age, years of experience, and annual salary were established from personnel records. From this information, it was possible to calculate the average annual salary by position and gender. There is nothing objectionable so far. But the investigators proceeded to calculate confidence intervals based on the average annual salary and the variance about this average, and to compare the confidence intervals for men and women at each position. Clearly, the available observations constituted the entire population of interest. No sample was involved, hence no justification could be given for the use of methods (such as confidence intervals) that apply to a sample drawn from an unknown population. The calculated averages and variances were population characteristics and could be compared purely as descriptive measures. The fact, for example, that for a certain position the average annual salary for women was higher than that for men should have been taken as the starting point for further investigation to determine if the difference could be explained by age, experience, or other sanctioned differentiating factors. Any variability in salaries after these factors are taken into account is evidence for discrimination.

10.3 IDENTIFYING THE POPULATION ELEMENTS

It has been assumed until now that the elements of the population of interest can be identified individually and labeled, so that they can be examined, measured, or questioned if selected for the sample.

Consider, however, estimating the quality of a truckload of potatoes. To select a simple random sample of potatoes, it is necessary to tag each potato with a label or identification number, a procedure that is laborious and expensive, and likely to cost more than a visual inspection of the entire truckload. Simple random sampling may not be feasible in this case. A possible alternative would be to sample as the potatoes are loaded or unloaded. For example, if the potatoes are unloaded down a funnel-like chute through which they pass one at a time, then a systematic sample of every, say, 50th potato could be treated as a simple random sample for all practical purposes.

A similar problem arises when some population elements are not physically present at the time the sample must be taken. Imagine, for example, having to select a simple random sample of pieces of mail from all such pieces handled by a courier company in a given month. Mail is delivered soon after it is received—there is no place where the entire month's mail is gathered. The number of pieces in a month's mail is not known in advance. If the company records such information as the sender, the recipient, the date and time of receipt and delivery, or the fee charged for each piece of mail handled, it is possible to select a random sample of a month's *records* in order to estimate the average time between receipt and delivery or other population characteristic. Of course, if the records are stored in a computer file, it is just as easy to calculate the population characteristics as to sample and then estimate. If records are not kept or the desired information (e.g., the weight) is not recorded, a sample of the month's mail may be necessary. A two-stage sample is one solution. For example, a number of one-minute intervals could be selected in advance from the known number of such intervals in the month. All pieces of mail processed during the selected interval would be inspected and measured, say, for weight. At the end of the month, the standard two-stage estimators could be calculated of the population average or proportion, and, assuming a count is kept of the number of pieces of mail processed during the month, of the corresponding population totals.

To select a simple or stratified random sample of *persons* from a given area it is necessary to have a list of all persons residing in the area. Lists of persons are not compiled frequently, especially lists of persons with specified characteristics—for example, female single parents, retired men, users of a particular product, etc.

Considerably easier to compile are lists of *residential dwellings*, each dwelling usually identified by its postal address. Indeed, there are firms that compile and sell such lists for regions where there is sufficient demand. Often these lists include additional information such as the name of the owners, the appraised value of the property, whether or not there are mortgages against the property, the number of residents, and so on.

A list of dwellings is often treated as a list of *households*, on the assumption that each household occupies one dwelling and each dwelling houses one household. The assumption is valid for most but not all dwellings and households, since some

dwellings are occupied by more than one household and some households reside in more than one dwelling.

A common method for selecting a sample of individuals or individuals with specified characteristics consists of first selecting a number of residential dwellings, contacting a member of the household, and obtaining from this member a list of the remaining members or of household members with the specified characteristics.

Occasionally, a list may be replaced by a *map*, and a random sample selected using a random sample of map coordinates.

Figure 10.1 shows a map of twelve residential lots, labeled A to L. The horizontal and vertical dimensions of the map can be scaled from 0 to 1, and a pair of random numbers in this range can be generated with the help of a table of, or computer program for, random numbers. For example, refer to Table 3.5, select a starting point, and form a 3-digit number, say, 123. Read this as the decimal 0.123, and treat it as the horizontal coordinate. The next 3-digit number will give the vertical coordinate, say, 0.654. The number of digits used depends on the desired accuracy. The point (0.123, 0.654) is located in Lot E, which becomes the first selected lot. Repeat the procedure to select the desired number of lots.

This procedure gives each lot a probability of selection proportional to its area. If the sample is to be without replacement and pairs of coordinates identifying previously selected lots are ignored, the probability that a given lot will be selected in a given draw is proportional to its area in relation to the total area of the as yet unselected lots.

It may happen that a list of population elements is not available, but another, closely related, list is. Think, for example, of an electrical utility serving a certain region. The utility has a list of its customers, the persons or organizations registered with the electricity meter and paying the bill. Customers may be classified as residential, commercial, industrial, government, and other. The residential customers could be homeowners or tenants. It is tempting to use the list of residential customers as a list of households in the region. The two lists, of course, are not

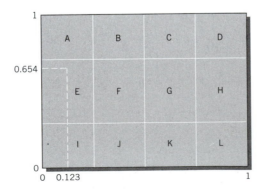

Figure 10.1 Land map

identical. A meter could serve more than one household or unrelated persons living together; some households do not have electricity. Whether or not the two lists can be considered identical for all practical purposes depends on the overlap, and that can be investigated. Apartments, for example, present a special problem. If they are individually metered, they can be treated as any other residences. If they are "block metered," that is, if one meter serves the entire apartment building, the list of meters/customers will be inappropriate for a sample of households when the region has a substantial number of apartments.

In general, the list from which the sample is actually drawn is called the *sampling frame* or, more simply, the *frame*. Ideally, the frame should be the target population. If that is not possible, then the frame should match as closely as possible the target. Occasionally, only a careful reading of the fine print will reveal the disparity between frame and target population.

Telephone surveys, which we examine in the following section, provide an interesting contrast of two lists used in practice to approximate populations of residential telephone numbers.

10.4 CASE: TELEPHONE SURVEYS

Mailed questionnaires, personal visits by field workers, or the telephone may be used for a survey of households. Although the telephone is not suited for all surveys (for example, it is not possible to show or demonstrate a product over the telephone), telephone surveys are frequently less costly and more conveniently administered.

Clearly, telephone numbers and households do not have a one-to-one correspondence. Telephone numbers are also used by businesses, government offices, and other organizations. Some households do not have a telephone number, others have more than one. Answering machines, modems, fax machines, and call forwarding may redirect telephone calls to the household.

Let us suppose, however, that the target population is that of all residential telephone numbers in a region, and let us consider how a random sample of these numbers can be selected. Two possible lists come immediately to mind: the telephone directories covering the region, and the list of all possible combinations of digits.

A complete *telephone directory* of all residential telephone numbers currently in use is the ideal list from which to select a random sample. The sample could be simple, stratified (as when, for example, a number of listings are selected at random from each page of the directory), two-stage (for example, when a number of pages are first selected at random and then a number of listings are selected from each selected page) systematic with a random start (for example, when the kth listing is selected in each page, or the listing located k inches from the top of each column), or of some other type.

Such an ideal list does not exist. Directories of residential-only telephone numbers are not available in all regions. The standard telephone directories also include

nonresidential listings, and these are sometimes not easily distinguishable from residential ones. Telephone directories are not updated frequently, and may be several months out of date. And, of course, directories do not include unpublished or unlisted numbers. This last deficiency is serious because it is estimated that about one-fourth to one-third of the numbers in service are unlisted.

In the United States and Canada, telephone numbers consist of 10 digits: a three-digit area code, a three-digit prefix, and a four-digit suffix. Thus, a list of all possible telephone numbers is the list of the 10 billion 10-digit numbers starting with 000-000-0000 and ending with 999-999-9999. Telephone surveys usually focus on geographical areas served by particular telephone exchanges within the broader area designated by the area code. These exchanges service telephone numbers with certain prefixes, which are not used elsewhere within the coded area. Telephone surveys, therefore, frequently employ a stratified sampling design where each stratum consists of telephone numbers having the same area code and prefix. There are 10,000 possible telephone numbers in each such stratum. It would appear that a simple random sample of these numbers could be easily obtained by generating 4-digit random numbers and appending these to the 6-digit area code and prefix combination.

This method, also known as *random digit dialing (RDD)*, solves the main problem of sampling from telephone directories because each number, whether listed or unlisted, now has the same chance of being generated. The method also does away with bulky, out-of-date, and inconvenient directory books. On the other hand, it is not possible to distinguish residential from nonresidential numbers except after they are dialed. And, of course, there remains the problem of nonworking numbers, for not all possible combinations of area code, prefix, or suffix are assigned and functioning telephone numbers. Completely random RDD can be expensive because many calls must be made in order to reach a given number of working telephone numbers. Detecting a nonworking number may not be easy because dialing such a number may produce an informative message (e.g., "the number you dialed is not in service"), but also a busy signal, unanswered rings, or no response at all, depending on the practices of the local telephone company. The sampling cost, however, can be reduced with some preparation. Lists of nonassigned area codes and prefixes are relatively easy to obtain from telephone companies; it is also possible to purchase from companies specializing in this task "banks" of unassigned four-digit suffixes.

Figure 10.2 shows the relationship between the target population of all residential telephone numbers on the one hand, and, on the other, the population of all numbers listed in the telephone directory and the list of all numbers that can be generated by random digit dialing.

Clearly, the three populations are not identical. Despite the efforts made in practice, it may not be possible to arrive exactly at the target population by screening elements of one of the other two. This does not mean that telephone surveys do not provide useful information. Estimates based on telephone surveys, however, must be interpreted for what they are: they are estimates of characteristics of

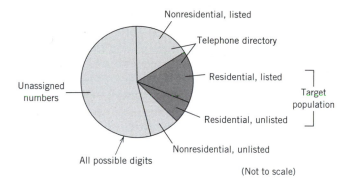

Figure 10.2 Residential telephone numbers, directories, and RDD

populations that are not identical but overlap, or have common elements with, the target population. On balance, telephone sampling remains a method with strong appeal for its cost and convenience.[1]

10.5 NONRESPONSE

Another assumption carried through all preceding chapters is that any selected element can be inspected or questioned, and the "true" values of the variables (or the categories of the attributes) of interest determined.

In some situations, this assumption is easily satisfied. Think, for example, of sampling for quality control: a sample is drawn from a lot of manufactured items, and the selected items are measured for width, height, hardness, flexibility, durability, or other variable relevant to the technical specifications. In most cases, there should be no problem carrying out these measurements or measuring correctly. There are, of course, exceptions: the selected item may not be accessible, the measurement may be difficult or subject to human or instrument error.

In surveys of people, however, it is not unusual for a selected person to be absent or unwilling to be questioned. Even when the person is willing to respond, he or she may not answer the question truthfully, whether intentionally or not. We examine the problem of nonresponse in this section, and that of inaccurate response in the next.

Certain sensible precautions can always be taken before the sample is drawn to reduce the nonresponse rate as much as possible. The questionnaire should be as short as possible. If the survey requires visits to private residences, these visits should be scheduled at times when persons to be interviewed are at home. The wording of

---

[1] For further information see, for example, Paul J. Lavrakas, *Telephone Survey Methods: Sampling, Selection and Supervision,* 2nd ed. (Newbury Park, CA: Sage Publications, 1993).

the introduction, the tone and manner in which it is delivered, even the appearance of the interviewer could be planned so as not to cause offense, dislike, or irritation.

An obvious consequence of nonresponse is that the actual sample size is less than the planned one. If those that respond can be assumed to have the same characteristics on average as those that do not, the smaller effective sample size is the only serious consequence. Such an assumption may be reasonable in a situation where the only nonrespondents are those that were absent when the interviewer visited their home. In other cases, however, estimates of the characteristics of the population based on the available responses may be misleading.

Consider, for example, a cable television company questioning randomly selected subscribers about the quality of the service they receive. It is likely that dissatisfied customers will feel strongly enough about their grievances to take the trouble to respond to the questionnaire with negative comments. Satisfied customers, on the other hand, may not be as motivated to respond to the questionnaire with favorable comments. Consequently, the proportion of dissatisfied subscribers may be greater, and the comments on average more negative, in the sample than in the population at large.

In some studies, it is possible to elicit responses from nonrespondents by contacting them a second time, perhaps offering an additional reward or incentive, or appealing to their sense of duty or responsibility. (Appeals, rewards, and incentives should be used with caution because they may alter what would otherwise have been the first-time response—for example, from a desire to please the donor.)

If *all* first-time nonrespondents now respond truthfully, the problem is solved: the ordinary sample average and proportion based on all responses received are unbiased estimators of the population mean and proportion. The first-round nonresponses have no other effect than to delay the gathering of the sample observations.

For example, if the objective is to estimate the population mean of a variable Y and n_1 of the n sampled persons respond, the information available after the first contact may be pictured as follows:

$$\underbrace{Y_1 \quad Y_2 \quad \cdots \quad Y_{n_1}}_{n_1 \text{ responses}} \underbrace{? \quad ? \quad \cdots \quad ?}_{n-n_1 \text{ nonresponses}}.$$

After the second contact, all the question marks are replaced by true responses:

$$\underbrace{Y_1 \quad Y_2 \quad \cdots \quad Y_{n_1}}_{\text{first contact}} \underbrace{Y_{n_1+1} \quad Y_{n_1+2} \quad \cdots \quad Y_n}_{\text{second contact}}.$$

Clearly, the average of the n sample Y values, $\bar{Y}$, *is* the familiar unbiased estimator of the population mean of Y, μ. For the same reasons, the proportion of sampled elements in a category, as calculated after the second contact, *is* the familiar unbiased estimator of the population proportion, π.

In the event that some persons do not respond to the second contact, it may be possible to contact them a third time, perhaps with an even better appeal or

inducement (provided, always, that these do not distort the true responses). Again, if all respond to this third contact, the average or proportion based on all responses received is the desired estimator. If there are still some that do not respond to the third contact, a fourth appeal may be considered, and perhaps a fifth, sixth, and so on.

It is obvious, however, that additional rounds of contacts are time-consuming, add to the cost of sampling, and increase the likelihood of distorted responses. Some persons will not respond for any reasonable inducement. Practical considerations, therefore, usually limit the number of recontacts, and after all appeals are exhausted there are likely to be some who have not responded.

Clearly, there can be no solution to the nonresponse problem, at least not without additional assumptions. Put plainly, there is no information about those elements of the population that do not respond, and estimates cannot be created without some information.

An assumption frequently made explicitly or implicitly is that those who do not respond to all appeals are similar on average to those that do. Whether or not this assumption is reasonable must be judged in each individual case. Among the factors to consider in making this judgment is the distribution of the reasons for the nonresponses (the percentages of not at home, busy at time of call, outright refusals, refusals after the introduction, etc.), the wording of the questionnaire and introduction, and the magnitude of the nonresponse rate.

In surveys of voting intentions, for example, a number of eligible voters are selected and usually asked the simple question: "Can you tell us which party you would vote for if an election was held today?" If the answer is "no," the voter is classified as "undecided"; if "yes," the voter is asked for the party and classified as, say, Democrat, Republican, or Other. The survey results are reported usually in the following format: "$x\%$ of the voters are undecided. Among decided voters, $y\%$ would vote Democrat and $z\%$ Republican." The undecided voters are, of course, the nonrespondents; any party that treats them as similar with respect to voting preferences to those that respond does so at its own peril.

10.6 MEASUREMENT ERROR

We now turn to the assumption that the selected elements can be measured without error. By "measurement" we understand determining the "true" value of a variable or category of an attribute of interest.

Think, for instance, of a survey of households and suppose that the head of each selected household is asked the simple question: "What is your household income?"

We can well anticipate some problems that this question will create. To begin with, is it annual, monthly, or weekly income that is requested? Some wage earners are paid weekly, others monthly, and still others are used to annual figures. Does the question refer to current income (for example, this year's annual income) or

to past incomes (for example, last year's)? In the former case, is the respondent asked to guess the entire year's income on the basis of realized income for a part of the year and expectations concerning the remaining part? In the latter case, and assuming past annual incomes varied, is it the most recent, average, median, or modal income that is to be reported? Is income understood to be before or after taxes and other deductions such as pension fund contributions? Which taxes (e.g., income, property, sales), if any, are to be excluded? The incomes of which members of the household are to be included? Is grandmother's income to be included or not? Which forms of income are to be considered? For example, how is a capital gain to be treated in the case where there was an appreciation in the value of stocks held but not sold? Is the imputed value of a housewife's services to be included? If so, how is it to be imputed? Let us also not overlook the fact that many people are secretive about their income, and either refuse to answer the question or give a random answer.

It is clear that the definition of what is to be measured may be quite difficult. Communicating this definition to the person interviewed may be even more difficult. Complicated definitions tend to make respondents impatient and irritated. Simplified definitions increase the likelihood of different interpretations from that assumed in the study.

Questions that invite respondents to recall a past event may also be answered incaccurately. For example, the question "How much did you spend on your last vacation?" may be difficult to answer accurately if expense records were not kept and the last vacation was several years ago.

Long questionnaires and interviews create fatigue and may force some respondents to hurry their responses without adequate reflection simply in order to have the questions done with.

Questions that invite a quantitative rating of an inherently nonquantitative attribute also invite trouble. Consider, for example, the request: On a scale from 1 to 5, rate the importance you attach to good friendship. What is friendship, what is good friendship, how can one measure the importance of good friendship, and of what possible use are such ratings from different people, each with his or her own perception of what the scale from 1 to 5 means?

The fact that it is often difficult to measure accurately and without error should not, of course, induce paralysis. After all, we are used to dealing with incomplete, inaccurate, or inappropriate information in everyday life. Common sense, a good grasp of the language, and a thorough testing of the questions on a crosssection of respondents before it is actually used will go a long way toward eliminating avoidable sources of confusion and misinterpretation.[2]

---

[2]Some excellent advice regarding problems related to question design and measurement can be found in A. N. Oppenheim, *Questionnaire Design, Interviewing, and Attitude Measurement*, New Ed. (New York: Pinter Publishers, 1992).

10.7 RANDOMIZED RESPONSE

One reason why people do not respond truthfully or altogether refuse to respond is their sensitivity to the question asked. For example, imagine a survey designed to estimate the proportion of adults who rent X-rated videos. To the question: Do you rent X-rated videos? an adult who does not rent will probably respond with a "No." The response of renters, however, could be "Yes," "No," or an outright refusal to answer this question and others that follow. Similar problems may be experienced with attempts to solicit information from individuals who have committed a crime, evaded taxes, used drugs, etc.

A reasonable precaution is to treat the individuals' responses confidentially, and provide assurances that the response cannot be traced back to the respondent. Such assurances can be given when the data are collected by means of a questionnaire, but not in a face-to-face interview, where the person interviewed may feel embarrassed, alarmed, or apprehensive about revealing the truth to the interviewer. Randomized response methods, introduced by S. L. Warner, aim at encouraging truthful responses by dissociating the question from the response, while still permitting unbiased estimates to be made of the population characteristics of interest. Let us explain.

Assume that interviews with selected adults are planned in order to estimate the proportion who use drugs. Instead of being asked directly: "Do you use drugs?," the person interviewed is invited to select at random *and in private* one of the following questions:

Q1: Do you use drugs?
Q2: Do you abstain from drugs?

and answer *the selected question* with a "Yes" or "No."

The interviewer does not know to which question the answer refers—a feature that is fully explained to the persons interviewed. Thus, a "Yes" could mean that the person interviewed either uses or abstains from drugs; likewise for a "No." Since the response is dissociated from the question, it can be hoped that all will respond truthfully to the selected question.

Random selection can be implemented, for example, by labeling a number of tags with "Q1" and "Q2," placing these tags in a box, and having the person interviewed mix the tags before selecting one at random. Alternatively, the person interviewed could be invited to spin a pointer and answer Q1 or Q2 depending on the position of the pointer when it comes to rest.

An important element of the randomized response technique is the probability that the question presented will be Q1. It can be controlled by varying the percentage of tags labeled Q1 in the first randomization scheme, or the pointer positions indicating Q1 in the second scheme. The probability q that the question will be Q1 is fixed in advance of the interviews and is known to the persons interviewed. The probability that the question will be Q2 is, of course, $1 - q$. As we shall now

demonstrate, q plays a role in forming an unbiased estimator of the proportion who use drugs, and influences the accuracy of this estimator.

Let π be the proportion of drug users in the population of N adults. Suppose that a simple random sample of n adults will be selected *with replacement*. If it can be assumed that all selected persons will answer truthfully the question they select, then the possible events and their probabilities can be determined with the help of a probability tree as shown in Figure 10.3.

The person interviewed will be either a user with probability π, or a nonuser with probability $1 - \pi$. Whether user or nonuser, the person interviewed will be presented either Q1 with probability q, or Q2 with probability $1 - q$. If the person is a drug user and the question is Q1, it is assumed that the response will be "Yes" with probability 1, or "No" with probability 0. The probability that the person interviewed will be a drug user, the question will be Q1, and the response "Yes" is the product of the probabilities along the corresponding branches of the tree: $(\pi)(q)(1)$, or πq. The probabilities of all other outcomes are similarly calculated.

The probability that the response will be "Yes" is the sum of the probabilities of all "Yes" outcomes:

$$Pr(\text{Yes}) = \pi q + (1 - \pi)(1 - q) = (2q - 1)\pi + (1 - q).$$

A reasonable estimator of this probability is the proportion of "Yes" responses in the sample of size n. Let us call this quantity Q. Bearing in mind that q is a given number, a reasonable estimator P of the proportion of drug users in the population can be found by solving for P the following equation:

$$Q = (2q - 1)P + (1 - q),$$

or, assuming $q \neq 0.5$,

$$P = \frac{Q - (1 - q)}{2q - 1}. \tag{10.1}$$

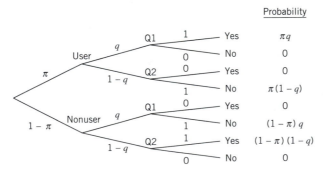

Figure 10.3   Probability tree, randomized response

For example, if $q = 0.1$ and 79% of 150 adults interviewed responded with "Yes," the estimate of the proportion of drug users would be:

$$P = \frac{0.79 - (1 - 0.1)}{2(0.1) - 1} = 0.1375.$$

It can be shown that P is an unbiased estimator of π, and that its variance is given by

$$Var(P) = \frac{\pi(1 - \pi)}{n} + \frac{q(1 - q)}{n(2q - 1)^2}. \tag{10.2}$$

Two conclusions may be drawn from an inspection of $Var(P)$. First, as can be expected, $Var(P)$ decreases as the sample size n increases. As for the second, recall that q can be controlled. The first term of Eq. 10.2 does not depend on q. The second term, which is always nonnegative, does. It equals 0 when q equals 0 or 1, and, as the following table shows, increases as q approaches 0.5 from either direction.

q	$\frac{q(1-q)}{(2q-1)^2}$
0.00 or 1.00	0.0000
0.05 or 0.95	0.0586
0.10 or 0.90	0.1406
0.15 or 0.85	0.2602
0.20 or 0.80	0.4444
0.25 or 0.75	0.7500
0.30 or 0.70	1.3125
0.35 or 0.65	2.5278
0.40 or 0.60	6.0000
0.45 or 0.55	24.7500

We conclude that, for given n, the unbiased estimator P of the proportion of drug users will tend to be more accurate the closer q is to 0 or 1. But, of course, such values of q may make the persons interviewed fear that their responses will reveal the truth with high probability, thereby defeating the objective of randomized response. A compromise solution may be to use moderate values of q—for example, about 0.25 or 0.75.

We have presented only the simplest randomized response technique. The method can accommodate sampling without replacement with equal or unequal inclusion probabilities, attributes with more than two categories, and variables as well as attributes.[3]

---

[3]In some versions, the person interviewed randomly selects and answers one of two questions, but one of the questions is now unrelated to the attribute being studied. For example, Q2 in the above illustration would be replaced with Q2′: "Is your grandmother alive?" For further information see, for example, A. Chaudhuri and R. Mukerjee, *Randomized Response: Theory and Techniques* (New York: Marcel Dekker, 1988).

10.8 CASE: RADIO AND TELEVISION MEASUREMENT

There is perhaps no application of sampling that generates more interest and can be better appreciated than the measurement of the size and characteristics of radio and television audiences. This application also provides an interesting illustration of the compromise that must often be made among the conflicting objectives of low cost, satisfactory coverage of the population of interest, high response rate, and accurate measurement.

Let us begin by reviewing some familiar features of the two media, television and radio. Watching television requires a television set. At any given time, the set is either on or off. If the set is on, it can be tuned to one of a finite number of channels through which it receives signals from stations broadcasting through the air, by means of cable, via satellite, or from a tape or disc. These signals carry programs: news, sports, movies, soap operas, documentaries, etc. The stations may be local or affiliated with networks. The channels can be changed with a set or remote control. The television set may be off, on with no one watching, or on with any number of viewers. A VCR may be taping a program for future viewing while the set is off or tuned to a different channel. A television set may be portable or a permanent fixture of the home or office. A home or office may have none, one, or more than one television set.

Radio has similar features. Instead of channels, a radio set is tuned to frequencies that, because of the regulation of the airwaves, are finite in number. All approaches for measuring television audiences can be applied to radio and vice versa. To avoid repetition, we concentrate on television in what follows.

Let us imagine it is possible to wave a magic wand at any instant, and immediately obtain a list of all persons in the country showing the channel, program, station, and network they watch—if, of course, they are watching television at all. Let us also imagine that this list also includes such personal characteristics as the gender, age, address, income, level of education, and so on, of the listed persons.

From such a list, one could calculate the percentage of people watching a particular program (the so-called *rating* of the program), the percentage of people who watch a program out of all people watching television at the time (the *share* of the program), and summary audience characteristics of a particular program (for example, the proportion of males among, and the average age and income of, viewers of the Rose Bowl game). Similar calculations can be made for particular stations, networks, specific locations and regions. And, of course, all this information can be examined over time to detect any trends and patterns.

Who would use such information? In many countries, the television industry is supported by advertising. The industry sells time slots in programs into which advertisers can insert their messages (the all-too-familiar "commercials"). The revenue from these sales is used to purchase, commission, create, and maintain programs. Advertisers are anxious to have their message seen, not necessarily

by as many people as possible, but by as many people as possible among those likely to buy their products or services. For example, it makes no sense for the advertiser of a ladies' perfume to purchase time in a hockey game telecast, even a popular one, because that audience is likely to be largely male, hence unlikely to buy the perfume (unless, of course, the aim of the commercial is to induce men to buy the perfume for women). Advertisers, therefore, are interested not only in the size of a program's audience but also in the characteristics of that audience. Television stations and networks, on the other hand, attempt to maximize their revenue by buying, or investing in the production of, programs having the greatest possible appeal among advertisers. It is not surprising, therefore, that advertisers, the advertising industry, the television industry, and the independent producers of programs all scrutinize the size and characteristics of program audiences in an effort to infer the reasons for the popularity of some programs and the demise of others.

Obviously, there does not exist a magic wand to help create this information. Equally obvious is the fact that the information will have to come from samples. It is interesting, therefore, to examine the different approaches that have been applied in the past to estimate radio and television audiences. We shall overlook the strictly sample design aspects of these approaches (for example, whether the samples are simple, stratified, or multistage) in order to concentrate on issues related to this chapter, concerning the nature and quality of the information that these approaches can provide.

The first systematic attempt to measure radio audiences made use of the so-called *telephone recall* approach. Applied to television, a number of persons would be interviewed over the telephone and asked to recall which programs they watched the evening or day before, or during the past so many days. A typical response may take the following form: "Last night, I watched program X on channel A from about 9:15 to 10, and program Y on channel B from 10 to 10:50, when I went to bed." The persons interviewed are also asked their gender, age, and other personal characteristics.

The population of persons who have a telephone is not, of course, identical to the target population of all individuals: the viewing habits and preferences of persons having a telephone are likely to be different from those that do not. The special shortcoming of the telephone recall approach, however, stems from the fact that people's recollections are not always accurate and tend to become less so the more distant the past event. As with other telephone surveys, the cost of telephone recall tends to be low and the response rate relatively high. This approach is currently used by two radio ratings services in the United States: Birch Radio and SRI.

An alternative to the telephone recall is the *telephone coincidental* approach. Again, the telephone would be used to interview selected individuals, but these are now asked to indicate which program they were watching at the time the call was made. The responses should tend to be more accurate than under the recall approach. However, the cost is likely to be higher because more calls are now necessary to estimate the audiences of a day's programs. Also, because calls

cannot disturb people when they rest or sleep, it is difficult to gather information about late night or early morning viewing.

The reliance on the telephone could, of course, be avoided by using *personal interviews* to inquire about the current or past viewing of the persons interviewed. Indeed, the personal interview approach dominated local radio measurement in the 1950s and early 1960s.

Nowadays, however, the most widely used approach employs *diaries*. Diaries are used for radio and for local area television measurements where the demand cannot offset the higher cost of more sophisticated approaches. Ratings services using diaries currently include A. C. Nielsen and Arbitron in the U.S., and A. C. Nielsen and BBM in Canada. Details of the diary approach vary with each service, but the measurement process begins with the selection of a number of households. Preliminary interviews establish whether or not the household will participate, the number of television sets, the members of the household, the preferred language, and other household characteristics. A separate diary is kept for each set. The diary covers usually one week of viewing and is kept by a designated member of the household. (Individual diaries are used for radio measurements.) The designated member is instructed to record which programs, channels, and stations members of the household were viewing at quarter-hour intervals. At the end of the week, the completed diaries are mailed back to a central location where they are edited and the information is processed. A nominal compensation is often given to households mailing completed diaries. Figure 10.4 outlines a typical completed television diary.

Accurately completed diaries can provide a wealth of information at relatively low cost. Accurate diary keeping, however, requires that the diary keeper be able to monitor the television set at all times and to observe who is watching. Otherwise, the diary keeper must rely on the recollections of the other members of the household. Large and busy households, a large number of available channels, remote controls that permit frequent channel changes, VCRs that shift the time at which a program is seen—these are some factors that make accurate diary keeping difficult.

The first successful *set meter* (or *household meter*), A. C. Nielsen's Audimeter, was a device in the shape of a cigar box that was attached to the television set. The meter could record at one-minute intervals whether or not the set was on and, if on, the channel to which the set was tuned. At first, these recordings had to be extracted and mailed for processing. Later, they could be automatically transmitted over telephone lines to a central processing station.

The set meter could supply information to subscribers at a speed that the diary could never match. Estimates of an evening's program audiences could be at a subscriber's desk early the following morning. The set meter, however, was expensive, and was used primarily for estimates of national audiences and in large population centers. The information provided by the meter was entirely accurate, as far as it went, but the set meter could not tell whether or not any one was watching while the set was on, or which members of the household were watching. Thus, the characteristics of a program's audience had to be inferred from other information.

DATE: TUESDAY EVENING, SEPT. 15

TIME FROM	TIME TO	TV SET OFF	TV SET ON	STATION CHANNEL NO.	STATION CALL LETTERS	PROGRAM NAME, OR MOVIE TITLE	JOHN 1 M 35	JANE 2 F 32	MARY 3 F 8	TOM 4 M 2	5	6	NAME / ID. NO. / SEX / AGE
...	...												
8:00P	8:14P	X											
8:15P	8:29P		￨										
8:30P	8:44P	X											
8:45P	8:59P		X	4	WKAN	BOSTON POPS	X	X					
9:00P	9:14P		￨	5	CDLX	TARZAN IN	￨		X	X			
9:15P	9:29P		￨	￨	￨	NEW YORK	￨		￨	￨			
9:30P	9:44P		￨				￨		￨	￨			
9:45P	9:59P		￨				￨		￨	￨			
10:00P	10:14P	X		5	CDLX		X		X	X			
...	...												

Figure 10.4 Sample television diary

The *people meter* has supplanted the set meter for the measurement of national audiences ever since it was first introduced by AGB and A. C. Nielsen in the 1980s. A people meter consists of four components: a television set meter, a VCR meter, the "people meter" proper, and a microprocessor that records the input from the first three components and automatically transmits the data over telephone lines to a central processing station at frequent intervals. The people meter proper is a device equipped with buttons or remote controls, one such button or control being assigned to each member of the household. A household member presses his or her button to register his or her arrival, and again to register his or her departure from the room. The people meter, in effect, combines the functions of the set meter and the diary, except that the diary is kept electronically. The only input required from each member of the household is the pressing of a button.

Even this apparently minimal cooperation is not always forthcoming: industry research suggests a compliance rate of about 70%. Two questions therefore arise. Is the manual diary more accurate? Can a more sophisticated device be designed that requires no human input to register without error who watches a program?

At the time of this writing, the industry is experimenting with various passive technologies. One system, for example, would require members of the household to wear or carry a device that would signal the member's near presence or absence.

Another system would store electronically the image of each member of the household. At frequent intervals, it would scan the room and compare the images of those present with the stored images, marking down a member of the household as present if a match is made. It is too early to tell how accurate these systems can be. Any improvement in accuracy may well be offset by an increase in the nonresponse rate if more households refuse to accept the invasion of privacy that these systems entail.[4]

Debates concerning the methodology of television audience measurements often tend to reflect vested interests. The faults of a particular approach tend to be magnified with little regard to its virtues or the faults of alternative approaches. It is interesting, therefore, to compare estimates of television audiences based on different approaches in order to get a sense of the magnitude of their differences.

Because space is limited, we can present only two comparisons.[5] The first is of estimates of the national audience of one network over one day from noon to 11 P.M.; one set of estimates is based on television set meters, the other on diaries. The second comparison is of audience estimates for one television station broadcasting in a relatively small region isolated from the signals of neighboring stations; the estimates are made by two competing ratings services, both using diaries and the same delineation of the area.

Details of the first comparison are given in Table 10.1, and the estimates are shown in Figure 10.5. Details concerning the second comparison can be found in Table 10.2; the estimates are shown in Figure 10.6. Of course, two comparisons cannot make a case, but it is interesting that despite the different sources of the estimates, sampling methods, sample size, and approach, the estimates have a strikingly similar pattern.

10.9 TO SUM UP

- Evidently, sampling methods should not be applied when available observations constitute the entire population. At the beginning of a study, therefore, it is useful to establish whether or not the situation calls for a sample.
- The list or "whole" from which the sample is selected (the "sampling frame") should be the target population. Users of the sample information should be aware of any deviations between the frame and the target population.
- Telephone surveys appear to be surveys of households, but, on closer examination, neither the telephone directory nor the list of all possible telephone numbers

---

[4]For additional information on radio and television audience measurement see, for example, James G. Webster and Lawrence W. Lichty, *Ratings Analysis: Theory and Practice* (Hillsdale, NJ: Lawrence Erlbaum Associates, Publishers, 1991), and Karen Buzzard, *Electronic Media Ratings: Turning Audiences into Dollars and Sense* (Stoneham, MA: Focal Press, 1992).

[5]These comparisons are due to Mr. Peter W. Foster.

TABLE 10.1	**Network Audience Comparison**	
Source:	**A. C. Nielsen**	**BBM**
Survey:	Nov. 4–10, 1991	Oct. 31–Nov. 6, 1991
Sample size:	1,400 households	14,381 adults 18+
Approach:	Metered television sets	Household diaries
Time:	Monday, Nov. 4, 1991, noon to 11 P.M.	Monday, Nov. 4, 1991, noon to 11 P.M.
Network:	CBC English Network	CBC English Network
Averaging:	One-minute intervals	Quarter-hour intervals

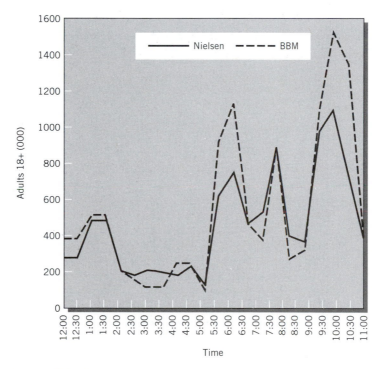

Figure 10.5   Network audience estimates

that can be produced by random digit dialing are identical to the target population of households. Still, telephone sampling is a method with strong appeal for its cost and convenience.

- Among the reasonable precautions that can always be taken before the sample is selected to reduce the nonresponse rate as much as possible are: short questionnaires, consideration of the respondents' convenience, avoidance of questions

TABLE 10.2 Local Station Audience Comparison		
Source:	**A. C. Nielsen**	**BBM**
Survey:	Oct. 28–Nov. 17, 1991	Oct. 31–Nov. 20, 1991
Sample size:	317 households	458 adults 18+
Approach:	Household diaries	Household diaries
Time:	Monday (3-week avg.), 4–11 P.M.	Monday (3-week avg.) 4–11 P.M.
Local area:	Thunder Bay, Ont.	Thunder Bay, Ont.
Station:	CKPR	CKPR
Averaging:	Quarter-hour intervals	Quarter-hour intervals

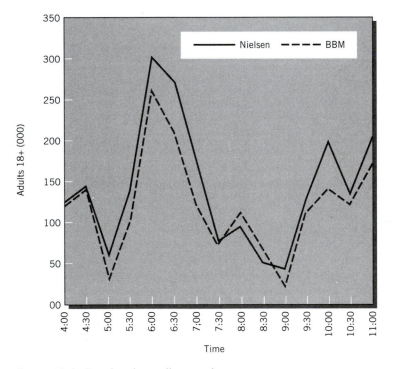

Figure 10.6   Local station audience estimates

that annoy or about which respondents feel sensitive, and a clear wording of the questions. The questioning of persons is an imposition and invasion of privacy, so it should not surprise that some persons do not respond, or that the nonresponse rate for some surveys is large, despite perhaps repeated attempts to contact nonrespondents. One issue that users of the sample information need to

address is whether or not those that did not respond can be regarded as having essentially the same characteristics as those that did. If the conclusion is negative, then users must consider what, if anything, useful was learned from the available responses.

- Common sense, a good grasp of the language, and a thorough testing of the questions on a pilot sample of respondents will go a long way toward eliminating avoidable sources of confusion and misinterpretation of the questionnaire.
- Randomized response methods attempt to encourage truthful responses and reduce measurement error by dissociating the question from the response, while still permitting unbiased estimates to be made and the variability of these estimators to be controlled.
- Radio and television audiences and their characteristics are currently estimated mainly using people meters and diaries. Other approaches used in the past, and for special purposes at present, employ set meters, personal interviews, and telephone surveys. This application of sampling provides an interesting illustration of the compromise that must often be made among the conflicting objectives of low cost, satisfactory coverage of the target population, high response rate, and accurate measurement.

PROBLEMS

10.1 United States Federal law[6] forbids discrimination against people 40 years of age and older with respect to employment decisions. Under the doctrine of adverse impact, it is not necessary to prove that the employer had discriminatory intent, but only to show that his or her actions had the actual effect of disadvantaging a disproportionate number of people of protected age. To overcome such a showing, an employer can demonstrate a nondiscriminatory business reason for the decisions.

The case under consideration here involves a company that was adversely affected by the downturn in the basic metals markets in the early 1980s. In that period the company reduced the size of its work force in a series of moves that were the occasion for the lawsuit. In a lawsuit brought against the company, it was alleged that the company had disproportionally fired people 40 and over. The case was a civil suit; such cases are decided on "the preponderance of the evidence."

(a) In civil litigation, each side is entitled to discovery, that is, to whatever records and analyses the other side intends to use in court and, additionally, all other "relevant" records. It was therefore possible for the plaintiff to assemble a list of all management employees during the period in question, their birth dates, and

---

[6]This problem draws from Joseph B. Kadane, "A Statistical Analysis of Adverse Impact of Employer Decisions," *Journal of the American Statistical Association* 85, no. 412 (December 1990), pp. 925–933.

the dates they left the company, if they had left involuntarily. The list revealed that employees had been fired in four "waves" on four specific dates: June 1982, November 1982, May 1983, and June 1984. The following table shows the ages of those fired and retained in these waves.

Age	Fired	Retained
June 1982 firings		
40+	18	129
39−	0	102
November 1982 firings		
40+	26	105
39−	10	83
May 1983 firings		
40+	13	92
39−	14	66
June 1984 firings		
40+	13	81
39−	2	52

Was the company discriminating?

(b) An expert witness for the plaintiff presented the results of chi-square tests applied to the above data. The null hypothesis is that age and firing decision are independent. (A description of the chi-square test for independence can be found in any introductory statistics text.) The test statistic, V, the critical value at the 5% level of significance, $V_{0.05}$, and the level of significance of the test statistic (that is, the probability of exceeding the test statistic if the null hypothesis is true) for each wave of firings and over all waves are given below.

Wave	V	$V_{0.05}$	Significance Level
June 1982	13.463	3.841	0.000
November 1982	3.335	3.841	0.068
May 1983	0.955	3.841	0.329
June 1984	3.861	3.841	0.049
All waves	8.511	3.841	0.004

On the basis of these results, the expert witness concluded that the firing decisions of the company were discriminatory.

You have been asked by the company for assistance in responding to this expert witness. Was the company indeed discriminating?

10.2 *"A little more than 50 per cent of all business software in use today is pirated"* (from an advertisement by the Business Software Alliance, *PC Magazine*, July 1994).

Design a sample to estimate the percentage of pirated software. Your report should describe in detail the population and its elements, the manner in which the elements should be selected and the estimates made, and the questionnaire to be used.

10.3 (a) The Post Office plans to estimate the distribution of the time between mailing and delivery of first class mail (that is, the percentages of letters delivered within 0, 1, 2, . . . days from the date of mailing) as well as the average of this time. These estimates will be made at frequent intervals so as to detect any improvement or deterioration in the speed of mail delivery.

Design the sample by which these estimates will be made. Your report should describe in detail the population and its elements, the manner in which the elements should be selected and the estimates made, and the time of mailing and delivery measured.

(b) Same as **(a)**, except that the study is commissioned by a nonprofit consumer organization.

10.4 Design a sample from which to estimate the composition and tonnage of a city's garbage in terms of such components as plastic, glass, paper, organic waste, etc.

Describe clearly the population and its elements, the variables and attributes of interest, the likely cost, advantages and disadvantages of at least two alternative sampling methods suitable for the problem, the manner in which the estimates will be calculated, and any problems of coverage, nonresponse, and measurement that you can anticipate.

10.5 A cable television company serving a metropolitan area of about two million people is considering the addition of a weather channel to its "extended package" of channels. The weather channel would be exclusively devoted to short- and long-term local and regional weather forecasts. The content of this channel will be produced by an independent company. The cable company will pay a monthly fee for this production and recover its costs by selling commercial time on this channel. Weather channels have been successful in some areas but not so in others.

Design a survey to estimate the potential size and characteristics of the audience of a weather channel. In doing so, you should pay appropriate attention to the definition of the population, to the sampling method, the sample size, the manner in which the selected elements will be questioned or interviewed, the wording of the questions, the likely response rate and the handling of nonrespondents, the estimation method, and any other aspects of the survey likely to influence its accuracy. Your report should be specific and readable, so that it could be implemented by people having less knowledge of sampling than yourself.

10.6 Consider the problem of estimating the number of fish in a lake. A commonly used approach is to first catch a number n_1 of fish, tag or mark them, and release them again into the lake. Some time later, after giving the tagged fish an opportunity

to mix well with the remaining fish in the lake, a second catch of n_2 fish is made, and the number of tagged and untagged fish observed:

$$
\begin{array}{cl}
m & \textbf{marked or tagged} \\
n_2 - m & \textbf{not marked or tagged} \\
n_2 & \textbf{total, second catch}
\end{array}
$$

Let N denote the number of fish in the lake. Assuming that the proportion of marked fish in the population is the same as in the second sample, we have

$$\frac{n_1}{N} = \frac{m}{n_2}.$$

Solving for N, we find

$$N = \frac{n_1 n_2}{m}.$$

For example, if $n_1 = 800$, $n_2 = 1{,}000$, and $m = 5$, then

$$N = \frac{(800)(1{,}000)}{5} = 160{,}000.$$

This is the simplest version of the *capture-recapture* model, frequently used to estimate wildlife populations.

(a) What precautions would you take when using this approach to measure the number of fish in a lake? What additional assumptions must be satisfied?

(b) How would you apply this approach to estimate the number of criminals from arrest records?

10.7 The National Association of Composers, Authors, and Music Publishers (NASCAMP) collects royalties from performances of music works, and distributes these royalties to publishers owning the rights to the works. Fees are paid by radio stations, television stations, concert halls, bars, taverns, and other organizations playing music. These organizations pay NASCAMP a flat quarterly royalty fee for a "blanket license" to play any music they wish to play. NASCAMP distributes the fees it collects to music publishers on the basis of the number of times the publisher's music was played during the quarter.

NASCAMP estimates the number of times a piece of music was played by sampling radio stations only; it assumes that the play frequency of a piece of music at concert halls, bars, taverns, and other media is proportional to that of radio. A number of radio stations are selected from all the stations in the country and these stations are monitored over selected time periods. In the past, this was done by inspecting the logs of the selected stations. Because the accuracy of the logs was questioned, NASCAMP began recently to record the stations off the air and to have its staff identify the music being played. The current monitoring practice, however, is also likely to be replaced: a company has recently announced successful tests of

a computer algorithm that is said to identify accurately a recorded piece of music from the power spectrum of any 48-millisecond-long sample of the music. This technology should make it possible to monitor the selected stations over longer periods and to identify the music items played accurately and without human intervention.

(a) Describe precisely the population of this case, the variables and attributes of interest, and the manner in which these determine the amount due to a publisher. Specify the population size, preferably after finding out the number of radio stations in the country. There are about 50,000 pieces of music the rights of which have not expired.

(b) Does the present system reward publishers fairly? Can you suggest a better alternative?

(c) Exactly how should the sample be taken? Describe clearly and in detail the sample elements, the manner in which they should be selected, and estimates of the population characteristics of interest made.

10.8 *The Monitor* is the country's leading music chart magazine. It is not sold at newstands but distributed to subscribers in the music recording and broadcasting industries. The magazine publishes weekly "charts" for each type of music, ranking records according to their national sales.

The Monitor collects data from a sample of 150 retail record stores, selected to "cover" the country and said to be "weighted" according to their size. Every week, the selected stores report over the phone the best-selling records, but not their actual dollar or unit sales. This is so because at present most record stores are equipped with simple cash registers and cannot track down sales of individual records. Consequently, *The Monitor* must "interpret" the raw data prior to ranking the best-selling records. The number of selected stores being so small, it is rumored that record companies occasionally make large purchases in order to influence the charts.

The magazine is actively considering having computerized inventory systems installed at selected stores at its own expense. This will certainly increase the cost of the information collected, but much of the additional cost is likely to be passed on to the magazine's clients as the price of improved accuracy.

You have been invited to act as a consultant to *The Monitor* and advise its management on the manner in which the new charts should be compiled. Write a clear report explaining in as much detail as possible how the necessary information should be collected and processed.

10.9 The Department of Transportation has just secured approval for the introduction of "photo radar." Unmarked vans, equipped with radar and camera, will be parked at varying locations along major highways. Operators in these vans will photograph speeding vehicles using telephoto lenses. The photographs will

be automatically imprinted with the date, time, and vehicle speed. When these photographs are later processed, it is hoped that a magnification of the portion containing the license plate will lead to the identification of the speeding vehicle. The owner of this vehicle will then receive the speeding citation in the mail. It is up to the owner to prove that another driver may have been the actual offender.

Photo radar has met strong opposition in the press, in part because of the difficulty in identifying the actual driver of the vehicle and in part because the press has chosen to emphasize the large revenue expected to materialize as a result of photo radar fines. The Department, on the other hand, has maintained that the only goal of photo radar is the reduction of highway speeds, which it believes have reached dangerous levels.

As a party neutral to the dispute, design a sampling plan by which to estimate the average speed of vehicles on major highways before and after the introduction of photo radar.

PART TWO

CASES

Canabag Manufacturing Company — Part I

Canabag is a manufacturing company that produces plastic bags, such as shopping and garbage bags. It has three divisions, located in Chicago, Boston and Atlanta, with the one in Boston being the major division and headquarters. The three divisions were originally three independent plastic bag manufacturers. Boston Canabag took over the Chicago company seven years ago, and the Atlanta company five years ago. Each division is responsible for manufacturing and marketing in its own region. Divisions maintain their own accounting information and send quarterly accounting reports to headquarters. Some operations of Canabag are centralized. These include purchasing of raw materials, advertising, discount policy, product research and development, and auditing of accounts receivable.

The firm maintains four product lines: (A) high-density shopping bags, (B) low-density shopping bags, (C) garbage bags, and (D) clothing bags. The two types of shopping bags are the major products of the firm. Different types of bags are usually sold to different kinds of customers. For example, the major customers for high density bags are department stores; for low-density bags, supermarkets; for clothing bags, clothing manufacturers; and for garbage bags, public institutions such as schools, hospitals, etc. The products are distributed either through commission-paid representatives or directly to the customers.

Canabag has experienced rapid growth over the past several years due to the rising cost of paper and paper products in general. Its regular customers are concentrated in the Boston, Chicago, and Atlanta regions. These customers average about 12 orders per year and the average order size is about $450. The term of payment is 30 days net. Sales amounted to about $5.5 million last year. All accounting operations are now computerized.

Shortage of current capital is a serious problem for many small and medium-sized manufacturing firms such as Canabag. Collection of accounts is a time-consuming process and suppliers usually give only short-term credit. Canabag accounts receivable comprises nearly one-third of total assets, and its control has become increasingly important.

In view of the importance of accounts receivable to Canabag's operations, the president ordered a thorough independent audit at the end of the first year of computerized operations.

At the president's request, the auditor spelled out the audit objectives as follows:

1. to determine the proportion of accounts receivable under control
2. to determine if the error rate on accounts receivable after computerization has changed significantly from that of previous years
3. to determine if accounts receivable are materially overstated or understated.

To determine if an account receivable is under control, the auditor will examine all transactions for the account and establish whether or not control procedures were properly followed for all these transactions. As a result, each audited account will be classified as either "in control" (that is, all control procedures were properly followed), or "not in control" (at least one control procedure was not complied with).

The control system for sales orders and accounts receivable is set up as follows. The sales orders department prepares a sales order upon receipt of a customer's purchase order. Copies of the sales order are sent to the credit department for a credit check. If credit is approved, the sales order is sent to the storeroom and the shipping department. A signed packing slip and a copy of the sales order are then sent to the billing department, where invoices are prepared, batched, and posted to the appropriate accounting records. The accounts receivable department receives a copy of the approved invoice for each sales transaction, debits the customer's account with each invoice issued, and credits the account with each payment received.

For some established customers, such as local hospitals and schools, sales orders are initiated by Canabag at fixed times unless otherwise instructed, and credit checking is usually ignored. On some occasions, however, deliveries have been returned because the customers did not need them.

The two types of shopping bags are usually distributed through commission-paid representatives. Due to high competition in the sales of these bags, representatives tend to provide biased information in favor of the customers' credit ratings in order to maximize

their sales volume. Collection of accounts receivable from these customers is traditionally more difficult than from others.

Each selected account will also be checked for the numerical accuracy of the recorded transactions, and classified as "in error" if the stated balance is incorrect, or as "correct" if otherwise. An account may be "in error" for a variety of reasons: a clerical error in calculating the sum of invoiced items, a failure to credit a payment from the customer, a credit to the wrong account, a discrepancy between goods actually shipped out and goods invoiced, and so on. In the past few years, about 25% of Canabag's accounts receivable were in error. The president was concerned about this high error rate and especially interested in finding out whether or not the computerized accounting system succeeded in reducing this rate.

The auditor must also estimate the total dollar error in accounts receivable, and indicate whether accounts receivable are "materially" overstated or understated. Based on the size of the accounts receivable balance in relation to total assets, expected income, and outstanding debit, the auditor feels that $20,000 would be an appropriate "material" amount. In other words, if the total dollar error amount exceeds $20,000, the difference will be judged substantial enough to warrant further investigation and action.

Since it would be impossible to examine all accounts receivable with the limited resources and staff available, the auditor plans to rely on a random sample of accounts in order to carry out the audit objectives. Samples were used in previous years as well, but they were judgment, not random, samples. Guided by professional experience, the auditor would subjectively select accounts that appeared suspicious, audit each for proper authorization and accuracy, and make adjustments whenever an error was found. The auditor would continue selecting accounts for audit until it was determined that the accounts checked were satisfactory or available time had run out.

The auditor estimates that no more than 100 accounts can be properly audited with the resources available.

A computer file is available at headquarters containing an alphabetical listing of all accounts receivable, and, for each account, the current book value, product line, and originating division. The information shown in Table 10.1 is extracted from this file.

QUESTIONS

1. What are the possible advantages and disadvantages of random and judgment samples? Which type of sample should the auditor choose?
2. Suppose that a simple random sample without replacement of ten accounts receivable was audited with the following results:

ACCT NO	LINE	DIVISION	BOOK VALUE	ERROR?	ERROR VALUE	CONTROL?
42	A	BOS	64.37	NO	0.00	YES
69	B	BOS	316.10	NO	0.00	NO
143	B	CHI	17.16	NO	0.00	YES
183	B	CHI	40.40	NO	0.00	YES
209	C	ATL	496.09	NO	0.00	YES
468	B	CHI	1865.19	NO	0.00	YES
598	A	BOS	2.17	NO	0.00	YES
867	B	CHI	309.14	YES	54.24	YES
917	A	CHI	60.83	NO	0.00	YES
970	B	BOS	484.19	NO	0.00	YES

The ERROR VALUE shown is the difference between the true account balance (as determined by the audit) and the book balance; a positive error means the book balance understates the true one, and vice versa.

In light of the sample results, estimate the proportion of accounts receivable not under control, the proportion in error, and the total dollar error. Has the error rate changed? Are the accounts receivable materially over- or understated?

3. Same as (2), except that a simple random sample of 100 accounts was selected and audited. The results are listed in the file CANABAG.DAT.

4. Is the sample size used adequate? If not, how many additional accounts should be selected?

5. Can the sample design be improved? In view of the additional information provided by the sample, should the next sample be stratified, two-stage, or another type of random sample?

TABLE 1	Supplementary Information				
Product Line	**Division**				
	Atlanta	**Boston**	**Chicago**	**Total**	
A	90,379	122,391	40,505	253,275	Total book value ($)
	105	201	97	403	Number of accounts
B	31,615	58,407	32,240	122,262	Total book value ($)
	85	125	88	298	Number of accounts
C	31,364	41,279	26,222	98,865	Total book value ($)
	61	92	55	208	Number of accounts
D	5,329	15,028	8,004	28,361	Total book value ($)
	14	41	24	79	Number of accounts
Total	158,687	237,105	106,971	502,763	Total book value ($)
	265	459	264	988	Number of accounts

6. A method of sample selection that is becoming increasingly popular in auditing is "dollar-unit sampling (DUS)," according to which an account is selected with probability proportional to its book value—not with equal probability as is the case with simple random sampling. Would you recommend DUS to Canabag?

7. Discuss any other relevant aspects of the case not specifically mentioned above.

Canabag Manufacturing Company — Part II

The sample estimates obtained as described in Question (3) of Part I alarmed the president, who, on further reflection, became convinced that *any* errors were unacceptable, as they indicated that the accounting department was not performing its job satisfactorily.

The accounting department responded by questioning the wisdom of relying on a sample for such an indictment of its performance and the accuracy of the sample itself.

After some discussion, the president decided to have a complete audit performed, that is, to have the remaining 888 accounts examined. The combined results of the audit of all 988 accounts receivable can be found in the file CANABAGP.DAT. A partial listing of this file is shown below.

ACCT	LINE	DIV	BOOK	ERROR	EFLAG	CFLAG
1	4	1	492.07	0.00	0	0
2	1	3	103.63	0.00	0	0
3	3	1	410.27	186.35	1	0
4	2	1	135.56	0.00	0	0
5	1	2	345.04	0.00	0	0
6	1	3	43.46	0.00	0	0
7	1	2	34.81	0.00	0	0
8	1	1	538.90	0.00	0	0
9	3	1	1118.51	0.00	0	0
10	3	2	240.76	0.00	0	0
⋮	⋮	⋮	⋮	⋮	⋮	⋮

The interpretation of the column headings is as follows:

ACCT	Account number
LINE	Product line (1=A, 2=B, 3=C)
DIV	Division (1=ATL, 2=BOS, 3=CHI)
BOOK	Book value of account ($)
ERROR	Error value (difference between audit and book value)
EFLAG	=1, if account has an error; =0, otherwise
CFLAG	=1, if account is not under control; =0, otherwise.

As a result of the complete audit, the population characteristics became known after one sample was taken. This is a fortuitous opportunity for exploring a number of additional issues requiring a known population.

QUESTIONS

1. Using any statistical or other program capable of performing statistical calculations, determine the population mean, total, variance, and standard deviation of book and error values. Likewise, calculate the population proportion and number of accounts in error and not under control.
2. Compare these population characteristics with your estimates in Question (3) of Part I. Are you surprised that differences exist?
3. The computer program CANABAG.EXE can simulate the selection of any number of simple random samples without replacement of given size from the population of accounts. The output of the program appears on the computer screen and is also stored in the file CANABAG.OUT. A partial listing of this file, produced after it was requested that 30 samples of size 100 be selected, appears below:

```
Sample size= 100, Number of samples= 30
  1      581.9087     1248.8658      35.9118     274.1400     0.140     0.020
  2      586.0149     1184.2287      16.6297      95.9974     0.240     0.020
  3      635.4372     1409.2759      26.8637     147.9628     0.210     0.000
  4      341.7148      524.4985      17.5358      89.0130     0.160     0.040
  5      419.5784      700.7640       0.6567      25.2954     0.120     0.010
  ⋮          ⋮            ⋮             ⋮           ⋮           ⋮         ⋮
```

The row numbers have the following interpretation:

Col. 1 Sample number

Col. 2 Sample mean book value

Col. 3 Sample standard deviation of book values

Col. 4 Sample mean error value

Col. 5 Sample standard deviation of error values

Col. 6 Sample proportion of accounts with an error

Col. 7 Sample proportion of accounts not under control.

(a) For each simulated sample listed above, calculate point estimates and 90% confidence intervals for the population mean book value, mean error value, proportion of accounts in error, and proportion of accounts not under control. Determine if each interval estimate "brackets" the corresponding population characteristic.

(b) Apply the program CANABAG.EXE to simulate the selection of 100 samples of size 50. With the help of a statistical or other program, read the file CANABAG.OUT and calculate the estimates and confidence intervals in (a) for each sample in the file. According to sampling theory, the average value of an unbiased estimator in a large number of samples should equal the target population characteristic, and the proportion of confidence intervals bracketing the target population characteristic should be equal to the confidence level (in this case, 90%). Are these expectations fulfilled in this experiment? If not, why?

(c) Following your instructor's guidelines, simulate the selection of a large number of samples of given size(s). For each simulated sample, calculate point estimates and confidence intervals of varying levels for the target population characteristics. Investigate the issues in (b) in order to see if there is better agreement between theory and experimental results as the sample size and the number of repetitions increase. Does this agreement depend on the values of the population characteristics themselves? If so, in what manner?

Valuation of Government Properties — Part I

The Department of Government Services is responsible for the acquisition, maintenance, and sale of real estate properties owned by the Federal Government. This case deals with a subset of 2,195 federally owned properties located throughout the country.

The precise definition of the subset of properties need not concern us here. These properties were acquired at various times over the last 100 years. Some are quite large and tend to be located in major metropolitan areas, but most are fairly small.

For reasons having to do with public accounting practices, there is no unified record of the cost of acquisition or construction, nor of that of any alterations or improvements made to the properties in the past.

A recent change in the law has made it necessary for the Department to arrive at an estimate of the total value of the 2,195 properties.

Such an estimate, of course, *could* be obtained by having each property individually appraised by real estate appraisers and adding up the individual appraisals, but the cost of this approach is so high as to make it impractical.

After some deliberation, the Department has chosen on a compromise approach in which 42 of the largest properties will be appraised individually by a real estate firm commissioned for this purpose. Their aggregate value will obviously be the sum of these appraisals. The value of the remaining 2,153 properties, however,

will be estimated on the basis of a sample. In other words, a number of these properties will be randomly selected, their values individually appraised by real estate consultants, and an estimate of the value of all 2,153 properties will be made on the basis of the appraised values of the sampled properties.

Leaving aside the appraisal of the 42 large properties which will be carried out in a straightforward way, you are asked to consider carefully the following questions related to the design and implementation of the sample.

(a) How large should the sample be?
(b) How should the sample be selected so that the estimate of the total value is as accurate as possible?
(c) Once a sample of a given size is taken according to the recommended method and the sampled properties appraised, precisely how is the aggregate estimate to be formed?

The Department has funds for appraising up to 200 properties, but would very much prefer to appraise 150 or even 100 properties, if the smaller samples are accurate enough. The *only* information available to you is in the file GOVPRO1.DAT, which contains the floor area of the building (SIZE, in m^2), the year of acquisition or construction of the building (YEAR, last two digits), and the location (LOCATION) of the property. A partial listing of this file is shown in the following table.

SIZE	YEAR	LOCATION
238	58	BEDFORD
720	53	HALIFAX
1397	14	DARTMOUTH
97	65	EASTERN
304	48	HALIFAX
93	67	WAVERLY
112	77	WINDSOR
109	78	PORTERS
2041	59	NORTH
73	66	BICKERTON
41	64	ASPEN
445	56	BADDECK
112	61	BRAS
50	65	BOYLESTON
50	66	CLEVELAND
50	66	CHRISTMAS
1379	61	GLACE
49	65	D'ESCOUSE
226	50	DOMINION
50	65	DINGWALL

Write a brief but clear report to the Department addressing the above questions and making specific recommendations concerning the size of the sample, the method of selection, and the manner in which the aggregate estimates will be calculated.

Valuation of Government Properties — Part II

A consultant retained by the Department of Government Services to make specific recommendations regarding the size and method of selection of the sample of properties to be appraised, submitted the report reprinted below.

(a) Comment critically on the consultant's recommendations.

(b) Verify the calculations in the consultant's report.

(c) Assuming the consultant's recommendations are adopted and a stratified sample of size 100 is approved, select the properties to be appraised using the data file GOVPRO1.DAT, described in Part I of this case.

CONSULTANT'S RECOMMENDATIONS FOR SAMPLING OF GOVERNMENT PROPERTIES

The purpose of the study is to obtain an estimate of the total value of properties administered by the Department of Government Services.

There are 2,195 such properties across the country, 42 of which will be appraised individually. From the remaining 2,153 properties, a random sample will be

selected, the sampled properties will be appraised, and an estimate of the total value of all 2,153 properties will be made on the basis of the sample.

The problem is how to select the sample so that the resulting estimate will be as accurate as possible, and, once the sample is taken, how to estimate the total appraised value.

SUMMARY OF RECOMMENDATIONS

(1) The appraised value of a property depends mainly on its size, age, and location. Practical considerations do not allow taking location into account.

A stratified sample is recommended, with categories or strata formed according to property size and age of the building, as listed in cols. (1) to (3) of Table 1. The codes are explained in Table 2.

A random sample will be selected from each category. The proportions of the total sample allocated to each category are listed in col. (4) of Table 1. The numbers of properties to be sampled from each category are listed in col. (5) of Table 1 for a sample of size 100.

(2) The selected properties will be appraised by real estate consultants. The estimate of the total value of all 2,153 properties (T) should be calculated according to the following formula:

$$T = N_1 \bar{X}_1 + N_2 \bar{X}_2 + \cdots + N_{30} \bar{X}_{30} = \sum_{i=1}^{30} N_i \bar{X}_i, \qquad (1)$$

where N_i represents the number of properties, and $\bar{X}_i$ the mean appraised value in the ith category.

(3) If desired, a confidence interval can be constructed in addition to the point estimate described above. For example, the probability is approximately 95% that the true value of all 2,153 properties lies between $(T - 1.96 N S_{\bar{X}})$ and $(T + 1.96 N S_{\bar{X}})$. In these expressions, $N = 2,153$, and $S_{\bar{X}}$ is calculated by the following formula:

$$S_{\bar{X}}^2 = \sum_{i=1}^{30} \left(\frac{N_i}{N}\right)^2 \frac{S_i^2}{n_i} \frac{N_i - n_i}{N_i - 1}, \qquad (2)$$

where S_i^2 is the variance of the appraised values of the n_i sampled properties in the ith category, and $N = 2,153$.

ILLUSTRATION

To illustrate as simply as possible these calculations, suppose there are only $N = 50$ properties and only 3 (rather than 30) categories formed according to size and age.

(a) Suppose that the sampling plan calls for appraising properties with identification numbers 5, 15, and 27 in category 1. The appraised value of these properties (in $000) is shown in the second column of the following table.

TABLE 1 Optimal Allocation

Category (1)	Code Size (2)	Code Age (3)	Sample Allocation (%) (4)	Sample Size (5)
1	1	1	1.2309	1
2	1	2	9.6263	9
3	1	3	0.4967	1
4	1	4	0.0627	1
5	1	5	0.0128	1
6	2	1	1.1679	1
7	2	2	0.8067	1
8	2	3	0.2070	1
9	2	4	0.0666	1
10	2	5	0.0248	1
11	3	1	0.2059	1
12	3	2	0.2153	1
13	3	3	0.1073	1
14	3	4	0.0626	1
15	3	5	0.0000	1
16	4	1	4.1876	4
17	4	2	3.7468	3
18	4	3	1.7424	2
19	4	4	1.3304	1
20	4	5	0.6277	1
21	5	1	3.1098	3
22	5	2	3.5074	3
23	5	3	3.6231	3
24	5	4	1.3248	1
25	5	5	1.2985	1
26	6	1	10.0276	9
27	6	2	16.8117	15
28	6	3	21.3319	19
29	6	4	5.6664	5
30	6	5	7.3706	7
Totals			100.0000	100

Property No.	Appraised Value, X	$X - \bar{X}$	$(X - \bar{X})^2$
5	57.5	$57.5 - 70.9 = -13.4$	179.56
15	30.1	$30.1 - 70.9 = -40.8$	1664.64
27	125.0	$125.0 - 70.9 = +54.1$	2926.81
Totals	212.6		4771.01

TABLE 2	Explanation of Codes		
Code	Size	Code	Age
1	100 m² or less	1	Built in 1970 or later
2	101 to 150 m²	2	Built between 1960 and 1969
3	151 to 200 m²	3	Built between 1950 and 1959
4	201 to 500 m²	4	Built between 1930 and 1949
5	501 to 1000 m²	5	Built before 1930
6	More than 1000 m²		

Therefore, the mean and variance of the appraised values in category 1 are:

$$\bar{X}_1 = 212.6/3 = 70.9,$$
$$S_1^2 = 4771.01/3 = 1590.34.$$

(b) Suppose further that similar calculations are made for categories 2 and 3, and the results summarized as in the following table.

	Total	Sample		
Category	Number of Properties, N_i	Number of Properties, n_i	Mean, $\bar{X}_i$	Variance, S_i^2
1	10	3	70.9	1590.34
2	20	5	120.3	675.21
3	20	2	56.4	712.11
	50	10		

The estimate of the total value of the 50 government properties is calculated according to Eq. 1:

$$T = (10)(70.9) + (20)(120.3) + (20)(56.4) = 4,243,$$

that is, the estimated total value is \$4,243,000.

If a confidence interval is desired, calculate first

$$S_{\bar{X}}^2 = (\frac{10}{50})^2(\frac{1590.34}{3})(\frac{10-3}{10-1}) + (\frac{20}{50})^2(\frac{675.21}{5})(\frac{20-5}{20-1})$$
$$+ (\frac{20}{50})^2(\frac{712.11}{2})(\frac{20-2}{20-1}) = 87.521,$$

and

$$S_{\bar{X}} = \sqrt{87.521} = 9.35.$$

Therefore, the desired 95% confidence interval is from

$$T - 1.96NS_{\bar{X}} = (4, 243) - (1.96)(50)(9.35)$$

to

$$T + 1.96NS_{\bar{X}} = (4, 243) + (1.96)(50)(9.35),$$

that is, between \$3,326,700 and \$5,159,300. (The numbers, of course, are fictitious. The range in this illustration is larger than can be expected.)

COMMENTS

The recommended stratified sample requires that a number of properties be selected at random from each category. By contrast, in a simple random sample, the selection is made from the entire population of properties without distinguishing strata. It can be shown that the accuracy of a stratified sample is likely to be greater than that of a simple random sample of the same total size, provided that the allocation of the total sample among categories is made according to the following formula:

$$n_i = n \frac{N_i \sigma_i}{\sum N_i \sigma_i} \tag{3}$$

In this expression, n is the total sample size, N_i is the number of properties, and σ_i the standard deviation of property values in the ith category.

Optimal allocation requires knowledge of the standard deviation (σ_i) of appraised values in each category. This information, of course, is not available. Had the government properties been appraised earlier, these appraisals could be used as a guide in forming the categories and in the calculation of the required statistics. The appraisal, however, is done for the first time.

The next best alternative, therefore, is to find another available variable, as closely related to appraised value as possible, and to use this variable in determining an approximate optimal allocation. One such variable is the size of the property. Obviously, it is not a perfect predictor of property value, because it does not take into account land and the varying values of land and size across the country. But it should be closely enough related to value to justify the expectation that a sample allocation based on size will not be far from the unknown optimal one based on value.

To determine this allocation, formula (3) is used, but with σ_i now representing the standard deviation of the *size* of all properties in the ith category. These standard deviations are listed in col. (4) of Table 3. The percentage allocation calculated in col. (5) and shown in the col. (6) of Table 3 is based on formula (3).

ACCURACY

Other things being equal, the larger the sample, the more accurate the estimates are likely to be. Of course, the larger the sample, the greater the cost. The sample

TABLE 3   Determination of Optimal Allocation

Category, i (1)	Number of Properties, N_i (2)	Average Size (m²) (3)	Standard Deviation of Size, σ_i (4)	$N_i\sigma_i$ (5)	Allocation (%) (6)
1	153	73.85	16.05	2455.6	1.2309
2	940	71.09	20.43	19204.2	9.6263
3	54	78.85	18.35	990.9	0.4967
4	5	57.60	25.01	125.1	0.0627
5	2	58.00	12.73	25.5	0.0128
6	154	114.78	15.13	2330.0	1.1679
7	121	117.00	13.30	1609.3	0.8067
8	29	127.38	14.24	413.0	0.2070
9	10	122.70	13.29	132.9	0.0666
10	3	132.00	16.52	49.6	0.0248
11	28	166.18	14.67	410.8	0.2059
12	28	173.79	15.34	429.5	0.2153
13	15	177.13	14.27	214.1	0.1073
14	7	169.29	17.84	124.9	0.0626
15	1	192.00	0.00	0.0	0.0000
16	86	307.57	97.14	8354.2	4.1876
17	91	291.64	82.14	7474.7	3.7468
18	41	329.93	84.78	3476.0	1.7424
19	33	289.79	80.43	2654.2	1.3304
20	16	367.81	78.26	1252.2	0.6277
21	49	677.94	126.61	6203.9	3.1098
22	52	710.77	134.56	6997.2	3.5074
23	46	740.48	157.13	7228.0	3.6231
24	18	702.67	146.83	2642.9	1.3248
25	18	703.55	143.92	2590.6	1.2985
26	33	1528.55	606.21	20004.9	10.0276
27	42	1808.81	798.55	33539.1	16.8117
28	48	2086.23	886.60	42556.7	21.3319
29	16	1825.69	706.52	11304.3	5.6664
30	14	2151.36	1050.30	14704.2	7.3706
Total	2153			199498.0	100.0000

size chosen is a compromise between cost and accuracy. To provide some idea of the likely accuracy of samples in the range from 100 to 200 properties, an index of approximate accuracy is calculated in Table 4 for samples of size 100, 150, and 200, and two sampling methods: simple and (the recommended) stratified sampling.

It can be easily seen that a stratified sample of size 100—allocated among groups according to col. (6) of Table 3—is more than 5.5 times as accurate as a simple

TABLE 4 Approximate Sample Accuracy, First Approach

Sampling Method	Sample Size		
	$n = 100$	$n = 150$	$n = 200$
Simple	100	124	145
Optimal stratified	555	753	987

random sample of the same size. Stratified samples of size 150 and 200 are about 36 and 78% more accurate than a stratified sample of size 100.

The accuracy index is based on the standard deviation of the estimate of total size. The stronger the relationship between size and appraised value, the more closely the above index approximates the unknown true index.

Another measure of the likely accuracy of a sampling plan is the probability that the estimate will be within $\pm c$ of the true total value of all properties, where c is a specified dollar amount. Unfortunately, to calculate this probability it is necessary that the variance of property values be known. As an approximation we may use instead the probability that the sample estimate of total *size* will be within $\pm c$ (m^2) of the known total. (The total size of all 2,153 properties is 630,890 m^2.) Table 5 shows the calculations.

For example, the probability is 95% that the estimate of the total size of all properties based on an optimal stratified sample of size 100 will be within $\pm$ 33,927 m^2. The other entries in Table 5 can be similarly interpreted.

NOTE ON CALCULATIONS

Because the recommended allocation (Table 1) requires that samples of only one property be selected from a number of categories, the categories must be grouped if estimates of group variances (S_i^2) and of the overall variance ($S_{\bar{X}}^2$) are to be made.

TABLE 5 Approximate Sample Accuracy, Optimal Stratified Sample

Probability	Sample Size		
	$n = 100$	$n = 150$	$n = 200$
95%	33,927	25,024	19,073
90%	28,562	21,065	16,057
80%	22,156	16,341	12,457

In reducing the 30 categories of Table 1, a number of choices are possible. It is recommended that the original five age classes be combined to form two new age classes, as follows:

A: built in 1960 or later (former classes 1 and 2)
B: build before 1960 (former 3, 4, and 5).

As a result, for the purposes of estimating S_i^2 and $S_{\bar{X}}^2$, and forming a confidence interval, the sampled properties should be classified into 12 categories as shown in Table 6.

The above choice ensures that: (a) in each new category, there will be at least two sampled properties; and (b) the stratification by size (the principal factor affecting value) is preserved.

TABLE 6   Revised Allocation

	Age		
Size	A	B	Total
1	10	3	13
2	2	3	5
3	2	3	5
4	7	4	11
5	6	5	11
6	24	31	55
Total	51	49	100

Valuation of Government Properties — Part III

One hundred government properties were randomly selected according to the sampling plan described in Part II, and appraised by real estate consultants.

The following table shows the building floor area (SIZE, in m²), year of acquisition or construction (YEAR, last two digits), building replacement value (REPL, in $000), depreciated building replacement value (DEPR, in $000), land value (LAND, $000), and total value (VALUE = DEPR + LAND, $000) of the first 10 sampled and appraised properties. The complete appraisal results are contained in the file GOVPRO2.DAT.

SIZE	YEAR	REPL	DEPR	LAND	VALUE
2478.9	52	2225.70	1391.06	293.00	1684.06
2405.3	7	2349.80	1174.90	56.00	1230.90
2041.2	59	2108.80	790.80	166.00	956.80
1904.7	61	2157.20	1186.46	44.70	1231.16
1899.0	60	2115.50	1586.63	55.00	1641.63
1295.2	74	1438.30	1150.64	305.00	1455.64
754.8	31	809.40	404.70	35.00	439.70
691.6	26	803.80	401.90	67.40	469.30
595.0	60	727.40	454.63	13.50	468.13
206.5	39	291.10	36.39	7.70	44.09

QUESTIONS

(a) Using these data, estimate the total value, total replacement value, total depreciated replacement value, and total land value of all 2,153 government properties.

(b) Again using these data, calculate 95% confidence intervals for the total value, total replacement value, total depreciated replacement value, and total land value of all 2,153 properties.

Valuation of Government Properties — Part IV

After the sampled 100 properties were appraised and the aggregate estimates described in Part III were made, the Department of Government Services approached the consultant with the following problem.

Occasionally, a government property is sold or transferred to the custody of another department. This does not happen very frequently (about once or twice a year on the average), but when such a transfer occurs, it is convenient to have an estimate of the total value of that property, as well as estimates of its replacement, depreciated replacement, and land value. These estimates will be used only for accounting purposes—to remove, so to speak, the property from the books.

If such estimates are made for all 2,153 properties, then the sum of these estimates must be consistent with the aggregate estimates arrived at in Part III. The latter may be taken as follows:

Total value	$543,663,000
Total replacement value	$731,550,000
Total depreciated replacement value	$457,274,000
Total land value	$86,390,000

As the consultant in this case, you are asked to write a report to the Department explaining how such estimates can be made and offering some evidence of their likely accuracy.

PackGoods Inc.

PackGoods Inc. is a market research firm specializing in estimating the retail sales of packaged food and drink products. Packaged products are sold in bottles, boxes, cans, and other containers, through supermarkets, grocery, convenience, drug and other stores. Figure 1 gives some idea of the voluminous information supplied both in printed form and electronically. It shows the national sales estimates for a few of the most popular brands of a leading manufacturer of ready-to-eat cereals, the estimated market shares of these brands by tonnage, number of units, and dollar sales, and the estimated average retail prices per brand unit. Shown are sales over the most recent 12- and 52-week periods, but similar information is available weekly as well.

These estimates are based on a sample of stores. Following is a description of the method used by PackGoods Inc. as it appears in a document sent to clients.

Our retail sales estimates are based on retail store level scanning data. The advantage of using store data over individual consumer data is that variables such as cultural and social biases, memory failure, etc. are eliminated as factors.

To produce high quality data for our clients, we require five basic elements:

1. A good universe estimate
2. A representative sample of stores
3. Excellent data collection work
4. Exact inside processing of the raw data
5. Quantitative and qualitative interpretation of the final data.

READY-TO-EAT CEREALS
KRAFT GENERAL FOODS
DECEMBER 12, 19X5
TONNAGE IN POUNDS

NATIONAL --- GROCERY SUPERMARKETS							
CATEGORY	BASIS		VOLUME	%CHNG	BASIS		VOLUME	%CHNG
TOTAL	12 WKS	TONNAGE	971	3	52 WKS	TONNAGE	4223	1
RTE		UNITS	867	2		UNITS	3815	−1
CEREALS		DOLLARS	2898	5		DOLLARS	12586	2

VOLUME IN 000,000

NATIONAL --- GROCERY SUPERMARKETS ITEMS RANKED BY 12 WKS TONNAGE SHARE								
					SHARES			AVGE PRICE /UNIT
ITEM/UNIT	PERIOD	RANK	VOLUME	%CHNG	TONN.	UNITS	DLRS	/UNIT
KELLOG'S CORN FLAKES	12 WKS	1	51885	−18	5.3	4.0	3.3	$2.73
675 GM	52 WKS	1	242080	−17	5.7	4.3	3.4	$2.64
KELLOG'S RICE KRISPIES	12 WKS	2	30278	−1	3.1	2.3	2.7	$3.96
700 GM	52 WKS	3	123181	8	2.9	2.1	2.5	$3.96
NABISCO SHREDDED WHEAT	12 WKS	3	25763	19	2.7	2.0	1.8	$3.06
675 GM	52 WKS	5	106449	−12	2.5	1.9	1.7	$3.06
KELLOG'S RAISIN BRAN	12 WKS	4	24666	6	2.5	1.6	2.2	$4.46
300 GM	52 WKS	2	129660	10	3.1	1.9	2.5	$4.25
KELLOG'S CORN FLAKES	12 WKS	5	22999	23	2.4	3.0	1.8	$1.99
400 GM	52 WKS	4	109910	16	2.6	3.3	2.0	$2.00

VOLUME IN 000

Figure 1 Sample report, PackGoods Inc.

UNIVERSE

The major portion of our universe data is supplied by local, state, and federal governments. Additional sources of universe data are field surveys and listings from chain organizations. Data obtained include number of stores, all commodity dollar sales, name/address of stores, number of checkouts.

SAMPLING

PackGoods uses a stratified, disproportionate, systematic random sample with ratio estimation.

1. *Stratified.*

The universe is split into smaller "cells" based on the following three criteria:

(A) *Geographical areas.* The country is split into:

(a) metropolitan areas with populations greater than one million

(b) states/provinces, excluding the metropolitan areas (a).

(B) *City size.* The following city sizes are distinguished:

"A" city: population over 120,000

"B" city: population 5,000 to 120,000

"C" city: population under 5,000.

(C) *Store types.* The following store type classifications are used:

1. Grocery supermarkets: food chains plus selected independent banners

2. Remaining grocery: all other food stores

3. Defined nongrocery: selected drug and mass merchandiser organizations.

All PackGoods' sales estimates are created on a cell by cell basis (e.g., Area 21, "A" city, Remaining grocery).

A sample of stores is selected from each cell. To ensure proper sample representation within a cell, the universe list is ordered based on geographical area (county/postal code).

The diagram in Figure 2 clearly illustrates the stratification concept in our sample selection.

2. *Disproportionate sample.*

A proportionate sample is a sample in direct proportion to the universe. The sample below is proportionate to the store count universe.

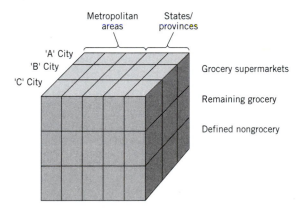

Figure 2   Stratification illustrated

	Number of Stores in Universe	10% Proportionate Sample
Grocery supermarkets	100	10 stores
Remaining grocery	200	20 stores
Defined nongrocery	400	40 stores
Total	700	70 stores

For a proportionate sample, all projection factors are the same.

However, a better sample can be achieved by an operation called optimum allocation of sample to strata. The optimum allocation will result in the best possible sample for the lowest possible cost.

	Number of Stores in Universe	Optimum Sample
Grocery supermarkets	100	31 stores
Remaining grocery	200	21 stores
Defined nongrocery	400	18 stores
Total	700	70 stores

3. *Systematic random.*

- Arrange all stores (universe) within a cell into a predetermined pattern (e.g., geography).
- Based on the sampling rate, break list into substrata. For example, if the sampling rate is 1 in 5 for a universe of 15 stores, break the list into three substrata of 5 stores each.
- Select a random number.
- If the random number is 2, then every second store in each substratum is sampled.

Store

Substratum #1

    #1
    #2   ⟸ Sample store
    #3
    #4
    #5

Substratum #2

    #6
    #7   ⟸ Sample store
    #8
    #9
    #10

Substratum #3

> #11
> #12   ⇐ Sample store
> #13
> #14
> #15

The PackGoods sample is maintained on a continuous basis through

(A) normal maintenance: replacement of canceled stores (e.g., store closed, no longer willing to cooperate)

(B) continuous sample improvement program, where sample changes are indicated to ensure that changes in the universe are reflected in the sample.

The sample is continuously improved and each new sample store is analyzed for atypical behavior prior to its inclusion.

4. *Ratio estimation.*

If a high correlation exists between all commodity dollar sales and product class dollar sales, the accuracy of projected data will improve if projection is based on all commodity dollars. This is called ratio estimation.

However, if the correlation between all commodity dollar sales and product class dollar sales is poor or nonexistent, the accuracy of projected data will decrease if a ratio estimation is used. Therefore, store count projection has to be used.

Projection factors are calculated on a cell by cell basis. This factor is applied to sales of all stores within that cell. The projection is applied to make the sales from the sample look like sales for the universe.

$$\text{Cell projection factor} = \frac{\text{Universe all commodity dollar sales in cell}}{\text{Sample all commodity dollar sales in cell}}.$$

Basic all commodity dollar sales information for the universe is obtained from government statistics.

All commodity sales information for the sample is obtained at the store level by data collection.

ERROR CONSIDERATIONS

Sources of errors are: standard error, bias, reporting, administrative, interpretation.

1. *Standard error.*

Disregarding the reporting and administrative errors, only a total universe measurement can be 100% accurate. It would be far too expensive, time consuming, and error prone a procedure to audit all stores in the universe. PackGoods data

are a sample estimate of the real world and, as a result, they have a measure of precision (standard error) associated with them.

The standard error gives a parameter within which the true value may differ from report data, e.g., ±4%. The standard error applies only to sample estimates, not to universe surveys. The two major factors in the calculation of the standard error are: (a) sample size, and (b) coefficient of variation, a measure of the variation among the sample stores.

The size of the universe is not important, unless it is small. Approximate standard errors (national):

Food	- brand totals	±4.2%
	- product class totals	±2.5%
Drug	- brand totals	±6.0%
	- product class totals	±4.2%

The standard errors for regional, city size, and store type data are higher. Some caution needs to be applied to the interpretation of regional data which are obviously based on smaller samples because:

- estimates from small samples are less precise
- brand totals are usually more precise than individual sizes
- small changes in volume or share are usually not significant for small samples, however, a series of small changes can result in a significant trend
- the existence of bias is more important for small samples
- other sources of error (e.g., reporting) are more important for small areas
- individual stores can exert much greater influence on the data for small areas.

2. *Bias.*
An inability to specify a representative sample due to noncooperation of some organizations may result in bias errors. This means, for example, that private labels of these organizations are not reflected in report data and, thus, affect shares of other data. Other organizations are used to represent the noncooperators, resulting in a possible distortion of the true picture.

3. *Reporting.*
Reporting errors are human errors that affect sample data as well as universe surveys. Possible field mistakes, such as incorrect code and wrong price readout, can be sources of these errors.

4. *Administrative.*
Administrative errors are also human errors that affect sample data as well as universe surveys. Errors in processing the data, such as wrong classification and wrong projection factors, are administrative errors.

5. *Interpretation.*
Errors in interpretation are usually the most critical and result from misinterpretation of occurrences in the marketplace, i.e., misinterpretation of sales estimates.

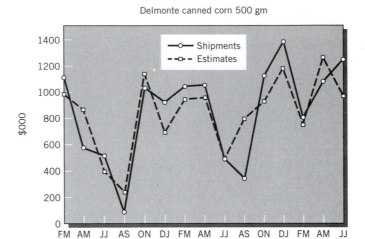

Figure 3   Coverage analysis illustrated

COVERAGE ANALYSIS

The quality of the estimates can be gauged by a coverage analysis, which is the mathematical relationship between client shipments and PackGoods sales estimates. As well, coverage analysis can be used to gauge developments in the market. Figure 3 compares PackGoods' monthly estimates and actual shipments for one of our client brands.

Coverage should be calculated on a bimonthly and 12-month moving basis.

Client data should consist of 2 to 3 years of monthly history on a converted basis, and should exclude shipments outside of the PackGoods universe.

Interpretation should consider:

- changes in warehouse stocks
- new items/dying items
- merchandise in the pipeline
- importance of nonmeasured outlets
- territorial differences.

QUESTIONS

1. *"Universe," of course, means "population." Given that the objective is to estimate sales of packaged food and drink products at the retail level, describe in detail the population and its elements.*
2. *In what sense is the "disproportionate sample" optimal?*

3. *Estimates are provided on a weekly basis. Do you think that a fresh sample of stores is selected every week? Does it matter?*

4. *"A sample of stores is selected from each cell. To ensure proper sample representation within a cell, the universe list is ordered based on geographical area (county/postal code)." What do you understand this statement to mean?*

5. *"For a proportionate sample, all projection factors are the same." What does this statement mean? Is it correct?*

6. *"If a high correlation exists between all commodity dollar sales and product class dollar sales, the accuracy of projected data will improve if projection is based on all commodity dollars. This is called ratio estimation. However, if the correlation between all commodity dollar sales and product class dollar sales is poor or nonexistent, the accuracy of projected data will decrease if a ratio estimation is used. Therefore, store count projection has to be used." What do these statements mean? Are they correct?*

7. *Using the terminology and notation of this book, explain as clearly as possible the ratio estimation method and "projection factors" used by PackGoods. Are the PackGoods' estimators unbiased?*

8. *What does PackGoods mean by "standard error"? Do you think the explanation provided is complete? If not, how would you revise the explanation?*

9. *Do you agree with PackGoods' definition of "bias"? If so, explain why. If not, how would you describe this "bias"?*

10. *How significant do you consider the other sources of error for this case?*

11. *Can the sampling method be improved? If so, describe how, and estimate the likely cost of your recommendations.*

12. *Comment on Figure 3, illustrating the "coverage analysis."*

13. *Comment on any other aspects of the case that were not specifically covered in the above questions.*

Pharmacom Research — Part I

Subscribers to Pharmacom Research's *Pharmaceutical Products Sales (PPS) Report* receive each month a 600-page publication listing the estimated sales of all pharmaceutical products distributed in the country.

Figure 1 illustrates the structure of the pharmaceutical industry and the flow of domestic sales from manufacturers to retailers.

There are over 100 manufacturers and over 3,200 pharmaceutical products. The *PPS Report* provides estimates of the sales of each product during the month, the

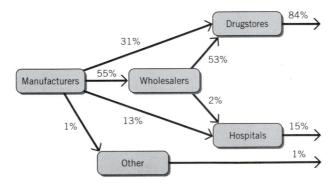

Figure 1 Flow of sales, pharmaceutical industry

year to date, and the most recent 12-month period. These sales are classified by the amount of sales, by manufacturer, and by therapeutic class. Examples of these listings are shown in Figures 2 to 4.

	Product	Manufacturer	Class	Sales (000$)	%Chng
		Leading Pharmaceutical Products			
1	Ventrolin	Granox	28	130987	−5.55
2	Kapoten	Streb	31	128951	9.89
3	Zantax	Granox	23	99025	3.1
4	Cardizon	Southern	31	89254	1.3
5	Vosotec	First	31	75509	8.03
6	Twophasil	Windham	33	72206	5.41
7	Edilat O	Morris	31	71558	5.59
8	Soltaren SR	Greenock	9	56338	−2.24
9	Seldine	Mersey	14	53677	−0.39
10	Enfilac	Reed	54	53518	−1.33
11	Mevakor	AC&E	32	53262	−3.05
12	Kardizem SR	Southern	31	52744	6.71
13	Soltaren	Greenock	9	49649	4.71
14	Ordec	Ordex	33	47899	4.35
15	Omnipast	Waters	40	47398	6.29
16	Ceclar	Larsen	15	45039	−6.62
17	Standimune	Zondac	30	40074	−9.03
18	Sulrate	Southern	23	39238	−0.28
19	Metoxin	AC&E	15	38541	−2.97
20	Tanormin	IFI	31	37348	5.01
21	Mintovral	Windham	33	36752	−7.41
22	Peptid	AC&E	23	36382	7.53
23	Modaret	AC&E	41	34291	2.82
24	Lopic	Daniels	32	32597	−3.88
25	Theordur	Strong	28	32184	−7.13
26	Isopton	Lerner	31	31410	2.27
27	Natrong SR	Rhinelab	31	31378	8.62
28	Diabita	Charet	39	30551	9.72
29	Minacin	Tanner	15	29323	2.01
30	Zevirax	Burwell	81	28987	−8.33
31	Amaxil	Armstrong	15	28148	8.17
32	Hismanol	Petersen	14	27807	−5.2
33	Atrovint	Mannheim	28	25649	−1.51
34	Premaron	Armstrong	52	25196	−9.02
35	Surgon	Rousseau	9	23959	−4.54
36	Bellovent	Granox	28	23894	8.67
37	Insure	Starr	60	23585	0.68
38	Cinemet	AC&E	20	23373	−8.5
39	Analatin	Morris	31	23015	−0.69
40	Motilion	Petersen	23	22770	1.65

Figure 2 *PPS Report*, sample listing 1

Product	Class	Sales ($000)		Total	% Of Total
		Drugstore	Hospital		
Seclor	15	43579	1488	45067	27.6
Humalin N	39	12144	1102	13246	8.11
Ceflex	15	5234	422	5656	3.46
Azid	23	5785	21	5806	3.56
Humalin R	39	6223	565	6788	4.16
Olosone	15	3900	72	3972	2.43
Aventol	64	4027	501	4528	2.77
Humalin L	39	3121	282	3404	2.08
Nafron	9	2067	12	2080	1.27
Dargon-N	2	1345	25	1370	0.84
Flotycin A	61	1405	1947	3352	2.05
Flotycin B	15	0	777	777	0.48
Dargon-N C	2	952	16	967	0.59
Vacillin K	15	907	21	929	0.57
Tulnal	67	463	9	471	0.29
Domelor	39	424	9	433	0.26
Ilecin L	39	222	34	256	0.16
Frenison	52	425	2	427	0.26
Tesatape	40	450	0	450	0.28
Vancotin	15	740	15661	16401	10.04
Nobcin	15	784	9770	10554	6.46
Seconil S	67	254	19	274	0.17
Ilecin R	39	226	34	260	0.16
Glufagon	78	270	1370	1640	1
Biotal S	4	249	108	357	0.22
Kesamet	17	231	300	531	0.33
Ceflin	15	97	1559	1656	1.01
Ondolin	30	380	2722	3102	1.9
Protamile S	78	4	1635	1638	1
Mandox	15	94	3923	4016	2.46
Eldirine	30	95	634	729	0.45
Delbex	30	62	514	576	0.35
Cefzol	15	109	4844	4954	3.03
Humatrone	52	0	10817	10817	6.62
Dobutril	31	364	5425	5789	3.55
Total		96633	66639	163271	100

Manufacturer: Larsen

Figure 3 *PPS Report*, sample listing 2

Of particular interest to subscribers is the classification according to therapeutic class or "market." More than 70 principal markets are distinguished, ranging from amoebacides and analgesics (classes 01 and 02) to thyroid preparations and vitamins (classes 72 and 76).

Obviously, manufacturers know accurately and with certainty the sales of *their own* products. They do not know the sales of their competitors' products, the total

Class 02: Analgesics					
Manufacturer	Product	Sales ($000)			Share
		Drugstore	Hospital	Total	(% of Total)
Bridge	Copercet	3947	189	4136	1.72
Bridge	Coperdan	2509	53	2562	1.07
Bridge	Endocin	411	25	436	0.18
Bridge	Nebalin	104	311	415	0.17
Bridge	Copercet-D	339	12	351	0.15
Carenis	Soniflex	245	2	247	0.10
Cerat	Borsin F	3082	11	3093	1.29
Cerat	Synidril	0	1938	1938	0.81
Cerat	Borsin-30	920	164	1084	0.45
Cerat	Borsin	967	109	1077	0.45
Cerat	Borsin-8	996	34	1029	0.43
Cerat	Ceragol	406	0	406	0.17
Cerat	Prosin-15	231	51	282	0.12
Chambers	Rolinex	879	507	1386	0.58
Chambers	Belfem	208	159	367	0.15
Coblenz	Tanrum	7439	134	7574	3.16
Crown	Arthrosol	1396	0	1396	0.58
Daniels	Steril	6520	92	6611	2.75
Daniels	Forcyl	288	0	288	0.12
Ergalion	Nigranil	524	0	524	0.22
First	Phenetrin	20487	69	20556	8.57
First	Acetolin comp w/	16741	78	16819	7.01
First	Dolomid	13029	244	13272	5.53
First	Lezitine	1531	770	2300	0.96
Zondac	Rinalin-C	14351	42	14393	6.00
Zondac	Rinalin	4687	26	4714	1.96
Zondac	Ferilin PB	2325	2	2327	0.97
Zondac	Migralin	2036	16	2051	0.85
Zondac	Ferilin	1840	12	1852	0.77
Zondac	Migralin DS	1516	0	1516	0.63
Zondac	Zondilex	720	5	726	0.30
Zondac	Atilin SR	473	0	473	0.20
TOTAL		211915	28072	239987	100.00

Figure 4   *PPS Report*, sample listing 3

sales in each market, or the market share of their own or competitors' products. It is for this reason that they are willing to pay the substantial cost of subscribing to the *PPS Report*. No other source for such information is available.

Pharmacom derives its estimates in the following manner. Once each month, field workers visit a number of selected drugstores and hospitals across the country, and photograph ("film") all invoices by drug manufacturers or wholesalers paid since the previous visit. Invoices show the manufacturer, the name of the product, the number of units sold, the unit price, and the dollar amount of the sale. Different sizes or strengths of the same product are treated as different products. The films are processed, and, after editing for accuracy and consistency, their contents are entered into a computer file. On the basis of this set of raw data from the selected

drugstores and hospitals, estimates are made of the total sales of pharmaceutical products for the entire country.

It is clear from the above description that Pharmacom monitors not manufacturers' *sales* but the *paid purchases* by drugstores and hospitals (exports, for example, are not monitored). However, because sales and purchases overlap to a very large extent, the tendency in the industry is to use the two terms interchangeably; this practice is followed in this case as well.

The sample of hospitals. The *Hospital Directory*, published annually, provides a list of hospitals classified according to region, type and size. Five regions are distinguished: East, South, Central, North, and West. Up to five size categories are formed according to the number of hospital beds, and three types of hospitals are distinguished: general, psychiatric, and specialty.

Table 1 shows the most recent classification of the population of hospitals. The entries show the number of hospitals and the total number of beds in each cell.

The sample design calls for a total of 300 hospitals, allocated approximately in proportion to the number of beds in each cell. The actual sample differs somewhat from that of the design for practical reasons having to do with the selection and "signing up" of a hospital. The number of hospitals called for by the sample design as well as the actual number are shown in Table 2.

Considerable time and effort is required to approach the hospital administration, make a formal request, and obtain permission to film the paid invoices of the hospital. Usually, the approval of the hospital Board is necessary, and this is not always forthcoming. It is estimated that it takes about seven months before approval or refusal is given. For these reasons, once a hospital is signed up, it remains in the sample until it decides to withdraw or changes in the characteristics of the hospital population make removal necessary. A very large proportion of the participating hospitals, therefore, have been in the sample for several years.

The sample of drugstores. Commercial directories are used to identify and classify drugstores according to region, type, and size. Two types are distinguished: independent and chain. Three size categories (small, medium, large) are formed according to the average dollar purchases of the drugstore. Table 3 shows the most recent classification of the population of drugstores.

The sample consists of 660 drugstores in total, and the allocation of this total according to region, size, and type is shown in Table 4.

The allocation of the sample by region is approximately proportional to the dollar sales of pharmaceutical products in the region, subject to a minimum of ten drugstores for each region. Within each region, the allocation by size is also proportional to dollar sales. Within each region/size category, the allocation is roughly proportional to the number of chain and independent drugstores in that category in the population.

Although the "signing up" of a drugstore for the sample involves less effort and time than that of a hospital, once a drugstore is selected, it tends to remain in

TABLE 1 Population of Hospitals

	East		South		Central		North		West		Total	
Type/Size	Hospitals	Beds	Hospitals	Beds	Hospitals	Beds	Hospitals	Beds	Hospitals	Beds	Hospitals	Beds
General:												
500+ beds	14	7721	47	32915	65	43355	29	22931	32	21380	187	128302
300–499 beds	23	8009	71	25517	107	40685	47	16964	35	11927	283	103102
200–299 beds	23	4772	68	16568	68	16079	23	4940	20	4376	202	46735
100–199 beds	68	9665	83	11297	98	13853	83	9962	47	6263	379	51040
under 100 beds	242	10358	122	6299	287	15443	839	25079	251	8369	1741	65548
Specialty:												
500+ beds	2	2	5	3254	20	11690	2	2	2	2	31	14950
300–499 beds	2	2	20	6989	11	3953	11	3368	11	3506	55	17818
200–299 beds	2	2	44	10544	17	3473	23	5015	8	1412	94	20446
100–199 beds	5	377	68	8549	29	3440	23	2642	14	1604	139	16612
under 100 beds	14	533	143	28214	44	6470	68	3014	29	2018	298	40249
Psychiatric:												
500+ beds	2	2	17	12815	5	1721	8	3515	5	3602	37	21655
under 500 beds	20	5420	35	7202	62	14531	23	3797	8	803	148	31753
Total	417	46863	723	170163	813	174693	1179	101229	462	65262	3594	558210

TABLE 2	Sample of Hospitals											
						Region						
	East		South		Central		North		West		Total	
Type/Size	Required	Actual	Required	Actual	Required	Actual	Required	Actual	Required	Actual	Required	Actual
General:												
500+ beds	4	7	16	16	25	16	13	13	10	13	68	65
300–499 beds	4	7	16	13	19	28	7	10	4	7	50	65
200–299 beds	4	4	7	4	7	7	4	7	4	7	26	29
100–199 beds	7	7	7	7	7	7	4	4	4	10	29	35
under 100 beds	7	4	4	1	7	7	13	4	4	4	35	20
Specialty:												
500+ beds	1	1	4	4	4	4	4	4	1	1	14	14
300–499 beds	1	1	4	7	1	1	1	1	4	1	11	11
200–299 beds	1	1	4	1	4	1	1	1	1	1	11	5
100–199 beds	1	1	7	7	1	1	4	1	1	1	14	11
under 100 beds	1	1	4	7	4	1	1	1	1	1	11	11
Psychiatric:												
500+ beds	1	1	4	7	4	4	4	1	1	1	14	14
under 500 beds	4	4	4	4	7	10	1	4	1	1	17	23
Total	36	39	81	78	90	87	57	51	36	48	300	303

TABLE 3 **Population of Drugstores**

Region	Small Indep.	Small Chain	Medium Indep.	Medium Chain	Large Indep.	Large Chain	Total
East	365	125	347	197	290	263	1587
South	1103	137	1355	122	1070	401	4188
Central	1334	443	1370	683	878	1172	5880
North	911	218	1040	218	689	521	3597
West	416	275	389	257	122	509	1968
Total	4129	1198	4501	1477	3049	2866	17220

TABLE 4 **Sample of Drugstores**

Region	Small Indep.	Small Chain	Medium Indep.	Medium Chain	Large Indep.	Large Chain	Total
East	4	1	10	7	19	19	60
South	10	1	34	4	76	28	153
Central	16	7	40	19	67	88	237
North	13	1	28	7	46	34	129
West	4	4	13	7	10	43	81
Total	47	14	125	44	218	212	660

the sample until it decides to withdraw. New drugstores are signed up only when necessary.

Estimation of hospital purchases. As noted earlier, the national purchases of pharmaceutical products are estimated on the basis of the observed purchases in a sample of drugstores and hospitals. To understand the method used for making these estimates ("projections," in the industry vernacular), consider hospital sales, and assume that a given classification cell contains five hospitals, of which two are in the sample. The purchases of three products by the sampled hospitals are shown in Table 5.

The number of beds for all hospitals in the cell is 1,535, and that for all hospitals in the sample is 565. The "projection factor" for the cell is 1535/565 or 2.72. The purchases of Product X by all hospitals in the cell are estimated as 519×2.72 or $1,412; similarly, those of products Y and Z are estimated to be $1,020 and $2,103, respectively.

TABLE 5	Hospital Estimation Method Illustrated				
			Purchases		
Hospital	Number of Beds	In Sample?	Product X	Product Y	Product Z
A	500				
B	490	Yes	$300	$200	$450
C	320				
D	150				
E	75	Yes	$219	$175	$323
Total	1,535	(565)	$519	$375	$773

The estimated national purchases of a given product are equal to the sum of the cell estimates.

Estimation of drugstore purchases. The estimation of purchases by drugstores follows quite a different procedure. Suppose that three drugstores are sampled in a given cell, and that their purchases of a given product in the current month and in the same month one year ago are as shown in Table 6.

The ratio of the two totals is $31,000/29,000$, or 1.069, indicating that in the given cell this month's purchases of Product X were 6.9% greater than those in the same month one year ago.

A weighted overall ratio is then calculated using as weights the cell shares of the total sales of pharmaceutical products.[1]

Finally, the current month's estimate of the national purchases of Product X by drugstores is obtained by multiplying the weighted overall ratio by the estimate of the total drugstore purchases of Product X in the same month last year.

TABLE 6	Drugstore Estimation Method Illustrated	
	Purchases of Product X ($)	
Drugstore	Current Month	Same Month Last Year
A	10,000	9,000
B	6,000	8,000
C	15,000	12,000
Total	31,000	29,000

---

[1] For the purpose of estimation, the classification "chain/independent" is ignored; as a result, 21 cells are used formed according to region and size.

QUESTIONS

1. As a newly hired analyst for Pharmacom, describe clearly and thoroughly the strengths and weaknesses of the current methodology, and explain how it may be improved. Estimate the additional cost of your recommendations.
2. Assume that the annual cost of a subscription to the *PPS Report* is $50,000. In your capacity as manager of Marketing Research of a pharmaceutical company, defend your recommendation to the President as to whether or not a subscription should be purchased.
3. Comment on any other aspects of the case not covered by the above questions.

Pharmacom
Research — Part II

From the beginning of its operations, Pharmacom has sponsored an annual "validation" designed to give subscribers some information regarding the accuracy of its sales estimates.

In March of each year, manufacturers of pharmaceuticals are invited to send confidentially to an independent consultant the actual sales of their products during the previous calendar year (January to December). The consultant is chosen jointly by Pharmacom and the manufacturers' market research association (MRA). The task of the MRA is to ensure a large participation; Pharmacom is responsible for the organization and cost of the validation.

Pharmacom prepares a form for each manufacturer, as outlined below, listing the manufacturer's products and the estimated drugstore and hospital sales of each product during the year.

Company: *Granox Pharmaceuticals*

Name of Product	Class	Drugstore		Hospital		Combined	
		Pharmacom Estimate ($000)	Actual Sales ($000)	Pharmacom Estimate ($000)	Actual Sales ($000)	Pharmacom Estimate ($000)	Actual Sales ($000)
⋮	⋮	⋮	⋮	⋮	⋮	⋮	⋮
Norbilon	60	1868		484		2352	
⋮	⋮	⋮	⋮	⋮	⋮	⋮	⋮

The participating manufacturer fills in its actual sales to drugstores and hospitals,[1] as shown in boldface below.

Company: *Granox Pharmaceuticals*

Name of Product	Class	Drugstore		Hospital		Combined	
		Pharmacom Estimate ($000)	Actual Sales ($000)	Pharmacom Estimate ($000)	Actual Sales ($000)	Pharmacom Estimate ($000)	Actual Sales ($000)
⋮	⋮	⋮	⋮	⋮	⋮	⋮	⋮
Norbilon	60	1868	**1827**	484	**529**	2352	
⋮	⋮	⋮	⋮	⋮	⋮	⋮	⋮

Nearly all manufacturers are able to distinguish drugstore from hospital sales, but a few who cannot do so complete the "Combined" column only.

From the contents of these forms, the independent consultant creates a data file, in which the drugstore and hospital sales of each product are treated separately as if they were sales of two different products. The typical data line illustrated in the preceding forms would thus become two lines in the validation file, and give rise to two so-called "R values" as follows:

Company	Name of Product	Class	Pharmacom Estimate	Actual Sales	R Value	Category
Granox	Norbilon	60	1868	1827	1.022	Drugstore
Granox	Norbilon	60	484	529	0.915	Hospital

The R value, it will be noted, is the ratio:

$$\frac{\text{Pharmacom estimate}}{\text{Actual sales}}.$$

These R values become the focus of the validation, and the independent consultant reports to Pharmacom and the MRA various aggregate statistics based on the R values. As the consultant observes in the introduction to the validation report,

"... if the estimates were perfectly accurate, all R values would equal 1. This is, of course, unlikely to occur. Several measures of accuracy are presented in this report: the average R value (should be close to 1), the standard deviation of R values (should be close to 0), and the 'proportion within 22.5%' (the proportion of R values in the range from 0.775 to 1.225; should be close to 1). These measures are calculated for various classifications and categories following the practice of the validation reports in earlier years."

---

[1] After an adjustment for sales to wholesalers, described in Appendix B below.

Interest in the annual validation appears to be growing: during the last six years, the number of participating manufacturers has increased steadily from 38 to 56. Figures 1 to 4 show the trends of selected statistics extracted from Table 1.

The terms "ethical" and "nonethical" should not be interpreted literally. Nonethical products are simply those falling into the therapeutic classes listed in Appendix A; all other products are called ethical. Products in the nonethical classes tend not to be conventional pharmaceuticals, and to be distributed to a large extent through channels other than retail pharmacies and hospitals. Pharmacom and the industry are primarily concerned with the accuracy of sales estimates of ethical products.

Volume categories are formed according to the actual sales of the product. The lower and upper limits of each volume category are adjusted annually ac-

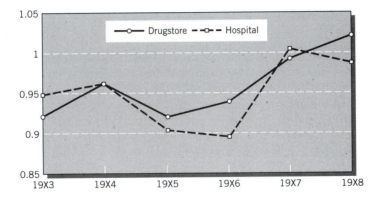

Figure 1   Mean R value

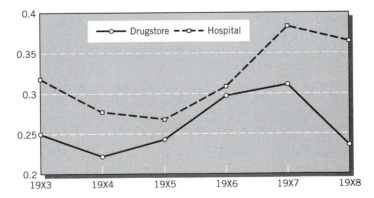

Figure 2   Standard deviation of R values

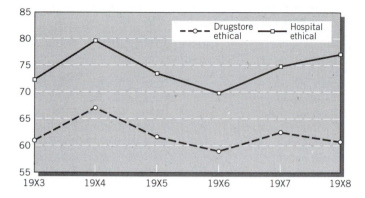

Figure 3   Proportion within 22.5%—ethical

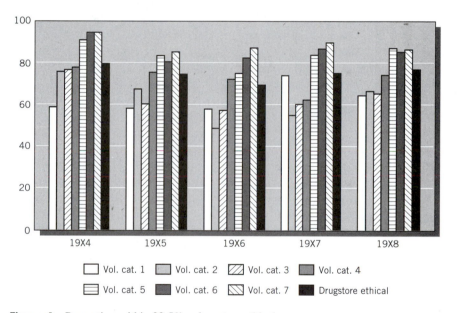

Figure 4   Proportion within 22.5%—drugstore ethical

cording to the government's price index for pharmaceutical products; the current limits are shown in Appendix C. The purpose of these volume categories is to allow comparisons of the accuracy of Pharmacom estimates for products of similar popularity. It is widely believed that accuracy tends to improve with volume.

Pharmacom takes considerable pride in the validation, pointing out that it is the *only* market research firm providing its customers with information on the accuracy of the estimates it sells.

TABLE 1	Selected Validation Results, 19X3–19X8					
Category	19X3	19X4	19X5	19X6	19X7	19X8
Mean R value:						
Drugstore, all products	0.921	0.963	0.920	0.938	0.991	1.019
Hospital, all products	0.948	0.962	0.903	0.894	1.003	0.985
Combined, all products	0.904	0.964	0.960	0.988	0.987	1.056
Drugstore, ethical	0.927	0.968	0.929	0.949	1.002	1.027
Hospital, ethical	0.968	0.992	0.912	0.904	1.011	0.988
Combined, ethical	0.901	0.969	0.961	0.993	0.987	1.074
Ethical, all products	0.935	0.959	0.927	0.942	1.003	1.019
Drugstore, ethical, taxable			0.854	0.852	0.982	1.010
Drugstore, ethical, nontaxable			0.954	0.977	1.011	1.071
Standard deviation of R values:						
Drugstore, all products	0.252	0.222	0.242	0.297	0.309	0.235
Hospital, all products	0.317	0.279	0.268	0.306	0.382	0.364
Combined, all products	0.203	0.194	0.214	0.237	0.218	0.215
Proportion within 22.5%:						
Drugstore, all products	71.3	78.3	71.4	68.3	73.8	76.2
Hospital, all products	59.5	63.7	60.3	58.1	61.8	59.9
Drugstore, ethical	72.4	79.4	73.0	69.7	74.9	76.8
Hospital, ethical	61.1	66.9	61.3	58.8	62.2	60.7
Drugstore, ethical, taxable			56.4	56.1	73.6	75.0
Drugstore, ethical, nontaxable			78.7	73.6	75.4	77.4
Drugstore, ethical, vol. cat. 1	63.8	57.4	57.0	57.3	74.0	65.6
Drugstore, ethical, vol. cat. 2	64.3	73.3	69.8	47.8	53.1	68.3
Drugstore, ethical, vol. cat. 3	67.9	74.2	59.7	55.8	60.3	67.5
Drugstore, ethical, vol. cat. 4	72.3	76.7	74.0	71.4	62.7	74.4
Drugstore, ethical, vol. cat. 5	81.5	90.6	82.1	76.0	86.1	87.5
Drugstore, ethical, vol. cat. 6	81.0	90.2	80.9	83.8	88.7	85.1
Drugstore, ethical, vol. cat. 7	86.7	94.8	84.7	89.9	90.7	86.6
Hospital, ethical, vol. cat. 1		56.5	60.6	54.0	51.5	41.3
Hospital, ethical, vol. cat. 2		73.1	46.1	59.0	48.6	64.0
Hospital, ethical, vol. cat. 3		47.4	46.1	48.1	64.1	56.2
Hospital, ethical, vol. cat. 4		66.7	54.2	46.4	65.2	61.5
Hospital, ethical, vol. cat. 5		64.5	71.7	66.7	54.9	67.9
Hospital, ethical, vol. cat. 6		85.7	69.6	75.0	71.9	66.7
Hospital, ethical, vol. cat. 7		82.4	77.5	61.8	80.5	68.4

"Take television ratings as an example," says Mr. John Taylor, a Pharmacom Vice President. "We read that Monday Night Football had a rating of 30 last week. What does this '30' mean? It is an *estimate* of the percentage of people who watched that program. Where does the estimate come from? It is based on a *sample* of people. For a given region, the sample size may be 300 or 200 out of

a population of millions. What is the *true* rating, the percentage of the *population* who watched the program? We do not know. It could be any number close to or far from 30—we will never know. The ratings firm will tell you, of course, that their sample is properly, statistically, designed, and that you should have confidence in well-designed samples. Pharmacom can give you the same assurance, and, *in addition*, some idea about how much confidence you can place on its estimates. To the best of my knowledge," concludes Mr. Taylor, "Pharmacom is the only market research firm giving its clients this kind of service."

For some time now, however, Mr. Taylor and Pharmacom have been concerned about the validation results. The concern began after the 19X4 and 19X5 validations, when the proportion within 22.5% reached record lows, and the deviation of the mean R value from 1 reached record highs. In 19X7, a thorough review was ordered of the sample design, the selection of drugstores and hospitals, the filming of invoices, the data editing and input, and the projection methodology, in order to discover any errors that may have crept in. Some errors were found and corrected.

Apparently, these efforts paid off because the 19X7 and 19X8 validation results generally improved. Nevertheless, Pharmacom is somewhat anxiously awaiting the 19X9 validation results. Sixty manufacturers (a new record) participated. The independent consultant has prepared the 19X9 validation file, a partial listing of which appears in Figure 5. The complete data set is in the file `PHARMAC2.DAT`.

The interpretation of the variables in Figure 5 is as follows:

COMPANY	Company code (1 to 60).
CLASS	Therapeutic class number.
TAX	YES if product is taxable; NO otherwise.
TYPE	DE if drugstore ethical;
	DNE if drugstore nonethical;
	HE if hospital ethical;
	HNE if hospital nonethical;
	CE if combined ethical;
	CNE if combined nonethical.
VOLUME	Volume category (1–7, as per Appendix C).
PHARMA	Pharmacom estimate ($000).
ACTUAL	Actual sales ($000), as reported by manufacturer.
R	R value (=PHARMA/ACTUAL).

QUESTIONS

1. Using the 19X9 validation data in the file `PHARMAC2.DAT`, calculate the 19X9 values of the statistics listed in Table 1. Update Figures 1 to 4.
2. Comment on the validation results of the last seven years.

COMPANY	CLASS	TAX	TYPE	VOLUME	PHARMA	ACTUAL	R
1	12	NO	DE	6	5647	5733	0.985
1	12	NO	HE	4	460	654	0.703
1	15	NO	DE	6	3731	4010	0.930
1	15	NO	DE	5	3297	2642	1.248
1	15	NO	DE	6	3175	3703	0.857
1	12	NO	DE	6	3090	2991	1.033
1	12	NO	HE	3	265	387	0.684
1	37	YES	DE	6	2210	2811	0.786
1	64	NO	DE	5	2018	1739	1.160
1	15	NO	DE	5	1908	1900	1.004
1	30	NO	DE	5	1719	1661	1.035
1	30	NO	HE	4	156	603	0.259
1	15	NO	DE	4	1288	1095	1.176
1	76	YES	DE	5	1903	1804	1.055
1	61	YES	DNE	4	1104	1030	1.072
1	15	NO	DE	4	1102	1156	0.953
1	37	YES	DE	4	1009	1207	0.836
1	52	NO	DE	4	798	837	0.953
1	15	NO	DE	4	680	772	0.881
1	15	NO	HE	7	2559	2068	1.237
1	64	NO	DE	2	354	351	1.008
1	61	YES	DNE	3	323	515	0.626
1	67	NO	DE	1	280	216	1.294
1	31	NO	HE	1	445	238	1.872
1	4	NO	DNE	1	52	255	0.203
1	4	NO	HNE	7	1915	2124	0.901
1	53	NO	HNE	5	1137	1131	1.006
1	15	NO	HE	5	1055	1244	0.848
1	53	NO	HNE	5	989	755	1.310
1	31	NO	HE	3	188	377	0.499
1	53	NO	HNE	6	1495	1379	1.084
2	2	NO	DE	5	1179	1831	0.644
2	56	NO	DE	2	301	392	0.768

Figure 5   Validation file PHARMAC2.DAT, partial listing

3. Until now, the annual validations have been used to provide only general information on the accuracy of the Pharmacom estimates (for example, that the proportion within 22.5% went up 4 points). Can you think of other uses?
4. In light of the data presented in this part of the case, what do you recommend Pharmacom do to improve the accuracy of its surveys? Estimate any additional costs associated with your recommendations.
5. Comment on any other aspects of the case not specifically covered by the above questions.

APPENDIX A

Following are the non-ethical product classes:

Class	Description
03	Analgesics, prop.
04	Anesthetics
06	Antacids, prop.
10	Antiarthritics, prop.
19	Anti-obesity, prop.
22	Antiseptics, prop.
24	Antiasthmatics, prop.
25	Baby care preps, prop.
35	Cough and cold preps, prop.
36	Denture preps, prop.
38	Dermatologicals, prop.
40	Diagnostic aids
42	Diuretics, prop.
46	Feminine hygiene preps, prop.
47	Foot preps, prop.
50	Antihemorrhoidal preps, prop.
53	Hospital solutions
54	Infant formulas
57	Laxatives, prop.
58	Lip protectives, prop.
61	Eye washes and lotions
66	Rubbing alcohol
68	Sedatives, prop.
70	Suntan preps, prop.
71	Sweetening agents
73	Tonics, prop.
77	Vitamins, prop.
79	Miscellaneous, prop.

APPENDIX B

As noted in Part I of this case, Pharmacom monitors the *purchases* by drugstores and hospitals. Direct sales by manufacturers to these outlets correspond to direct purchases and require no adjustment. However, 55% of manufacturers' sales are to wholesalers, who resell to drugstores and hospitals at a markup. Consequently,

purchases from wholesalers by drugstores and hospitals will differ from manufacturers' sales by the amount of the markup. The average markup in the industry is 9%.

For the purpose of the validation, manufacturers are requested to distinguish direct sales to drugstores and hospitals from those to wholesalers, and to multiply the latter by 1.09. For example, if the direct sales of Norbilon by Granox Pharmaceuticals were $508 (000), and the sales to wholesalers were $1,210, for the purpose of the validation the total drugstore sales of Norbilon should be reported as $(1210 \times 1.09) + 508$, or $1,827.

APPENDIX C

The volume categories are as follows:

Volume Category	Actual Sales ($000)	
	Drugstore	**Hospital**
1	175–250	175–250
2	250–425	250–350
3	425–675	350–425
4	675–1275	425–675
5	1275–2550	675–1275
6	2550–5575	1275–1950
7	5575+	1950+

For sales reported on a combined basis, the drugstore volume categories apply. If actual sales are less than the lower limit of the first volume category, the corresponding R value is excluded from the validation.

Tenderdent Toothpaste

Mr. Brian Conti, the new chief executive officer of Medlabs Incorporated, had a feeling that the pharmaceutical company was at a crossroads with respect to one of its products, the Tenderdent toothpaste.

Medlabs was established fifty years ago to produce and market specialty pharmaceutical products to which it held the patent. The company prospered modestly over the years, first by exploiting the original patents and then, when these expired, by introducing successful modifications, acquiring additional patents, and developing new products.

One product in the latter category is Tenderdent toothpaste, designed to relieve the pain or discomfort produced by sensitive teeth. Tenderdent performs the cleaning duties of a regular toothpaste and, in addition, contains an ingredient that dulls sensitive tooth nerves, thereby helping to relieve the pain and discomfort some people experience with cold or hot drinks and food, or tooth brushing.

Tenderdent has never been marketed directly to the consumer, only to dentists, drugstores, and hospitals as part of a promotional package that covers all the products of the company.

The promotional strategy of the major competitor, Sensodyne, has been the opposite. Relying mainly on television advertising, Sensodyne has succeeded in capturing the lion's share of the market. It is estimated that Sensodyne accounts for more than 80% of the sales of toothpaste for sensitive teeth; Tenderdent's market share is thought to be about 15%.

A standard-sized (75 mL) tube of regular toothpaste retails for about $1. The retail price of a tube of Sensodyne or Tenderdent toothpaste of the same size is about $5. Medlabs' wholesale price is $3.50; the cost of production and normal marketing

is $1.75 per tube. One standard-sized tube lasts for about one month of normal usage. Tenderdent's current sales are fairly steady at about 150,000 tubes per month.

Mr. Conti believes that Tenderdent is by no means inferior to Sensodyne. Laboratory tests have shown Tenderdent to be slightly more effective than Sensodyne in cleaning performance and protection against tartar formation. In clinical tests, subjects found both brands equally effective in relieving pain, but reported a mild preference for the taste of Sensodyne.

Mr. Conti believes, therefore, that the company has a good product, the potential of which has not been fully exploited. The head of Medlabs' laboratory assures Mr. Conti that the problem of taste can be resolved, and is currently carrying out experiments with this objective in mind.

By far the most important reason for Mr. Conti's optimism has to do with the company's promotional policy—or the lack of it, as Mr. Conti perceives it.

"We have two problems," says Mr. Conti. "First, the focus of our promotion has been the dentist and drugstore. But people are not buying toothpaste for sensitive teeth as a medicine and on the advice of their doctor, dentist, or friendly druggist. They buy it as an alternative to their regular toothpaste, off the open shelves. That is the first mistake. The second," he continues, "is we have not promoted it enough. I am convinced that there are many people out there who are not aware they have sensitive teeth, don't know there is something they can do about it, and don't think 'Tenderdent' when they look for a solution to their problem.

"What we've got to do," says Mr. Conti, "is two things. First we must find out the size of the potential market for our product. If this market is large enough, we go after it with an aggressive promotional campaign. The campaign should be aimed at people who are bothered by sensitive teeth. It should make them aware of their problem, and convince them to buy Tenderdent to solve this problem. The message should play up the good lab results, and Tenderdent's new improved taste (assuming the lab finally gets this little problem licked)."

Mr. Conti's enthusiasm is not shared widely among Medlabs' "old guard" (the term is Mr. Conti's). They tend to see Tenderdent as a minor element of the company's product portfolio, and its sales potential as small. About three months ago, however, they were willing to approve a proposal by Mr. Conti to commission a market research survey to shed some light on the issue.

PMR, a highly respected market research firm, was given the commission to carry out the survey. PMR advised, and Medlabs agreed, that the survey be limited to one region that could also serve as a test market for the launch of the advertising campaign in the event the survey results are favorable.

The selected region is the Nortown Metropolitan Area, with a population of approximately 3.4 million. This area is divided into 724 districts. For the first few districts, Table 1 shows the number of households, the average household income, the percentage of households owning their home, the percentage of persons with some university education, and the percentage of persons who are employed. The complete data for all districts can be found in the file FCENSUS.DAT.

There are altogether 1,199,490 households in the Nortown Metropolitan Area. It was decided by Medlabs and PMR that a number of households will be selected

TABLE 1 District Information, Nortown Metropolitan Area (file FCENSUS.DAT)

District	Number of Households	Average Household Income ($000)	Percent Owning	Percent with Some University Education	Percent Employed
1	200	32	62	4	42
2	230	32	89	18	53
3	2900	23	7	10	59
4	2050	21	11	11	55
5	1645	24	13	10	62
⋮	⋮	⋮	⋮	⋮	⋮

and all adults (persons 20 years old and older) in the selected households will be interviewed. The most recent census shows 1,859,473 adults residing in Nortown.

PMR will not design the sample, but will implement it, interview the selected households and adults, and provide to Medlabs the raw survey results in the form of a computer file. The cost of the survey will be $50 per selected household.

There is abundant documentation available to PMR concerning the Nortown Metropolitan Area, including lists of households in each district. For the purpose of this case it can be assumed that households in each district are numbered consecutively, and that they can be identified and located simply from their district and household identification numbers.

The selected adults will be interviewed along the lines of the questionnaire shown in Figure 1.

FOR PRACTICE

1. In order to become gradually acquainted with the case, assume that the Nortown Metropolitan Area consists of only three districts, as shown in Table 2 (the data can also be found in the file SCENSUS.DAT).

TABLE 2 Fictional District Information, Nortown Metropolitan Area (file SCENSUS.DAT)

District	Number of Households	Average Household Income ($000)	Percent Owning	Percent with Some University Education	Percent Employed
1	500	15	30	10	85
2	1100	20	40	15	80
3	800	30	50	17	90
	2400				

A. Household information:
  1. District number: ___
  2. Household number: ___
  3. Number of adults in household: ___
  4. Number of nonadults in household: ___
  5. Household income: ___ ($000)
B. Personal information:
  6. Adult identification number: ___
  7. What is your age? ___ (years)
  8. What is your gender? Male ___ Female ___
  9. Have you attended a university? Yes ___ No ___
  10. Are your teeth sensitive to hot or cold food or drink? Yes ___ No ___
     If your answer to Question 10 was No, please go to Question 17.
     Otherwise,
  11. How would you describe this sensitivity?
     ___ Mild, can live with it.
     ___ Bothers me, I wish I could do something about it.
  12. Are you aware there are special toothpastes designed to relieve the
     pain and discomfort from sensitive teeth? Yes ___ No ___
     If your answer to Question 12 was No, please go to Question 17.
     Otherwise,
  13. Do you buy toothpaste for sensitive teeth? Yes ___ No ___
     If your answer to Question 13 was No, please go to Question 17.
     Otherwise,
14–16. Which brand of toothpaste for sensitive teeth do you buy?
     14. Sensodyne ___ 15. Tenderdent ___ 16. Other ___
  17. On average, how many hours a day do you watch television during
     the period from 6 a.m. to 7 p.m.? ___ (hours)
  18. On average, how many hours a day do you watch television during
     the period from 7 p.m. to 11 p.m.? ___ (hours)

Figure 1   Survey questionnaire

The latest census shows that 4,000 adults reside in the area. Assume that a *simple random sample* of five households yielded the following survey results:

1	2	3	4	5	6	7	8	9	10	11	12	13	14	15	16	17	18
1	99	2	0	7.7	1	25	0	0	0	0	0	0	0	0	0	1.1	1.4
1	99	2	0	7.7	2	26	1	0	0	0	0	0	0	0	0	2.1	1.6
1	324	1	0	10.8	1	28	0	0	1	1	1	0	0	0	0	1.5	1.2
2	487	2	1	7.0	1	60	0	0	1	1	0	0	0	0	0	2.7	2.2
2	487	2	1	7.0	2	36	1	0	0	0	0	0	0	0	0	1.8	1.4
2	664	2	3	18.4	1	32	0	0	0	0	0	0	0	0	0	1.5	1.4
2	664	2	3	18.4	2	34	1	0	1	1	0	0	0	0	0	2.0	1.5
3	657	2	3	50.4	1	24	0	0	0	0	0	0	0	0	0	1.0	1.0
3	657	2	3	50.4	2	21	1	0	0	0	0	0	0	0	0	1.3	1.3

The sample of five households resulted in interviews with nine adults. The column numbers correspond to the question numbers in Figure 1. In col. 8, "female" is coded 1, "male" coded 0. Bothersome sensitivity is coded 1 in col. 11, mild sensitivity is coded 0. In other questions, "yes" is coded 1, "no" is coded 0. These data can also be found in the file TENDER11.DAT.

The totals of the personal variables by household were as follows (see also file TENDER12.DAT):

1	2	3	4	5	7	8	9	10	11	12	13	14	15	16	17	18
1	99	2	0	7.7	51	1	0	0	0	0	0	0	0	0	3.2	3.0
1	324	1	0	10.8	28	0	0	1	1	1	0	0	0	0	1.5	1.2
2	487	2	1	7.0	96	1	0	1	1	0	0	0	0	0	4.6	3.6
2	664	2	3	18.4	66	1	0	1	1	0	0	0	0	0	3.5	2.9
3	657	2	3	50.4	45	1	0	0	0	0	0	0	0	0	2.3	2.2

Estimate the population characteristics of interest in this case, including: the proportion of adults in the Nortown Metropolitan Area who have sensitive teeth; the proportion of adults in the Nortown Metropolitan Area who have sensitive teeth and are aware of special toothpastes; the proportion of adults in the Nortown Metropolitan Area who are sensitive, aware, and buy a toothpaste for sensitive teeth; the average time adults in the Nortown Metropolitan Area watch television during prime time[1]; the average time adults with sensitive teeth in the Nortown Metropolitan Area watch television during prime time; the average household income; and the proportion of households with no nonadults (roughly, the proportion of households without children).

2. Same as in (1), except that a *stratified random sample* of two households from each district gave the following results (see also file TENDER21.DAT):

1	2	3	4	5	6	7	8	9	10	11	12	13	14	15	16	17	18
1	21	2	0	8.1	1	32	0	0	1	1	0	0	0	0	0	1.6	1.5
1	21	2	0	8.1	2	27	1	0	0	0	0	0	0	0	0	2.2	1.6
1	68	1	1	10.3	1	42	1	0	0	0	0	0	0	0	0	1.6	1.5
2	453	2	1	7.0	1	58	0	0	1	1	1	0	0	0	0	1.9	1.1
2	453	2	1	7.0	2	58	1	0	1	1	1	1	1	0	0	2.1	2.2
2	534	2	3	3.5	1	22	1	0	0	0	0	0	0	0	0	1.3	1.4
2	534	2	3	3.5	2	44	0	0	0	0	0	0	0	0	0	1.4	1.7
3	216	2	0	12.2	1	40	0	0	0	0	0	0	0	0	0	1.6	1.1
3	216	2	0	12.2	2	21	1	1	0	0	0	0	0	0	0	1.2	0.9
3	261	1	0	8.4	1	45	0	0	1	1	1	1	1	0	0	1.6	0.9

---

[1]*Prime time* is the period from 7 to 11 P.M.; for the purpose of this case, *off-prime* is the period from 6 A.M. to 7 P.M.

The sums of the personal variables by household were as follows (see also file TENDER22.DAT):

1	2	3	4	5	7	8	9	10	11	12	13	14	15	16	17	18
1	21	2	0	8.1	59	1	0	1	1	0	0	0	0	0	3.8	3.1
1	68	1	1	10.3	42	1	0	0	0	0	0	0	0	0	1.6	1.5
2	453	2	1	7.0	116	1	0	2	2	2	1	1	0	0	4.0	3.4
2	534	2	3	3.5	66	1	0	0	0	0	0	0	0	0	2.6	3.1
3	216	2	0	12.2	61	1	1	0	0	0	0	0	0	0	2.7	2.1
3	261	1	0	8.4	45	0	0	1	1	1	1	1	0	0	1.6	0.9

3. Same as in (1), except that a *stratified random sample* was selected in the following manner. The districts were stratified according to average household income into two groups: the first consisting of districts with average household income less than or equal to 25 ($000), and the second consisting of districts with average household income greater than 25 ($000). The total sample size of six households was allocated to these groups in proportion to the number of districts. The results were as follows (see also file TENDER31.DAT):

1	2	3	4	5	6	7	8	9	10	11	12	13	14	15	16	17	18
1	107	2	3	7.7	1	60	0	0	1	1	1	1	1	0	0	2.7	2.5
1	107	2	3	7.7	2	56	1	0	1	1	1	1	1	0	0	2.3	1.9
2	23	2	0	7.8	1	30	1	0	1	1	0	0	0	0	0	1.9	2.1
2	23	2	0	7.8	2	55	0	0	0	0	0	0	0	0	0	1.8	1.4
2	133	1	2	56.5	1	59	1	0	0	0	0	0	0	0	0	2.3	1.7
2	796	2	0	35.2	1	33	0	0	0	0	0	0	0	0	0	1.1	1.3
2	796	2	0	35.2	2	23	1	1	0	0	0	0	0	0	0	1.1	1.1
3	339	2	1	37.6	1	24	0	0	0	0	0	0	0	0	0	1.4	0.9
3	339	2	1	37.6	2	23	1	1	0	0	0	0	0	0	0	0.9	1.4
3	536	3	1	23.4	1	47	0	0	0	0	0	0	0	0	0	1.4	1.6
3	536	3	1	23.4	2	39	1	0	1	1	1	0	0	0	0	1.6	1.9
3	536	3	1	23.4	3	52	1	1	1	1	1	1	1	0	0	1.7	1.5

The sums of the personal variables by household were as follows (see also file TENDER32.DAT):

1	2	3	4	5	7	8	9	10	11	12	13	14	15	16	17	18
1	107	2	3	7.7	116	1	0	2	2	2	2	2	0	0	4.9	4.4
2	23	2	0	7.8	85	1	0	1	1	0	0	0	0	0	3.7	3.5
2	133	1	2	56.5	59	1	0	0	0	0	0	0	0	0	2.3	1.7
2	796	2	0	35.2	56	1	1	0	0	0	0	0	0	0	2.2	2.4
3	339	2	1	37.6	47	1	1	0	0	0	0	0	0	0	2.2	2.3
3	536	3	1	23.4	138	2	1	2	2	2	1	1	0	0	4.7	5.0

4. Same as in (1), except that a *two-stage random sample* of two districts and three households from each selected district yielded the following results (see also file TENDER41.DAT):

1	2	3	4	5	6	7	8	9	10	11	12	13	14	15	16	17	18
1	85	2	0	9.6	1	33	1	0	0	0	0	0	0	0	0	1.9	1.9
1	85	2	0	9.6	2	50	1	0	0	0	0	0	0	0	0	1.7	2.1
1	163	2	0	7.1	1	26	0	0	0	0	0	0	0	0	0	1.3	1.6
1	163	2	0	7.1	2	25	1	0	0	0	0	0	0	0	0	1.3	2.1
1	243	1	1	9.0	1	26	1	0	0	0	0	0	0	0	0	1.7	1.9
3	442	2	0	7.5	1	81	0	0	1	1	1	1	1	0	0	2.9	2.2
3	442	2	0	7.5	2	49	1	0	1	1	0	0	0	0	0	1.8	1.6
3	638	2	0	10.1	1	21	0	0	0	0	0	0	0	0	0	0.8	1.0
3	638	2	0	10.1	2	22	1	0	0	0	0	0	0	0	0	0.9	1.3
3	778	2	1	10.3	1	51	0	0	0	0	0	0	0	0	0	1.7	1.1
3	778	2	1	10.3	2	44	1	0	1	1	1	1	1	0	0	1.4	2.2

The sums of the personal variables by household were as follows (see also file TENDER42.DAT):

1	2	3	4	5	7	8	9	10	11	12	13	14	15	16	17	18
1	85	2	0	9.6	83	2	0	0	0	0	0	0	0	0	3.6	4.0
1	163	2	0	7.1	51	1	0	0	0	0	0	0	0	0	2.6	3.7
1	243	1	1	9.0	26	1	0	0	0	0	0	0	0	0	1.7	1.9
3	442	2	0	7.5	130	1	0	2	2	1	1	1	0	0	4.6	3.8
3	638	2	0	10.1	43	1	0	0	0	0	0	0	0	0	1.8	2.3
3	778	2	1	10.3	95	1	0	1	1	1	1	1	0	0	3.1	3.2

FOR REAL

5. The files TENDER51.DAT and TENDER52.DAT contain the individual and household data obtained from a simple random sample of 300 households selected from the Nortown Metropolitan Area. The format of these files is the same as for the data listed above. The indication NRA in these files stands for "No Response/Answer," and means there is no information for that household.

Estimate the population characteristics of interest in this case, including: the proportion of adults in the Nortown Metropolitan Area who have sensitive teeth; the proportion of adults in the Nortown Metropolitan Area who have sensitive teeth and are aware of special toothpastes; the proportion of adults in the Nortown Metropolitan Area who are sensitive, aware, and buy a toothpaste for sensitive teeth; the average time adults in the Nortown Metropolitan Area

watch television during prime time; the average time adults with sensitive teeth in the Nortown Metropolitan Area watch television during prime time; the average household income; and the proportion of households with no nonadults (roughly, the proportion of households without children).

6. Same as Question (5), except that the files `TENDER61.DAT` and `TENDER62.DAT` contain the individual and household data obtained from a two-stage random sample of 100 districts and three households from each selected district in the Nortown Metropolitan Area.

7. Same as Question (5), except that a stratified random sample was selected as follows. The districts were stratified into four groups according to the average household income. The first group consisted of districts with income in the range 0 to 35; the second, third, and fourth groups of districts with average household income in the ranges 35 to 43, 43 to 51, and over 51 ($000), respectively. The total sample size of 300 households was allocated to the four groups in proportion to the number of households in each group. The individual and household data can be found in the files `TENDER71.DAT` and `TENDER72.DAT`, respectively.

8. Same as Question (5), except that a stratified random sample was selected as follows. The districts were stratified into four groups according to the percentage of households owning their home. The four groups consisted of districts with percentage owning in the ranges 0 to 46, 46 to 68, 68 to 84, and over 84, respectively. From each group, 75 households were randomly selected. The individual and household data can be found in the files `TENDER81.DAT` and `TENDER82.DAT`, respectively.

9. Same as Question (5), except that a stratified random sample was selected as follows. The districts were stratified into four groups according to the percentage of persons with university education. The groups consisted of districts where this percentage was in the ranges 0 to 5, 5 to 9, 9 to 14, and over 14, respectively. The total sample size of 300 households was allocated to the four groups arbitrarily in the proportions 0.15, 0.30, 0.25, and 0.30. The individual and household data can be found in the files `TENDER91.DAT` and `TENDER92.DAT`, respectively.

10. Select *one* of the five sampling methods described in Questions (5) to (9) as the best for the purposes of the Nortown survey. Analyze the data files produced by this method and estimate the relevant characteristics of the Nortown Metropolitan Area population. *You may not combine data produced by different methods.* In a detailed report addressed to the management of Medlabs, explain what can be usefully learned from the survey.

11. Comment on any other aspects of the case that were not specifically covered in the preceding questions.

APPENDIX A

Basic Concepts

A.1 THE SUMMATION NOTATION

The summation notation is a convenient shorthand, very helpful in reducing the size of certain mathematical expressions.

For example, suppose we wish to indicate the sum of n observations $Y_1, Y_2, \ldots, Y_n$. We can write this sum either as

$$Y_1 + Y_2 + \cdots + Y_n, \tag{A.1}$$

or, in summation notation, more simply as

$$\sum_{i=1}^{n} Y_i. \tag{A.2}$$

$\sum$ is the summation symbol, i is the index of summation, and (A.2) is read as "the sum of the Y_i for i taking values from 1 to n." Clearly, (A.2) is a much more compact way of writing a sum than (A.1).

When there is no danger of misunderstanding, even simpler notation is sometimes used in place of (A.2):

$$\sum_{1}^{n} Y_i, \qquad \sum Y_i, \qquad \sum Y.$$

All these expressions translate into (A.1).

The summation notation can be used for writing the sums of any expressions identified by subscripts. Some examples are shown below.

$$\sum_{i=1}^{n} Y_i^2 = Y_1^2 + Y_2^2 + \cdots + Y_n^2$$

$$\sum_{i=1}^{n} (Y_i - c)^2 = (Y_1 - c)^2 + (Y_2 - c)^2 + \cdots + (Y_n - c)^2$$

$$\sum_{i=1}^{n} X_i Y_i = X_1 Y_1 + X_2 Y_2 + \cdots + X_n Y_n$$

$$\sum_{i=1}^{m} f(y_i) = f(y_1) + f(y_2) + \cdots + f(y_m)$$

$$\sum_{i=1}^{m} y_i f(y_i) = y_1 f(y_1) + y_2 f(y_2) + \cdots + y_m f(y_m).$$

The particular index used does not affect the sense of the summation:

$$\sum_{i=1}^{n} Y_i = \sum_{j=1}^{n} Y_j = \sum_{k=1}^{n} Y_k = Y_1 + Y_2 + \cdots + Y_n.$$

The sum from 1 to n of a constant is simply n times that constant:

$$\sum_{i=1}^{n} c = \underbrace{c + c + \cdots + c}_{n \ terms} = nc.$$

The sum of terms each multiplied by the same constant equals the constant times the sum of terms:

$$\sum_{i=1}^{n} cY_i = cY_1 + cY_2 + \cdots + cY_n = c \sum_{i=1}^{n} Y_i.$$

The summation of a sum of terms is the sum of the summation of the terms:

$$\sum_{i=1}^{n} (X_i + Y_i) = (X_1 + Y_1) + (X_2 + Y_2) + \cdots + (X_n + Y_n) = \sum_{i=1}^{n} X_i + \sum_{i=1}^{n} Y_i,$$

and more generally,

$$\sum_{i=1}^{n} (X_i + Y_i + \cdots Z_i) = \sum_{i=1}^{n} X_i + \sum_{i=1}^{n} Y_i + \cdots + \sum_{i=1}^{n} Z_i.$$

From the last results we see, for instance, that

$$\sum_{i=1}^{n} (X_i - c)^2 = \sum_{i=1}^{n} (X_i^2 - 2cX_i + c^2) = \sum_{i=1}^{n} X_i^2 - 2c \sum_{i=1}^{n} X_i + nc^2.$$

Double summation is shorthand for sums of expressions which can be identified by two subscripts. For example, suppose we wish to write the sum of mn observations arranged in a table with m rows and n columns, as follows:

		Columns		
Rows	**1**	**2**	$\cdots$	n
1	Y_{11}	Y_{12}	$\cdots$	Y_{1n}
2	Y_{21}	Y_{22}	$\cdots$	Y_{2n}
$\vdots$	$\vdots$	$\vdots$	$\vdots$	$\vdots$
m	Y_{m1}	Y_{m2}	$\cdots$	Y_{mn}

The first subscript identifies the row and the second the column. Thus, Y_{ij} represents the observation in the ith row and jth column (or, in the ijth *cell*) of the table.

The sum of all mn observations can be written in full as

$$(Y_{11} + Y_{12} + \cdots + Y_{1n}) + (Y_{21} + Y_{22} + \cdots + Y_{2n}) + \cdots + (Y_{m1} + Y_{m2} + \cdots + Y_{mn}),$$

or, more compactly, as

$$\sum_{i=1}^{m}(Y_{i1} + Y_{i2} + \cdots + Y_{in}),$$

or, even more compactly, as

$$\sum_{i=1}^{m}\sum_{j=1}^{n} Y_{ij}.$$

We make use of double summation only briefly in Chapter 8.

A.2 DISTRIBUTIONS

Almost every type of analysis begins with a set of observations. When the number of observations is large, it may be difficult to consider them in the raw form in which they were obtained—hence the need for various measures that summarize the data in a meaningful way.

A first step in this reduction is usually the classification of the observations into a number of categories, together with a count of the number ("frequency") of observations falling into each category. It is convenient always to construct these categories so that they form a mutually exclusive and collectively exhaustive set.[1]

A list of mutually exclusive and collectively exhaustive categories and of the corresponding frequencies of observations is called a *frequency distribution*. If the frequencies are divided by the total number of observations, a *relative frequency distribution* is obtained.

---

[1] A set of categories is called *mutually exclusive* if no observation can be classified into more than one category, and *collectively exhaustive* if every observation can be classified into one of the categories.

◪ EXAMPLE A.1   A questionnaire sent to a group of five adults included the questions: What is your gender? How many children do you have? The responses were as follows:

ID No.	Number of Children	Gender
1	1	F
2	2	M
3	1	F
4	2	M
5	1	F
Total	7	

The responses concerning the number of children can be summarized in the form of a frequency or relative frequency distribution:

Number of Children (1)	Frequency (2)	Relative Frequency (3)
1	3	0.6
2	2	0.4
Total	5	1.0

Columns (1) and (2) show the frequency distribution of the number of children; cols. (1) and (3) show the relative frequency distribution of the number of children. For example, 3 of the 5 adults (60% of the number of adults questioned) had 1 child each; etc.

The frequency and relative frequency distributions of adults by gender are as follows:

Gender	Frequency	Relative Frequency
Male	2	0.4
Female	3	0.6
Total	5	1.0	◪

With respect to the number of children, the observations assume numerical values and the categories into which they are classified are also in numerical form; we say that the number of children is a *variable*. For gender, however, there is no natural numerical description of the observations. A person is either male or female; gender is an *attribute*. For example, age, temperature, distance, weight are variables. Sex (male, female), marital status (single, married, divorced, other), nationality (German, French, other) are attributes.

A.3 MEASURES OF CENTRAL TENDENCY AND DISPERSION

For a *variable* of interest, it is often desirable to further reduce the observations to a single number which is, in a certain sense, "representative" of the observations. Two types of such measures are frequently encountered: (a) measures of location or central tendency; and (b) measures of dispersion, indicating the degree of variation or dispersion of the observations around a measure of central tendency or location.

Perhaps the most familiar and most widely used measure of central tendency is the arithmetic average (more simply, the *average* or *mean*) of the variable. If there are n observations of a variable Y with values $Y_1, Y_2, \ldots, Y_n$, their average or mean ($\bar{Y}$) is defined as

$$\begin{aligned}
\bar{Y} &= \frac{1}{n}(Y_1 + Y_2 + \cdots + Y_n) \\
&= \frac{1}{n}\sum_{I=1}^{n} Y_i.
\end{aligned} \tag{A.3}$$

In Example A.1, the average number of children per adult is 7/5, or 1.4.

When the observations have been grouped in the form of a frequency or relative frequency distribution, the calculations can be shortened. If a variable Y takes values $y_1, y_2, \ldots, y_m$ with respective frequencies $f(y_1), f(y_2), \ldots, f(y_m)$, the mean, $\bar{Y}$, is

$$\begin{aligned}
\bar{Y} &= \frac{1}{n}[y_1 f(y_1) + y_2 f(y_2) + \cdots + y_m f(y_m)] \\
&= \frac{1}{n}\sum_{i=1}^{m} y_i f(y_i),
\end{aligned} \tag{A.4}$$

where $n = f(y_1) + f(y_2) + \cdots + f(y_m) = \sum_{i=1}^{m} f(y_i)$.

It should be clear that Eq. A.4 follows from A.3, since there are $f(y_1)$ observations having the value y_1, $f(y_2)$ observations having the value y_2, and so on. Equation A.4 can also be written in terms of relative frequencies:

$$\begin{aligned}
\bar{Y} &= y_1 \frac{f(y_1)}{n} + y_2 \frac{f(y_2)}{n} + \cdots + y_m \frac{f(y_m)}{n} \\
&= y_1 r(y_1) + y_2 r(y_2) + \cdots + y_m r(y_m) \\
&= \sum_{i=1}^{m} y_i r(y_i),
\end{aligned} \tag{A.5}$$

where $r(y_i) = f(y_i)/n$ are the relative frequencies of the variable.

◪ **EXAMPLE A.1**
(Continued)

The average number of children per adult can also be calculated using either the frequency or the relative frequency distribution:

y_i	$f(y_i)$	$r(y_i)$	$y_i f(y_i)$	$y_i r(y_i)$
1	3	0.6	3	0.6
2	2	0.4	4	0.8
	5	1.0	7	1.4

The average number of chidren per adult can be calculated using (A.4),

$$\bar{Y} = \frac{1}{n} \sum_{i=1}^{m} y_i f(y_i) = \frac{7}{5} = 1.4,$$

or (A.5),

$$\bar{Y} = \sum_{i=1}^{m} y_i r(y_i) = 1.4.$$

The result is the same, and agrees with the figure calculated using the raw data. ▨

Measures of dispersion describe the "scatter," "variation," or "spread" of the observations around a measure of central tendency (a "central value"), usually the mean. The most widely used measure of dispersion is the *variance* or its square root, the *standard deviation.*

Suppose that n (ungrouped) observations on a variable Y: Y_1, Y_2, ..., Y_n are available. The *variance*, S^2, is the average squared deviation from the mean:

$$S^2 = \frac{1}{n} \sum_{i=1}^{n} (Y_i - \bar{Y})^2. \tag{A.6}$$

The (positive) square root of the variance is the *standard deviation, S:*

$$S = +\sqrt{S^2}. \tag{A.7}$$

▨ **EXAMPLE A.1**
(Continued)

For example, the variance of the number of children can be calculated as follows:

Y_i	$Y_i - \bar{Y}$	$(Y_i - \bar{Y})^2$
1	−0.4	0.16
2	+0.6	0.36
1	−0.4	0.16
2	+0.6	0.36
1	−0.4	0.16
Total	0	1.20

Therefore, since $n = 5$,

$$S^2 = (\frac{1}{5})(1.20) = 0.24.$$

The standard deviation is $\sqrt{0.24}$, or about 0.49. ◪

When the observations have been grouped in the form of a frequency or relative frequency distribution, the calculation of the variance can be simplified. If a variable Y takes values $y_1, y_2, \ldots, y_m$, with frequencies $f(y_1), f(y_2), \ldots, f(y_m)$, then the variance of Y is given by:

$$S^2 = \frac{1}{n}[(y_1 - \bar{Y})^2 f(y_1) + (y_2 - \bar{Y})^2 f(y_2) + \cdots + (y_m - \bar{Y})^2 f(y_m)]$$
$$= \frac{1}{n} \sum_{i=1}^{m} (y_i - \bar{Y})^2 f(y_i). \tag{A.8}$$

This formula follows from Eq. A.6 because $f(y_1)$ observations have the value y_1 and deviation $(y_1 - \bar{Y})$, $f(y_2)$ observations have deviation $(y_2 - \bar{Y})$, and so on. The variance can also be calculated using the relative frequencies. Since $r(y_i) = f(y_i)/n$,

$$S^2 = (y_1 - \bar{Y})^2 \frac{f(y_1)}{n} + (y_2 - \bar{Y})^2 \frac{f(y_2)}{n} + \cdots + (y_m - \bar{Y})^2 \frac{f(y_m)}{n}$$
$$= \sum_{i=1}^{m} (y_i - \bar{Y})^2 r(y_i). \tag{A.9}$$

◪ **EXAMPLE A.1**
(Continued)

The average number of children was earlier found to be $\bar{Y} = 1.4$. The frequency and relative frequency distributions of the number of children are shown in cols. (1) to (3) of the following table.

y_i (1)	$f(y_i)$ (2)	$r(y_i)$ (3)	$(y_i - \bar{Y})$ (4)	$(y_i - \bar{Y})^2$ (5)	$(y_i - \bar{Y})^2 f(y_i)$ (6)	$(y_i - \bar{Y})^2 r(y_i)$ (7)
1	3	0.6	-0.4	0.16	0.48	0.096
$m = 2$	2	0.4	$+0.6$	0.36	0.72	0.144
	$n = 5$	1.0			1.20	0.240

The variance of the number of children can be calculated using the frequencies in col. (2), as shown in cols. (4) to (6):

$$S^2 = \frac{1}{n} \sum_{i=1}^{m} (y_i - \bar{Y})^2 f(y_i) = \frac{1}{5} 1.20 = 0.24.$$

Alternatively, the variance may be calculated using the relative frequencies of col. (3), as shown in cols. (4), (5) and (7):

$$S^2 = \sum_{i=1}^{m} (y_i - \bar{Y})^2 r(y_i) = 0.24.$$

The results are the same.[2] ☑

The variance of a distribution can also be written in a form more convenient for hand calculations. If raw data are used,

$$S^2 = \frac{1}{n} \sum_{i=1}^{n} Y_i^2 - \bar{Y}^2.$$

If the data are grouped in the form of a frequency distribution, then

$$S^2 = \frac{1}{n} \sum_{i=1}^{m} y_i^2 f(y_i) - \bar{Y}^2.$$

In terms of relative frequencies, this is

$$S^2 = \sum_{i=1}^{m} y_i^2 r(y_i) - \bar{Y}^2.$$

These expressions can be shown to be equivalent to (A.6), (A.8), and (A.9), respectively.

The mean and variance are used primarily to compare variables measured in comparable units. For example, Figure A.1 shows the starting salaries of three accountants and four economists (the data are, of course, fictitious).

It is clear from a visual inspection of Figure A.1 that accountants, on average, earn more than economists; also, that their salaries tend to vary less around their average salary than do the salaries of economists. We would have reached exactly the same conclusions had we compared instead the means (35 v. 31.5) and variances (4.667 v. 30.250) of the two sets of observations.

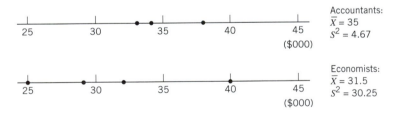

Figure A.1 Comparison of starting salaries

---

[2]When the number of distinct values of Y in the raw data is large, it may be more convenient to form a distribution using intervals—for instance, by listing the number or percentage of observations with Y values in the intervals, say, 0 to 9.99, 10 to 19.99, and so on. In such a case, the mean and variance of Y can be calculated exactly only from the raw data; they may, however, be approximated by assuming that in each interval all the observations have a Y value equal to the midpoint of the interval.

Similar comparisons can be based on grouped data. To illustrate, suppose that the distributions of starting salary of accountants and economists are as depicted in Figure A.2. The same scale is used in both panels.

Accountants' salaries tend to vary less around their average salary than do economists' salaries; the variance of accountants' salaries will therefore be smaller than that of economists' salaries.

A.4   PROBABILITY

The term "probability" is part of our everyday speech. Probability describes the likelihood of occurrence of a particular event. It has certain familiar properties: it is expressed as a number between 0 and 1, with a 0 indicating an impossible event and a 1 indicating that an event is certain to occur; probabilities between 0 and 1 indicate various degrees of likelihood, ranging from "very unlikely" to "very likely." (We make no attempt here to define precisely the terms in quotation marks.)

In assessing the likelihood of the occurrence of an event, at least two approaches may be distinguished.

The equal likelihood approach: If a "process" results in one of k mutually exclusive and collectively exhaustive events, and if it is reasonable to consider these events equally likely, then, according to this approach, the probability that one of these events will occur is $1/k$. For example, if the events "heads" and "tails" of a toss of a coin are considered equally likely, then the probability that tails will occur should be $1/2$; the probability of heads should also be $1/2$. The probability that a six will occur when an ordinary die is rolled should be $1/6$, if each of the six faces of the die is considered equally likely to show up.

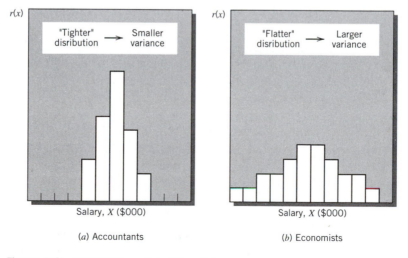

Figure A.2   Distributions of starting salaries

When we say that the events are equally likely, we express the belief that were we to observe the process (e.g., the flipping of a coin) a large number of times, we should expect to find the events (heads, tails) occurring with about equal relative frequencies.

The relative frequency approach: Under this approach, the probability of an event is set equal to the relative frequency of its occurrence in a large number of past observations. Implied in this approach is the assumption that the process remains stable—an essential condition for using past experience as a guide to future action.

Consider, for example, assessing the probability that a 40-year-old man applying for life insurance coverage will die during or survive the following five-year period. There are two possible events, death and survival, but these are not equally likely, as we know from intuition and from numerous studies. Indeed, survival is much more likely than death. Recent studies indicate that about 1.75% of men who reached age 40 died before they turned 45, and the remaining 98.25% survived. If the insurance company has no reason to believe that the mortality rate in the near future will differ from that observed in the near past, it may assume—as indeed it does—that, in the future, 1.75% of 40-year-old men will die before they turn 45. It may treat the number 0.0175 as the probability that any one of these persons will die before turning 45.

In this book, we rely mainly on the principle of equal likelihood for the assessment of probabilities. We interpret the probability of an event as the expected relative frequency of the event in a large number of future repetitions of the process, if such repetitions are possible.

A.5 PROBABILITY DISTRIBUTIONS

A *probability distribution* is a list showing the possible values of a variable (or the possible categories of an attribute) and the associated probabilities.

◢ EXAMPLE A.2 A box contains three balls; one is black, one is yellow, one is red. The balls will be mixed thoroughly and one of them will be drawn blindly. The probability distribution of its color is as follows.

Color	Probability
Black	1/3
Yellow	1/3
Red	1/3
Total	1

Because of the thorough mixing that is assumed to precede the draw, it is reasonable to consider the three events equally likely. If this process were to be repeated a large number of times (replacing the previously selected ball in the box

and mixing thoroughly prior to each draw), it is reasonable to expect that the color of the drawn ball will be black in one-third of the draws in the long run. ◪

Provided that the events are mutually exclusive and collectively exhaustive, the probability of this *or* that event occurring is simply equal to the sum of their probabilities. In the above example, for instance, the probability that the ball will be red *or* yellow is 1/3 plus 1/3, or 2/3. This "addition rule" for probabilities is eminently reasonable if we think of probabilities as expected relative frequencies, and it can be extended in a straightforward way.

In view of the similarity between relative frequencies and probabilities, it is not surprising that the mean, variance, and standard deviation of a probability distribution are defined and calculated in exactly the same way as for a relative frequency distribution.

For the sake of completeness we shall briefly restate these definitions, but it can be observed that *all the expressions of this section are identical to those in Section A.3,* the only difference being that probabilities, denoted by $p(x)$, replace relative frequencies, denoted by $r(x)$.

If the possible values of the variable Y are $y_1, y_2, \ldots, y_m$ with probabilities $p(y_1), p(y_2), \ldots, p(y_m)$, the *mean* or *expected value* of Y is

$$E(Y) = y_1 p(y_1) + y_2 p(y_2) + \cdots + y_m p(y_m) = \sum_{i=1}^{m} y_i p(y_i).$$

$E(Y)$ can be interpreted as the expected average value of Y in a large number of future repetitions of the process.

The *variance* of a probability distribution is defined exactly as in relative frequency distributions:

$$Var(Y) = \sum_{i=1}^{m} [y_i - E(Y)]^2 p(y_i).$$

An alternative expression is often more convenient for calculations by hand:

$$Var(Y) = \sum_{i=1}^{m} y_i^2 p(y_i) - [E(Y)]^2.$$

$Var(Y)$ can be interpreted as the expected average squared deviation of the values of the variable from its mean, $E(Y)$, in a large number of future repetitions of the process.

The *standard deviation* of Y is simply the square root of the variance:

$$Sd(Y) = +\sqrt{Var(Y)}.$$

◪ **EXAMPLE A.3**   The probability distribution of the variable Y is given in cols. (1) and (2) of the following table.

y_i (1)	$p(y_i)$ (2)	$y_i p(y_i)$ (3)	$y_i - E(Y)$ (4)	$[y_i - E(Y)]^2$ (5)	$[y_i - E(Y)]^2 p(y_i)$ (6)	y_i^2 (7)	$y_i^2 p(y_i)$ (8)
0	0.5	0.0	−0.9	0.81	0.405	0	0.0
1	0.2	0.2	0.1	0.01	0.002	1	0.2
2	0.2	0.4	1.1	1.21	0.242	4	0.8
3	0.1	0.3	2.1	4.41	0.441	9	0.9
	1.0	$E(Y) = 0.9$			$Var(Y) = 1.09$		1.9

The expected value of Y is 0.9, as calculated in col. (3). The variance may be calculated either as shown in col. (6), or, using the sum in col. (8), as

$$Var(Y) = \sum_{i=1}^{m} y_i^2 p(y_i) - [E(Y)]^2 = (1.9) - (0.9)^2 = 1.09.$$

It follows that the standard deviation of Y is $\sqrt{1.09}$ or 1.04. ◾

A.6 CORRELATION

We turn now to the measurement of the direction and strength of the relationship between two variables.

◾ **EXAMPLE A.4** Let us suppose that a questionnaire sent to a group of five adults included the following questions: What is your age? What is your income? The responses were as follows.

ID No.	Age (years), X	Income ($000), Y
1	35	29
2	50	46
3	20	15
4	60	54
5	40	36

Figure A.3 is a *scatter diagram* of age and income. Each point in this diagram represents one pair of values of the two variables.

Figure A.3 suggests that the two variables are related. The relationship can be described roughly by the straight line in the figure. It appears that as age increases, income increases. It also seems that the relationship is reasonably "strong," in the sense that the points tend not to deviate much from that line. ◾

The scatter diagram is a useful tool for obtaining an impression of the nature and extent of the relationship between two variables. Graphs, however, also have certain limitations: their construction is time-consuming, and sometimes appearances can

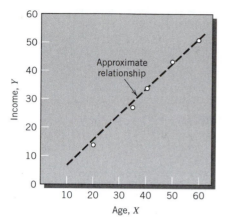

Figure A.3 Scatter diagram, Example A.4

be deceptive. The question then may be posed: Can the direction and strength of the relationship between two variables be summarized into a single number? With certain qualifications, the answer is yes. A widely used measure for this purpose is the *correlation coefficient*, which we shall now describe.

Suppose there are n pairs (X_i, Y_i) of values of two variables, X and Y. The *correlation coefficient* of X and Y is defined as:

$$r_{xy} = \frac{\sum_{i=1}^{n}(X_i - \bar{X})(Y_i - \bar{Y})}{\sqrt{\sum_{i=1}^{n}(X_i - \bar{X})^2}\sqrt{\sum_{i=1}^{n}(Y_i - \bar{Y})^2}}$$

$$= \frac{\sum_{i=1}^{n} X_i Y_i - n\bar{X}\bar{Y}}{\sqrt{\sum_{i=1}^{n} X_i^2 - n\bar{X}^2}\sqrt{\sum_{i=1}^{n} Y_i^2 - n\bar{Y}^2}}. \tag{A.10}$$

Before interpreting the correlation coefficient, we illustrate its calculation.

◼ EXAMPLE A.4
(Continued)

The preliminary calculations for the correlation coefficient of age and income are shown in Table A.1.

We find $\bar{X} = 205/5 = 41$ and $\bar{Y} = 180/5 = 36$. Also,

$$\sum_{i=1}^{n} X_i^2 - n\bar{X}^2 = (9325) - (5)(41)^2 = 920,$$

$$\sum_{i=1}^{n} Y_i^2 - n\bar{Y}^2 = (7394) - (5)(36)^2 = 914,$$

$$\sum_{i=1}^{n} X_i Y_i - n\bar{X}\bar{Y} = (8295) - (5)(41)(36) = 915.$$

TABLE A.1	Calculation of r_{xy}	Illustrated			
ID No.	X_i	Y_i	$X_i Y_i$	X_i^2	Y_i^2
1	35	29	1015	1225	841
2	50	46	2300	2500	2116
3	20	15	300	400	225
4	60	54	3240	3600	2916
5	40	36	1440	1600	1296
	205	180	8295	9325	7394

It follows that the correlation coefficient of age and income is

$$r_{xy} = \frac{915}{\sqrt{920}\sqrt{914}} = 0.998.$$

The properties of the correlation coefficient can be best understood with the help of Figure A.4, which shows six types of scatter diagrams and the associated approximate value of the correlation coefficient:

(a) The value of the correlation coefficient is always between -1 and $+1$.
(b) When all the pairs of values of X and Y lie on a straight line, r_{xy} is equal to $+1$ if the line is upward-sloping (Figure A.4a) or to -1 if the line is downward-sloping (Figure A.4d).
(c) When the pairs of (X, Y) values tend to cluster along an upward-sloping line, the value of r_{xy} will be a positive number between 0 and 1; the closer the points cluster around the line, the closer r_{xy} will be to $+1$ (Figure A.4b). Similarly, the closer the points cluster around a downward-sloping line, the closer will r_{xy} be to -1 (Figure A.4c).
(d) When, as shown in Figure A.4e, there is no relationship between X and Y, r_{xy} will tend to be near 0. Note, however, that r_{xy} may be near 0 also for certain types of curvilinear relationships, as, for example, in Figure A.4f.

The correlation coefficient may therefore be described as a standardized measure of the direction and strength of a *linear* relationship between two variables.
Equation A.10 can also be written as

$$r_{xy} = \frac{\frac{1}{n}\sum_{i=1}^{n}(X_i - \bar{X})(Y_i - \bar{Y})}{\sqrt{\frac{1}{n}\sum_{i=1}^{n}(X_i - \bar{X})^2}\sqrt{\frac{1}{n}\sum_{i=1}^{n}(Y_i - \bar{Y})^2}} = \frac{S_{xy}}{S_x S_y}.$$

The numerator is called the "covariance" of X and Y; the denominator is the product of the standard deviations of X and Y.

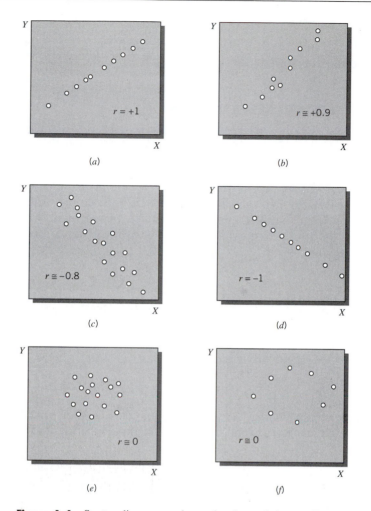

Figure A.4   Scatter diagrams and associated correlation coefficients

Thus, the *covariance* of X and Y, S_{xy}, is

$$S_{xy} = \frac{1}{n} \sum_{i=1}^{n} (X_i - \bar{X})(Y_i - \bar{Y}) = \frac{1}{n} \sum_{i=1}^{n} X_i Y_i - \bar{X}\bar{Y}. \qquad \text{(A.11)}$$

In Example A.4, the covariance of age and income is

$$S_{xy} = \frac{1}{5}(8295) - (41)(36) = 183.$$

The covariance is best thought of as an nonstandardized correlation coefficient: it has the same sign as the correlation coefficient, but does not necessarily lie between -1 and $+1$.

A.7 LEAST SQUARES ESTIMATION

An exact linear relationship between two variables X and Y can be written as

$$Y = a + bX,$$

where a and b are the *parameters* of the relationship. "Exact" means that all pairs of (X, Y) values lie on the line. Figure A.5 illustrates an exact linear relationship.

When, as in Figure A.3, the relationship is *approximately* linear, it is reasonable to look for a line that best fits the scatter of the observations. The expression for that line would be

$$\hat{Y} = a + bX,$$

where $\hat{Y}$ denotes the *estimated* or *predicted* value of Y.

Figure A.6 shows a typical point in the scatter diagram, (X_i, Y_i), a candidate line, $\hat{Y} = a + bX$; the predicted value of Y when $X = X_i$, $\hat{Y}_i$; and the *deviation* between actual and predicted value:

$$U_i = Y_i - \hat{Y}_i = Y_i - (a + bX_i).$$

Under *the method of least squares* (or *regression*, as it is also referred to), the parameters a and b are chosen so as to minimize the sum of the squared deviations

$$\sum_{i=1}^{n} U_i^2 = \sum_{i=1}^{n} (Y_i - a - bX_i)^2.$$

These values are known as the *least squares estimates* of the parameters.

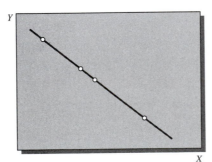

X **Figure A.5**   An exact linear relationship

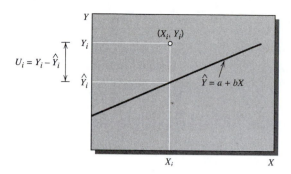

Figure A.6 Notation illustrated

With the help of calculus, it can be shown that the least squares estimates are given by

$$b = \frac{\sum_{i=1}^{n}(X_i - \bar{X})(Y_i - \bar{Y})}{\sum_{i=1}^{n}(X_i - \bar{X})^2} = \frac{\sum_{i=1}^{n} X_i Y_i - n\bar{X}\bar{Y}}{\sum_{i=1}^{n} X_i^2 - n\bar{X}^2}$$

$$a = \bar{Y} - b\bar{X}$$

◪ EXAMPLE A.4
(Continued)

Using the data of Table A.1, we calculate

$$b = \frac{(8295) - (5)(41)(36)}{(9325) - (5)(41)^2} = 0.9946$$

$$a = (36) - (0.9946)(41) = -4.7786$$

The estimated best fitting line is

$$\hat{Y} = -4.7786 + 0.9946X.$$

For example, the predicted value of Y when $X = 30$ is

$$\hat{Y} = (-4.7786) + (0.9946)(30) = 25.0594.$$

The best fitting line is plotted in Figure A.7. ◪

A special case of an approximate linear relationship between X and Y is one in which $a = 0$, that is, $\hat{Y} = bX$. In this case, $\hat{Y} = 0$ when $X = 0$, that is, the line is constrained to go through the origin ($X = 0$, $Y = 0$). The least squares estimate is that value of b which minimizes

$$\sum_{i=1}^{n} U_i^2 = \sum_{i=1}^{n}(Y_i - bX_i)^2$$

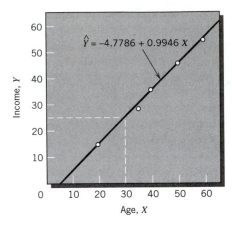

Figure A.7   Best fitting line, Example A.4

and can be shown to be

$$b = \frac{\sum_{i=1}^{n} X_i Y_i}{\sum_{i=1}^{n} X_i^2}.$$

Using the data of Table A.1 for illustration, we find

$$b = \frac{8295}{9325} = 0.889.$$

In general, if Y is approximately a linear function of k variables X_1, X_2, ... , X_k, that is,

$$\hat{Y} = a + b_1 X_1 + b_2 X_2 + \cdots + b_k X_k,$$

the least squares estimates are those values of a, b_1, b_2, ... , b_k which minimize

$$\sum_{i=1}^{n} U_i^2 = \sum_{i=1}^{n} (Y_i - a - b_1 X_{1i} - b_2 X_{2i} - \cdots - b_k X_{ki})^2,$$

and are invariably calculated with the help of special computer programs for regression. Examples of the output of such programs can be found in Chapters 6 and 9.

A.8 A NOTE ON VARIANCES AND COVARIANCES

Some sampling texts prefer to *define* the population variance of a variable Y as

$$\frac{1}{N-1} \sum_{i=1}^{N} (Y_i - \mu)^2, \tag{A.12}$$

and the sample variance of the same variable as

$$\frac{1}{n-1}\sum_{i=1}^{n}(Y_i - \bar{Y})^2. \tag{A.13}$$

Likewise, the covariance of Y and X is defined with denominator $N - 1$ or $n - 1$, rather than N or n as in this text.

It should be clear that all the expressions in this text could be written in terms of these definitions. For example, if we denote Eqs. A.12 and A.13 by $\tilde{\sigma}^2$ and $\tilde{S}^2$, respectively, then

$$\sigma^2 = \frac{N-1}{N}\tilde{\sigma}^2,$$

and $S^2 = (n-1)\tilde{S}^2/n$.

These definitions produce slightly simpler expressions and make their mathematical derivation easier. But they are confusing for nontechnical readers, appearing to suggest that the variances and covariances in sampling are different from those encountered in data analysis, probability theory, and other statistical subjects.

In this text, we have chosen to preserve the intuitively appealing definition of the variance as the average squared deviation from the mean. The reader, however, should rest assured that despite apparent differences, the results presented in this text are consistent with those of other texts.

APPENDIX B

Glossary and Technical Summary

B.1 NOTATION

Conventions:

Y Unknown variable of interest
X (and X_1, X_2, ...) Known auxiliary variable(s)
C Category of interest

Undivided population:

N Population size (number of elements in population)
μ Population mean of variable Y
τ $(= N\mu)$ Population total of variable Y
σ^2 Population variance of variable Y (or σ_y^2)
σ Population standard deviation of variable Y (or σ_y)
π Population proportion of elements in category C
τ' $(= N\pi)$ Population number of elements in category C
σ_{xy} Population covariance of variables X and Y
ρ_{xy} Population correlation coefficient of X and Y $(= \sigma_{xy}/\sigma_x\sigma_y)$
μ_r Ratio of population means of variables Y and X $(= \mu_y/\mu_x)$

Population divided into groups:

M	Number of groups
N_i	Size of (number of elements in) group i
w_i	$(= N_i/N)$ Proportion of elements in group i
μ_i	Mean of variable Y in group i
σ_i^2	Variance of variable Y in group i
π_i	Proportion of elements in category C in group i

Simple random sampling, simple estimators:

n	Sample size
$\bar{Y}$	Sample mean of variable Y, simple estimator of population mean of variable Y
T	$(= N\bar{Y})$ simple estimator of population total of variable Y
S^2	Sample variance of variable Y
S	Sample standard deviation of variable Y
P	Sample proportion in category C, simple estimator of population proportion in category C
T'	$(= NP)$ simple estimator of population number of elements in category C

Probability distributions of estimators:

$E(Z)$	Expected value of Z (Z being any estimator; e.g., $E(\bar{Y})$)
$Var(Z)$	Variance of Z (e.g., $Var(\bar{Y})$)
$\widehat{Var(Z)}$	Estimator of $Var(Z)$ (e.g., $\widehat{Var(\bar{Y})}$)

Stratified and two-stage sampling:

m	Number of selected groups
n_i	Size of sample from group i
$\bar{Y}_i$	Sample mean of variable Y in group i
S_i^2	Sample variance of variable Y in group i
P_i	Sample proportion in category C in group i
$\bar{Y}_s$	Stratified estimator of μ
P_s	Stratified estimator of π
T_s	Stratified estimator of τ
T'_s	Stratified estimator of τ'
$\bar{Y}_{ts}$	Two-stage estimator of μ
P_{ts}	Two-stage estimator of π
T_{ts}	Two-stage estimator of τ
T'_{ts}	Two-stage estimator of τ'

Other sampling methods and estimators:

$\bar{Y}_r$	Ratio estimator of μ
T_r	Ratio estimator of τ
$\bar{Y}_{lr}$	Regression estimator of μ
T_{lr}	Regression estimator of τ
$\bar{Y}_{ht}$	Horvitz–Thompson estimator of μ
T_{ht}	Horvitz–Thompson estimator of τ
P_{ht}	Horvitz–Thompson estimator of π
T'_{ht}	Horvitz–Thompson estimator of τ'
$\bar{Y}_{hh}$	Hansen–Hurwitz estimator of μ
T_{hh}	Hansen–Hurwitz estimator of τ
P_{hh}	Hansen–Hurwitz estimator of π
T'_{hh}	Hansen–Hurwitz estimator of τ'

Inclusion and selection probabilities:

p_j	Inclusion probability for element j
q_j	Probability that element j will be selected in any draw of sample with replacement

Prediction approach:

σ^2	Variance of deviations ($= Var(U_i)$)
S^2	Estimator of σ^2
$\tilde{E}$	($= T - \tau$) Prediction error
T^*	Best linear unbiased estimator (BLUE) of τ
$\tilde{E}^*$	Prediction error using BLUE T^*

$Z_{\alpha/2}$ for selected $1 - \alpha$

$1 - \alpha$	$Z_{\alpha/2}$	$1 - \alpha$	$Z_{\alpha/2}$
0.99	2.576	0.80	1.282
0.95	1.960	0.60	0.842
0.90	1.645	0.50	0.674

B.2 SIMPLE RANDOM SAMPLING (CHAPTER 3)

- An unbiased estimator of μ, the population mean of a variable Y, is the ordinary average of the n Y values in the sample:

$$\bar{Y} = \frac{1}{n}(Y_1 + Y_2 + \cdots + Y_n), \tag{3.1}$$

that is,

$$E(\bar{Y}) = \mu. \tag{3.2}$$

The variance of $\bar{Y}$ is given by

$$Var(\bar{Y}) = \frac{\sigma^2}{n}\frac{N-n}{N-1}. \tag{3.3}$$

- An unbiased estimator of π, the proportion of elements in the population that belong to a given category, is

$$P = \text{Proportion of sample elements in the category}, \tag{3.4}$$

that is,

$$E(P) = \pi. \tag{3.5}$$

The variance of P is given by

$$Var(P) = \frac{\pi(1-\pi)}{n}\frac{N-n}{N-1}. \tag{3.6}$$

- Provided that the accuracy requirements are demanding enough, the size (n) of a simple random sample needed to estimate μ or π within $\pm c$ with probability $(1-\alpha)$ is

$$n = \frac{NA}{(N-1)D^2 + A}, \tag{3.13}$$

where $D = (c/Z_{\alpha/2})$, and $A = \pi(1-\pi)$ when estimating π, or $A = \sigma^2$ when estimating μ. For estimating $\tau' = N\pi$ or $\tau = N\mu$ within $\pm c'$ with probability $(1-\alpha)$, apply Eq. 3.15 with $c = c'/N$.
- When n and $N - n$ are large, an approximate $100(1-\alpha)\%$ *confidence interval* for μ, the population mean of a variable Y, is

$$\bar{Y} \pm Z_{\alpha/2}\sqrt{\widehat{Var(\bar{Y})}}, \tag{3.14}$$

while one for π, the population proportion in a given category, is

$$P \pm Z_{\alpha/2}\sqrt{\widehat{Var(P)}}. \tag{3.15}$$

In the preceding expressions, $\widehat{Var(\bar{Y})}$ is an unbiased estimator of $Var(\bar{Y})$ given by

$$\widehat{Var(\bar{Y})} = \frac{S^2}{n-1} \frac{N-n}{N}, \tag{3.16}$$

and $\widehat{Var(P)}$ is an unbiased estimator of $Var(P)$ given by

$$\widehat{Var(P)} = \frac{P(1-P)}{n-1} \frac{N-n}{N}. \tag{3.17}$$

In Eq. 3.18 S^2 is the sample variance of the variable Y:

$$S^2 = \frac{1}{n} \sum_{i=1}^{n} (Y_i - \bar{Y})^2 = \frac{1}{n} \sum_{i=1}^{n} Y_i^2 - \bar{Y}^2. \tag{3.18}$$

- For large n and $N-n$, approximate $100(1-\alpha)\%$ confidence intervals for the total of a variable ($\tau = N\mu$) and the total in a category ($\tau' = N\pi$) are obtained by multiplying by N the limits of the above intervals, to get

$$N\left[\bar{Y} \pm Z_{\alpha/2} \sqrt{\widehat{Var(\bar{Y})}}\right] \tag{3.19}$$

and

$$N\left[P \pm Z_{\alpha/2} \sqrt{\widehat{Var(P)}}\right], \tag{3.20}$$

respectively.

B.3 STRATIFIED RANDOM SAMPLING (CHAPTER 4)

- An unbiased estimator of the population mean, μ of a variable Y is the *stratified estimator of μ*:

$$\bar{Y}_s = w_1 \bar{Y}_1 + w_2 \bar{Y}_2 + \cdots + w_M \bar{Y}_M = \sum_{i=1}^{M} w_i \bar{Y}_i, \tag{4.1}$$

and its variance is given by

$$Var(\bar{Y}_s) = \sum_{i=1}^{M} w_i^2 \frac{\sigma_i^2}{n_i} \frac{N_i - n_i}{N_i - 1}. \tag{4.2}$$

- An unbiased estimator of the proportion, π, of population elements belonging to a given category C is the *stratified estimator of π*

$$P_s = w_1 P_1 + w_2 P_2 + \cdots + w_M P_M = \sum_{i=1}^{M} w_i P_i, \tag{4.3}$$

and its variance is given by

$$Var(P_s) = \sum_{i=1}^{M} w_i^2 \frac{\pi_i(1 - \pi_i)}{n_i} \frac{N_i - n_i}{N_i - 1}. \tag{4.4}$$

- Provided that the accuracy requirements are demanding enough, the total size (n) of a stratified sample needed to estimate μ or π within $\pm c$ with probability $(1 - \alpha)$ and to be allocated to the groups according to $n_i = nv_i$ can be shown to be

$$n = \frac{\sum_{i=1}^{M}(N_i^2 A_i / v_i)}{N^2 D^2 + \sum_{i=1}^{M} N_i A_i} \tag{4.14}$$

where $D = (c/Z_{\alpha/2})$, and $A_i = \pi_i(1 - \pi_i)$ when estimating π, or $A_i = \sigma_i^2$ when estimating μ. For estimating $\tau' = N\pi$ or $\tau = N\mu$ within $\pm c'$ with probability $(1 - \alpha)$, apply Eq. 4.14 with $c = c'/N$.
- When all n_i and $N_i - n_i$ are large, an approximate $100(1 - \alpha)\%$ confidence interval for μ, the population mean of a variable Y, is

$$\bar{Y}_s \pm Z_{\alpha/2} \sqrt{\widehat{Var(\bar{Y}_s)}}, \tag{4.15}$$

and that for π, the population proportion in a given category, is

$$P_s \pm Z_{\alpha/2} \sqrt{\widehat{Var(P_s)}}. \tag{4.16}$$

$\widehat{Var(\bar{Y}_s)}$ is an unbiased estimator of $Var(\bar{Y}_s)$ given by

$$\widehat{Var(\bar{Y}_s)} = \sum_{i=1}^{M} w_i^2 \frac{S_i^2}{n_i - 1} \frac{N_i - n_i}{N_i}, \tag{4.17}$$

and $\widehat{Var(P_s)}$ is an unbiased estimator of $Var(P_s)$ given by

$$\widehat{Var(P_s)} = \sum_{i=1}^{M} w_i^2 \frac{P_i(1 - P_i)}{n_i - 1} \frac{N_i - n_i}{N_i}. \tag{4.18}$$

In Eq. 4.17, S_i^2 is the variance of Y in the sample from group i.
- Approximate $100(1 - \alpha)\%$ confidence intervals for the population totals $\tau = N\mu$ and $\tau' = N\pi$ are obtained by multiplying the upper and lower limits of (4.15) and (4.16), respectively, by N.

B.4 TWO-STAGE RANDOM SAMPLING (CHAPTER 5)

- An unbiased estimator of the population mean of a variable Y is the *two-stage estimator of μ*

$$\bar{Y}_{ts} = \frac{M}{m}(w_1\bar{Y}_1 + w_2\bar{Y}_2 + \cdots + w_m\bar{Y}_m) = \frac{M}{m}\sum_{i=1}^{m} w_i\bar{Y}_i, \qquad (5.1)$$

and its variance is given by

$$Var(\bar{Y}_{ts}) = \left(\frac{M}{N}\right)^2 \frac{\sigma_{01}^2}{m} \frac{M-m}{M-1} + \frac{M}{m}\sum_{i=1}^{M} w_i^2 \frac{\sigma_i^2}{n_i} \frac{N_i-n_i}{N_i-1}. \qquad (5.2)$$

- An unbiased estimator of the proportion of elements in the population belonging to a category C is the *two-stage estimator of π*

$$P_{ts} = \frac{M}{m}(w_1 P_1 + w_2 P_2 + \cdots + w_m P_m) = \frac{M}{m}\sum_{i=1}^{m} w_i P_i, \qquad (5.3)$$

and its variance is given by

$$Var(P_{ts}) = \left(\frac{M}{N}\right)^2 \frac{\sigma_{02}^2}{m} \frac{M-m}{M-1} + \frac{M}{m}\sum_{i=1}^{M} w_i^2 \frac{\pi_i(1-\pi_i)}{n_i} \frac{N_i-n_i}{N_i-1}. \qquad (5.4)$$

In Eqs. 5.2 and 5.4,

$$\sigma_{01}^2 = \frac{1}{M}\sum_{i=1}^{M}\left(N_i\mu_i - \frac{N\mu}{M}\right)^2$$

and

$$\sigma_{02}^2 = \frac{1}{M}\sum_{i=1}^{M}\left(N_i\pi_i - \frac{N\pi}{M}\right)^2.$$

- When m, $M-m$, and all n_i and N_i-n_i are large, an approximate $100(1-\alpha)\%$ confidence interval for μ, the population mean of a variable Y, is

$$\bar{Y}_{ts} \pm Z_{\alpha/2}\sqrt{\widehat{Var(\bar{Y}_{ts})}}, \qquad (5.7)$$

and that for π, the population proportion in a category, is

$$P_{ts} \pm Z_{\alpha/2}\sqrt{\widehat{Var(P_{ts})}}. \qquad (5.8)$$

In Eq. 5.7, $\widehat{Var(\bar{Y}_{ts})}$ is an unbiased estimator of $Var(\bar{Y}_{ts})$:

$$\widehat{Var(\bar{Y}_{ts})} = \left(\frac{M}{N}\right)^2 \frac{S_{01}^2}{m-1} \frac{M-m}{M} + \frac{M}{m}\sum_{i=1}^{m} w_i^2 \frac{S_i^2}{n_i-1} \frac{N_i-n_i}{N_i}$$

and

$$S_{01}^2 = \frac{1}{m} \sum_{i=1}^{m} (N_i \bar{Y}_i - \bar{Y}_0)^2,$$

where $\bar{Y}_0 = (\sum_{i=1}^{m} N_i \bar{Y}_i)/m$. In Eq. 5.8, $\widehat{Var(P_{ts})}$ is an unbiased estimator of $Var(P_{ts})$:

$$\widehat{Var(P_{ts})} = \left(\frac{M}{N}\right)^2 \frac{S_{02}^2}{m-1} \frac{M-m}{M} + \frac{M}{m} \sum_{i=1}^{m} w_i^2 \frac{P_i(1-P_i)}{n_i-1} \frac{N_i - n_i}{N_i}$$

and

$$S_{02}^2 = \frac{1}{m} \sum_{i=1}^{m} (N_i P_i - P_0)^2,$$

where $P_0 = (\sum_{i=1}^{m} N_i P_i)/m$.

- $100(1-\alpha)\%$ confidence intervals for the totals $\tau = N\mu$ and $\tau' = N\pi$ are calculated by multiplying the limits of (5.7) and (5.8), respectively, by N.

B.5 RATIO AND REGRESSION ESTIMATORS (CHAPTER 6)

- The *ratio estimator* of the population mean of a variable Y is

$$\bar{Y}_r = \frac{\bar{Y}}{\bar{X}} \mu_x, \tag{6.9}$$

where μ_x is the known population mean of an auxiliary variable X.
- The ratio estimator is unbiased under the modified simple random sampling described in Section 6.3.
- The ratio estimator is approximately unbiased under simple random sampling when n and $N - n$ are large.
- For large n and $N - n$, under either simple or modified simple random sampling, the approximate variance of $\bar{Y}_r$ is

$$Var(\bar{Y}_r) \approx \frac{1}{n} \frac{N-n}{N-1} \sigma_r^2, \tag{6.10}$$

where

$$\sigma_r^2 = \sigma_y^2 + \mu_r^2 \sigma_x^2 - 2\mu_r \rho_{xy} \sigma_x \sigma_y. \tag{6.11}$$

- The *regression estimator* of the population mean of a variable Y is

$$\bar{Y}_{lr} = \bar{Y} + b(\mu_x - \bar{X}), \tag{6.12}$$

where μ_x is the known population mean of an auxiliary variable X, and

$$
\begin{aligned}
b &= \frac{\sum_{i=1}^{n}(X_i - \bar{X})(Y_i - \bar{Y})}{\sum_{i=1}^{n}(X_i - \bar{X})^2} \\
&= \frac{\sum_{i=1}^{n} X_i Y_i - n\bar{X}\bar{Y}}{\sum_{i=1}^{n} X_i^2 - n\bar{X}^2}.
\end{aligned}
\tag{6.13}
$$

- The regression estimator is unbiased under the modified simple random sampling described in Section 6.5.
- The regression estimator is approximately unbiased under simple random sampling when n and $N - n$ are large.
- For large n and $N - n$, under either simple or modified simple random sampling, the approximate variance of $\bar{Y}_{lr}$ is

$$
Var(\bar{Y}_{lr}) \approx \frac{1}{n}\frac{N-n}{N-1}\sigma_{lr}^2,
\tag{6.14}
$$

where

$$
\sigma_{lr}^2 = \sigma_y^2(1 - \rho_{xy}^2).
\tag{6.15}
$$

- Provided the accuracy requirements are stringent enough, the size of a simple or modified simple random sample needed for the *ratio estimate* to be in the interval $\mu_y \pm c$ with probability $1 - \alpha$ is given by

$$
n = \frac{N\sigma_r^2}{(N-1)D^2 + \sigma_r^2},
$$

where $D = (c/Z_{\alpha/2})$ and $\sigma_r^2 = \sigma_y^2 + \mu_r^2\sigma_x^2 - 2\mu_r\rho_{xy}\sigma_x\sigma_y$.
- Provided the accuracy requirements are stringent enough, the size of a simple or modified simple random sample needed for the *regression estimate* to be in the interval $\mu_y \pm c$ with probability $1 - \alpha$ is given by

$$
n = \frac{N\sigma_{lr}^2}{(N-1)D^2 + \sigma_{lr}^2},
\tag{6.16}
$$

where $D = (c/Z_{\alpha/2})$ and $\sigma_{lr}^2 = \sigma_y^2(1 - \rho_{xy}^2)$.
- For large n and $N - n$, an approximate $100(1 - \alpha)\%$ confidence interval for μ_y, the population mean of a variable Y, based on the *ratio estimator* under either simple or modified simple random sampling, is

$$
\bar{Y}_r \pm Z_{\alpha/2}\sqrt{\widehat{Var(\bar{Y}_r)}},
\tag{6.19}
$$

where

$$
\widehat{Var(\bar{Y}_r)} = \frac{S_r^2}{n}\frac{N-n}{N}
$$

and

$$S_r^2 = \frac{1}{n-1} \sum_{i=1}^{n} (Y_i - \hat{Y}_i)^2 = \frac{1}{n-1} \sum_{i=1}^{n} \left(Y_i - \frac{\bar{Y}}{\bar{X}} X_i\right)^2. \qquad (6.20)$$

A $100(1 - \alpha)\%$ confidence interval for $\tau_y = N\mu_y$, the population total of Y, is calculated by multiplying the limits in (6.19) by N.

- For large n and $N - n$, an approximate $100(1 - \alpha)\%$ confidence interval for μ_y, the population mean of a variable Y, based on the *regression estimator* under either simple or modified simple random sampling, is

$$\bar{Y}_{lr} \pm Z_{\alpha/2} \sqrt{\widehat{Var(\bar{Y}_{lr})}}, \qquad (6.17)$$

where

$$\widehat{Var(\bar{Y}_{lr})} = \frac{S_{lr}^2}{n} \frac{N-n}{N}$$

and

$$S_{lr}^2 = \frac{1}{n-2} \sum_{i=1}^{n} (Y_i - \hat{Y}_i)^2 = \frac{1}{n-2} \sum_{i=1}^{n} (Y_i - a - bX_i)^2. \qquad (6.18)$$

A $100(1 - \alpha)\%$ confidence interval for $\tau_y = N\mu_y$, the population total of Y, is calculated by multiplying the limits in (6.17) by N.

B.6 SOME SPECIAL TOPICS (CHAPTER 7)

The following results apply to simple random sampling only.

- For large n, an unbiased estimator of the ratio of population totals (τ_1/τ_2) or means (μ_1/μ_2) of any two variables Y_1 and Y_2 is

$$R = \frac{T_1}{T_2} = \frac{\bar{Y}_1}{\bar{Y}_2}, \qquad (7.1)$$

and its approximate variance is given by

$$Var(R) \approx \frac{1}{\mu_2^2} \frac{1}{n} \frac{N-n}{N-1} \frac{1}{N} \sum_{i=1}^{N} \left(Y_{1i} - \frac{\mu_1}{\mu_2} Y_{2i}\right)^2. \qquad (7.2)$$

- For large n and $N - n$, an approximate $100(1 - \alpha)\%$ confidence interval for τ_1/τ_2 or μ_1/μ_2 is

$$R \pm Z_{\alpha/2} \sqrt{\widehat{Var(R)}}, \qquad (7.3)$$

where

$$\widehat{Var(R)} = \frac{1}{\bar{Y}_2^2} \frac{1}{n} \frac{N-n}{N} \frac{1}{n-1} \sum_{i=1}^{n} (Y_{1i} - RY_{2i})^2. \qquad (7.4)$$

In the preceding expressions, it is assumed that $\mu_2 \neq 0$ and $\bar{Y}_2 \neq 0$.

- For large n, the poststratified estimator of the population mean of a variable Y is approximately unbiased, and its approximate variance is given by

$$Var(\bar{Y}_{ps}) \approx \frac{1}{n}(1 - \frac{n}{N}) \sum_{i=1}^{M} w_i \sigma_i^2 + \frac{1}{n^2} \sum_{i=1}^{M}(1 - w_i)\sigma_i^2, \qquad (7.6)$$

where σ_i^2 is the variance of the Y values of all elements in the ith group.

- Again for large n, an approximate $100(1 - \alpha)\%$ confidence interval for the population mean of the variable Y based on the poststratified estimator is given by

$$\bar{Y}_{ps} \pm Z_{\alpha/2}\sqrt{\widehat{Var(\bar{Y}_{ps})}}, \qquad (7.7)$$

where $\widehat{Var(\bar{Y}_{ps})}$ is Eq. 7.6 with the σ_i^2 replaced by

$$\hat{S}_i^2 = \frac{n_i}{n_i - 1} \frac{N_i - 1}{N_i} S_i^2.$$

S_i^2 is the variance of the Y values of the sampled elements falling into group i. When n_i and N_i are large, $\hat{S}_i^2 \approx S_i^2$.

B.7 SAMPLING WITHOUT REPLACEMENT AND UNEQUAL PROBABILITIES (CHAPTER 8)

- For any sample selection method with known inclusion probabilities, an unbiased estimator of the population mean of a variable Y is the *Horvitz–Thompson estimator (HTE) of* μ_y, given by

$$\bar{Y}_{ht} = \frac{1}{N} \sum_{i=1}^{d} \frac{Y_i}{p_i}, \qquad (8.1)$$

where d is the number of distinct elements in the sample, p_i is the probability that the ith distinct sample element will be included in the sample under a given sampling method (the *inclusion probability* of the element), and Y_i is its Y value. An unbiased estimator of the population total of Y is $T_{ht} = N\bar{Y}_{ht}$.

- The variance of $\bar{Y}_{ht}$ is

$$Var(\bar{Y}_{ht}) = \frac{1}{N^2} \sum_{i=1}^{N} \sum_{j>i}^{N}(p_i p_j - p_{ij})\left[\frac{Y_i}{p_i} - \frac{Y_j}{p_j}\right]^2, \qquad (8.2)$$

where p_{ij} denotes the joint inclusion probability of elements i and j.

- An unbiased estimator of the proportion of elements in the population that fall into a given category, the *Horvitz–Thompson estimator of* π, is

$$P_{ht} = \frac{1}{N} \sum_{i=1}^{d} \frac{Y_i'}{p_i} \tag{8.3}$$

and its variance is

$$Var(P_{ht}) = \frac{1}{N^2} \sum_{i=1}^{N} \sum_{j>i}^{N} (p_i p_j - p_{ij}) \left[\frac{Y_i'}{p_i} - \frac{Y_j'}{p_j} \right]^2, \tag{8.4}$$

where Y_i' (and Y_j') is equal to 1 if the ith (or jth) element belongs to the category, and to 0 if it does not. An unbiased estimator of the number of elements in the population that fall into the category is $T_{ht}' = N P_{ht}$.

- For large n and $N - n$, an approximate $100(1 - \alpha)\%$ confidence interval for μ based on the HTE is

$$\bar{Y}_{ht} \pm Z_{\alpha/2} \sqrt{\widehat{Var(\bar{Y}_{ht})}}, \tag{8.7}$$

where $\widehat{Var(\bar{Y}_{ht})}$ is an estimator of $Var(\bar{Y}_{ht})$ given by

$$\widehat{Var(\bar{Y}_{ht})} = \frac{1}{n} \frac{N-n}{N-1} S_{ht}^2$$

and

$$S_{ht}^2 = \frac{1}{d-1} \sum_{i=1}^{d} \left(\frac{d Y_i}{N p_i} - \bar{Y}_{ht} \right)^2. \tag{8.8}$$

- For large n and $N - n$, an approximate $100(1 - \alpha)\%$ confidence interval for π based on the HTE is

$$P_{ht} \pm Z_{\alpha/2} \sqrt{\widehat{Var(P_{ht})}}, \tag{8.9}$$

where $\widehat{Var(P_{ht})}$ is an estimator of $Var(P_{ht})$ given by

$$\widehat{Var(P_{ht})} = \frac{1}{n} \frac{N-n}{N-1} S_{ht}'^2$$

and

$$S_{ht}'^2 = \frac{1}{d-1} \sum_{i=1}^{d} \left(\frac{d Y_i'}{N p_i} - P_{ht} \right)^2. \tag{8.10}$$

In the last expression, Y_i' is 1 if the sample element falls into the category, or 0 if it does not.

- The corresponding confidence intervals for the population totals, $\tau = N\mu$ and $\tau' = N\pi$, are formed by multiplying by N the limits of (8.7) and (8.9), respectively.

B.8 SAMPLING WITH REPLACEMENT AND UNEQUAL PROBABILITIES (CHAPTER 8)

- If sampling is *with replacement* and known selection probabilities, an unbiased estimator of the population mean of a variable Y is the *Hansen–Hurwitz estimator (HHE) of* μ:

$$\bar{Y}_{hh} = \frac{1}{nN} \sum_{i=1}^{n} \frac{Y_i}{q_i}, \qquad (8.11)$$

where q_i is the selection probability of element i in any given draw. An unbiased estimator of the population total of Y is $T_{hh} = N\bar{Y}_{hh}$.

- The variance of $\bar{Y}_{hh}$ is

$$Var(\bar{Y}_{hh}) = \frac{1}{n} \sum_{i=1}^{N} \left[\frac{Y_i}{Nq_i} - \mu\right]^2 q_i = \frac{1}{n}\left[\frac{1}{N^2} \sum_{i=1}^{N} \frac{Y_i^2}{q_i} - \mu^2\right]. \qquad (8.12)$$

- The same expressions give an unbiased estimator, P_{hh}, of the proportion of elements in the population that belong to a given category (π), and its variance, except that μ in Eq. 8.12 is replaced by π, and that Y_i in Eqs. 8.11 and 8.12 is equal to 1 if the ith element belongs to the category, or to 0 if it does not. An unbiased estimator of the number of elements in the population that belong to the category is $T'_{hh} = NP_{hh}$.

- In the special case of sampling with PPS, $q_i = X_i / \sum_{j=1}^{N} X_j$.

- For large n, an approximate $100(1 - \alpha)\%$ confidence interval for μ based on the HHE is

$$\bar{Y}_{hh} \pm Z_{\alpha/2} \sqrt{\widehat{Var(\bar{Y}_{hh})}}, \qquad (8.13)$$

where $\widehat{Var(\bar{Y}_{hh})}$ is an estimator of $Var(\bar{Y}_{hh})$ given by

$$\widehat{Var(\bar{Y}_{hh})} = \frac{1}{n(n-1)} \sum_{i=1}^{n} \left(\frac{Y_i}{Nq_i} - \bar{Y}_{hh}\right)^2. \qquad (8.14)$$

- For large n, an approximate $100(1 - \alpha)\%$ confidence interval for π, the proportion of elements in the population that belong to a given category, is

$$P_{hh} \pm Z_{\alpha/2} \sqrt{\widehat{Var(P_{hh})}}, \qquad (8.15)$$

where $\widehat{Var(P_{hh})}$ is an estimator of $Var(P_{hh})$ given by

$$\widehat{Var(P_{hh})} = \frac{1}{n(n-1)} \sum_{i=1}^{n} \left(\frac{Y_i'}{Nq_i} - P_{hh}\right)^2. \qquad (8.16)$$

In the last expression, Y_i' is 1 if the sample element falls into the category, or 0 if it does not.

- The corresponding confidence intervals for the population totals, $\tau = N\mu$ and $\tau' = N\pi$, are formed by multiplying the limits of (8.13) and (8.15), respectively, by N.

B.9 SAMPLING FROM A PROCESS (CHAPTER 9)

- The best linear unbiased estimator, T^*, of the population total of a variable Y under Model A,

$$\text{Model A:} \begin{cases} Y_i = \beta X_i + U_i, \\ U_i \text{ independent, with } E(U_i) = 0 \text{ and } Var(U_i) = \sigma^2, \end{cases}$$

is

$$T^* = \sum_{i=1}^{n} Y_i + b \sum_{j \in R} X_j, \tag{9.7}$$

where b is the least squares estimator of β:

$$b = \frac{\sum_{i=1}^{n} X_i Y_i}{\sum_{i=1}^{n} X_i^2}. \tag{9.8}$$

The error variance associated with this estimator is

$$Var(\tilde{E}^*) = \sigma^2 \Big[\frac{(\sum_{j \in R} X_j)^2}{(\sum_{i=1}^{n} X_i^2)} + N - n \Big]. \tag{9.9}$$

- The best linear unbiased estimator, T^*, of the population total of a variable Y under Model B,

$$\text{Model B:} \begin{cases} Y_i = \beta X_i + U_i, \\ U_i \text{ independent, with } E(U_i) = 0, \text{ and } Var(U_i) = \sigma^2 X_i, \end{cases}$$

is

$$T^* = \sum_{i=1}^{n} Y_i + b \sum_{j \in R} X_j,$$

where b is the ratio estimator of β:

$$b = \frac{\sum_{i=1}^{n} Y_i}{\sum_{i=1}^{n} X_i} = \frac{\bar{Y}}{\bar{X}}.$$

The error variance associated with this estimator is

$$Var(\tilde{E}^*) = \sigma^2 \Big[\frac{N^2}{n} (1 - \frac{n}{N}) \frac{\bar{X}_R}{\bar{X}} \mu_x \Big].$$

- The best linear unbiased estimator, T^*, of the population total of a variable Y under Model C,

$$\text{Model C:} \begin{cases} Y_i = \beta X_i + U_i, \\ U_i \text{ independent, with } E(U_i) = 0, \text{ and } Var(U_i) = \sigma^2 X_i^2, \end{cases}$$

is

$$T^* = \sum_{i=1}^{n} Y_i + b \sum_{j \in R} X_j,$$

where b is the average of the ratios Y_i / X_i:

$$b = \frac{1}{n} \sum_{i=1}^{n} \frac{Y_i}{X_i}.$$

The error variance associated with this estimator is

$$Var(\tilde{E}^*) = \sigma^2 \left[\frac{1}{n} (\sum_{j \in R} X_j)^2 + (\sum_{j \in R} X_j^2) \right].$$

- The best estimator of the population mean of variable Y under Models A, B, or C is T^*/N.
- When the sample size, n, and the population remainder, $N - n$, are large, an approximate $100(1 - \alpha)\%$ confidence interval for the population total of a variable Y under Models A, B, or C is

$$T^* \pm Z_{\alpha/2} \sqrt{\widehat{Var(\tilde{E}^*)}}, \tag{9.10}$$

where $\widehat{Var(\tilde{E}^*)}$ is an estimator of $Var(\tilde{E}^*)$, obtained by replacing σ^2 in the expression for $Var(\tilde{E}^*)$ with

$$S^2 = \frac{1}{n-1} \sum_{i=1}^{n} \frac{(Y_i - bX_i)^2}{v_i},$$

where $v_i = 1$ for Model A, $v_i = X_i$ for Model B, and $v_i = X_i^2$ for Model C. A confidence interval for the population mean of the variable is calculated by dividing the limits of (9.10) by N.

- The best linear unbiased estimator of the population total of a variable Y under Model D,

$$\text{Model D:} \begin{cases} Y_i = \alpha + \beta X_i + U_i, \\ U_i \text{ independent, with } E(U_i) = 0, \text{ and } Var(U_i) = \sigma^2, \end{cases}$$

is

$$T^* = \sum_{i=1}^{n} Y_i + \sum_{j \in R} (a + bX_j) = N[\bar{Y} + b(\mu_x - \bar{X})], \tag{9.11}$$

where a and b are the least squares estimators:

$$b = \frac{\sum_{i=1}^{n} X_i Y_i - n\bar{X}\bar{Y}}{\sum_{i=1}^{n} X_i^2 - n\bar{X}^2}$$ (9.12)

$$a = \bar{Y} - b\bar{X}.$$

The error variance associated with this estimator is

$$Var(\tilde{E}^*) = \sigma^2 N^2 \Big[(\frac{1}{n} - \frac{1}{N}) + \frac{(\bar{X} - \mu_x)^2}{\sum_{i=1}^{n}(X_i - \bar{X})^2} \Big].$$ (9.13)

The best estimator of the population mean of the variable is T^*/N.

- For large n and $N - n$, an approximate $100(1 - \alpha)\%$ confidence interval for the population total of a variable Y under Model D is given by

$$T^* \pm Z_{\alpha/2} \sqrt{\widehat{Var(\tilde{E}^*)}},$$ (9.14)

where $\widehat{Var(\tilde{E}^*)}$ is an estimate of $Var(\tilde{E}^*)$, obtained by replacing σ^2 in Eq. 9.13 with

$$S^2 = \frac{1}{n-2} \sum_{i=1}^{n} (Y_i - \hat{Y}_i)^2.$$

The corresponding confidence interval for the population mean of the variable is obtained by dividing the limits of (9.14) by N.

- The best linear unbiased estimator of the population total of a variable Y under Model E,

Model E: $\begin{cases} Y_i = \alpha + \beta_1 X_{1i} + \beta_2 X_{2i} + \cdots + \beta_k X_{ki} + U_i, \\ U_i \text{ independent, with } E(U_i) = 0, \text{ and } Var(U_i) = \sigma^2, \end{cases}$

is

$$T^* = N[\bar{Y} + \sum_{j=1}^{k} b_j (\mu_j - \bar{X}_j)],$$ (9.15)

where b_j is the least squares estimator of β_j, μ_j is the population mean and $\bar{X}_j$ the sample mean of explanatory variable X_j. The best estimator of the population mean of variable Y is T^*/N.

APPENDIX C

Computing Instructions

C.1 GENERAL

The diskette attached to this text contains (a) the data files referred to in the text and cases, and (b) two computer programs.

The first program, CANABAG, is described in the case *Canabag Manufacturing Company—Part II*.

The second, SCALC, is an elementary, "no frills" sampling calculator. It is an interactive DOS program designed to assist the calculation of complicated sampling formulas.

The files in the diskette may be copied into a hard-disk directory. The sequence of DOS commands (assuming the directory is STEXT and the diskette is in drive A) is:

```
CD\
MD STEXT
CD \STEXT
XCOPY A:*.*
```

To delete these files and directory when they are no longer needed use:

```
CD\STEXT
DEL *.*
CD\
RD STEXT
```

The two DOS programs, SCALC and CANABAG, can be run either from the diskette or from the hard disk directory. The file README.TXT should be read for additional instructions.

C.2   THE CANABAG PROGRAM

To execute the program, type CANABAG and press the return (enter) key. The program requires the file CANABAGP.DAT, which it assumes to be located in the same directory as the program itself. The program stores automatically the displayed results in the file CANABAG.OUT. This file (also located in the current directory) may then be edited using any text editor or printed in the usual way. Figure C.1 shows a sample "conversation" with the computer using this program. The user's input is underlined.

The contents of the file CANABAG.OUT are displayed in Figure C.2.

```
Program will select the desired number of random samples without
replacement of a size to be specified from the CAMABAGP.DAT file
(see Canabag Mfg Co. -- Part II), and will store sample statistics
in the file CANABAG.OUT.


Enter the sample size (1 to 988): 10
Enter the number of samples to be selected (1 to 1000): 5
Enter a positive integer up to 9 digits long: 123456789
You may wish to make a note of this integer 123456789 ...
It can be used to replicate the simulation results.

Press any key to continue...
Sample size= 10, Number of samples= 5
     1      257.0210      342.2298      2.0220      6.0660   0.100   0.000
     2     1096.4610     2510.5311     -1.3020      3.9060   0.100   0.000
     3      526.0210      710.7504      6.5700     19.7100   0.100   0.000
     4      211.5890      250.9470     35.8790     98.3817   0.200   0.100
     5      879.3030      727.2220      0.0000      0.0000   0.000   0.000
```

Figure C.1   Program CANABAG, sample conversation

```
Sample size= 10, Number of samples= 5
     1      257.0210      342.2298      2.0220      6.0660   0.100   0.000
     2     1096.4610     2510.5311     -1.3020      3.9060   0.100   0.000
     3      526.0210      710.7504      6.5700     19.7100   0.100   0.000
     4      211.5890      250.9470     35.8790     98.3817   0.200   0.100
     5      879.3030      727.2220      0.0000      0.0000   0.000   0.000
```

Figure C.2   Listing of file CANABAG.OUT

C.3 THE SCALC PROGRAM

To execute the program, type SCALC and press enter. Then, follow the prompts.

In what follows can be found a sample conversation with the program to verify some of the calculations appearing in Chapter 3. The user's input is underlined.

```
*********************************************************
This is an elementary Sampling CALCulator, to accompany
P. Tryfos, Sampling for Applied Research, Wiley, 1996.
*********************************************************
N.B. If you are using this program to check calculations
done by hand, remember that they are subject to rounding
error at all intermediate stages.
*********************************************************


Specify approach or exit program:
     1. Randomization (random sampling)
     2. Prediction
     0. To exit program
Enter 1 or 2 or 0: 1
Specify sampling method:
          1. Simple random sampling
          2. Stratified random sampling
          3. Cluster random sampling
          4. Two-stage random sampling
          5. Sampling with unequal probabilities,
               without replacement.
          6. Sampling with unequal probabilities,
               with replacement.
Enter 1 to 6: 1
Specify estimator:
          1. Simple
          2. Ratio
          3. Regression
Enter 1, 2, or 3: 1
Specify target population characteristic:
          1. Mean (mu) or total (tau) of a variable
          2. Proportion (pi) or number(tau') in a category
          3. Ratio of totals or means of two variables
Enter 1, 2, or 3: 1
Specify objective of calculations:
          1. Point estimate
          2. Point and interval estimate
          3. Variance of estimator (theoretical)
          4. Sample size determination
Enter 1 to 4: 3
You will have to provide the following:
          - the population size (N)
```

```
                - the sample size (n)
                - the population variance of the variable (sigma^2)
Do you have this information? If not, press ESC to abort
the program.  Press any other key to continue... [Key Pressed]
Enter the population size (N): 6  [See Section 3.4]
Enter the sample size (n): 2
Enter the population variance of the variable (sigma^2): 8.472
N= 6, n= 2, sigma^2= 8.472000
Variance of barY is 3.388800. Standard deviation of barY is 1.840869
Variance of T is 121.997. Standard deviation of T is 11.0452


Specify approach or exit program:
     1. Randomization (random sampling)
     2. Prediction
     0. To exit program
Enter 1 or 2 or 0: 1
Specify sampling method:
              1. Simple random sampling
              2. Stratified random sampling
              3. Cluster random sampling
              4. Two-stage random sampling
              5. Sampling with unequal probabilities,
                   without replacement.
              6. Sampling with unequal probabilities,
                   with replacement.
Enter 1 to 6: 1
Specify estimator:
              1. Simple
              2. Ratio
              3. Regression
Enter 1, 2, or 3: 1
Specify target population characteristic:
              1. Mean (mu) or total (tau) of a variable
              2. Proportion (pi) or number(tau') in a category
              3. Ratio of totals or means of two variables
Enter 1, 2, or 3: 2
Specify objective of calculations:
              1. Point estimate
              2. Point and interval estimate
              3. Variance of estimator (theoretical)
              4. Sample size determination
Enter 1 to 4: 4
You will have to provide the following:
              - the population size (N)
              - the accuracy requirements (1-alpha, c)
              - the population proportion in the category (pi)
Do you have this information? If not, press ESC to abort
```

```
the program.  Press any other key to continue...
Enter the population size (N): 50000
Enter the population proportion in category (pi): 0.40
Specify accuracy requirements:
    1. specify desired probability (1-alpha):
      1=0.99 2=0.95 3=0.90 4=0.80 5=0.60 6=0.50
    Enter 1 to 6: 2
    2. specify desired margin (c or c'): 0.01
Specify the target population characteristic:
        1. a population mean or proportion
        2. a population total or number
Enter 1 or 2: 1
N= 50000, A= 0.240000, 1-alpha= 0.950000, c= 0.010000
Required sample size is 7785. [See Example 3.1]


Specify approach or exit program:
    1. Randomization (random sampling)
    2. Prediction
    0. To exit program
Enter 1 or 2 or 0: 1
Specify sampling method:
        1. Simple random sampling
        2. Stratified random sampling
        3. Cluster random sampling
        4. Two-stage random sampling
        5. Sampling with unequal probabilities,
            without replacement.
        6. Sampling with unequal probabilities,
            with replacement.
Enter 1 to 6: 1
Specify estimator:
        1. Simple
        2. Ratio
        3. Regression
Enter 1, 2, or 3: 1
Specify target population characteristic:
        1. Mean (mu) or total (tau) of a variable
        2. Proportion (pi) or number(tau') in a category
        3. Ratio of totals or means of two variables
Enter 1, 2, or 3: 1
Specify objective of calculations:
        1. Point estimate
        2. Point and interval estimate
        3. Variance of estimator (theoretical)
        4. Sample size determination
Enter 1 to 4: 2
You will have to provide the following:
```

```
                    - the population size (N)
                    - the confidence level (1-alpha)
                    - the sample size (n)
                    - the sample average of the variable (bar Y)
                    - the sample variance of the variable (S^2)
Do you have this information? If not, press ESC to abort
the program.  Press any other key to continue...
Specify confidence level:
 1=0.99 2=0.95 3=0.90 4=0.80 5=0.60 6=0.50
Enter 1 to 6: 2
Enter the population size (N): 10000 [See Example 3.3]
Enter the sample size (n): 800
Enter the sample average of the variable (bar Y): 95
Enter the sample variance of the variable (S^2): 1156
N= 10000, n= 800, barY= 95.000000, s2= 1156.000000, hatSdY= 1.153717
Estimate of mu is: 95.000000
Estimate of tau is: 950000.000000
0.95 confidence interval for mu is from 92.738714 to 97.261286.
0.95 confidence interval for tau is from 927387.138599 to 972612.861401.


Specify approach or exit program:
    1. Randomization (random sampling)
    2. Prediction
    0. To exit program
Enter 1 or 2 or 0: 0
```

APPENDIX D

Solutions to Selected Problems

D.1 CHAPTER 3

3.2 (a) $\bar{Y} = 1078.33$, $T = 107,833$; (b) $P = 0.5$, $T' = 50$.

3.4 (a) $\mu = 22$, $\sigma^2 = 16$; (c) $Var(\bar{Y}) = 2.667$; (g) $Var(P) = 0.0267$.

3.12 (a) 117; (b) cannot be calculated.

3.13 147.92 to 152.08, 147,920 to 152,080.

3.14 0.36 to 0.50, 644 to 904.

D.2 CHAPTER 4

4.3 (a) $\bar{Y}_s = 418.75$, $P_s = 0.291$; (b) $T_s = 3,350,000$, $T'_s = 2,330$.

4.5 (a) $Var(\bar{Y}_s) = 0.398$, $Var(P_s) = 0.037$; (b) $Var(\bar{Y}_s) = 1.111$, $Var(P_s) = 0.111$.

4.11 (a) 7,663 with $\pi_1 = 0.17$ and $\pi_2 = 0.12$; (b) 2,136 with $\sigma_1^2 = 2,140$ and $\sigma_2^2 = 86$.

4.13 (a) 21.78 to 24.22; 1,088,822 to 1,211,177; (b) 0.082 to 0.146, 4,091 to 7,309.

D.3 CHAPTER 5

5.2 $\bar{Y}_{ts} = 117.04$, $P_{ts} = 0.259$.

5.5 (b) $Var(\bar{Y}_{ts}) = 2.805$, $Var(P_{ts}) = 0.055$.

5.9 (a) $\bar{Y}_{ts} = 144.37$, $P_{ts} = 0.359$; (b) 110.15 to 178.60, 1,101,536 to 1,785,964, 0.280 to 0.439, and 2,802 to 4,388.

D.4 CHAPTER 6

6.2 (a) 0.8101, 0.7857, 0.7689; (b) T_r is 150,703, 146,164 and 143,038 respectively; $\bar{Y}_r$ is 130.14, 126.22, and 123.52 respectively.

6.7 (b) $a = 51.613$, $b = 0.6475$; (d) $\bar{Y}_{lr} = 155.62$; (e) $T_{lr} = 180,208$.

6.14 (a) 243; (b) 408; (c) 642; (d) 1,228.

6.16 (a) $S_r^2 = 1.3945$, $\widehat{Var}(\bar{Y}_r) = 0.2092$; (b) $S_{lr}^2 = 0.1826$, $\widehat{Var}(\bar{Y}_{lr}) = 0.0274$; (c) $\bar{Y}_{lr} = 6.66$, 6.34 to 6.99, 126.8 to 139.8; (d) $\bar{Y}_r = 4.96$, 4.07 to 5.86, 81.3 to 117.2.

D.5 CHAPTER 7

7.2 (b) 30.56, -6.4 to 67.5.

7.4 (a) 660,000; (b) 3,000; (c) 110; (d) 0.5.

7.8 (a) 0.790; (b) cannot be calculated; (c) 6.356; (d) 0.376.

7.10 (a) 0.3, 420,000; (b) 2.84; (c) 0.14 to 0.46, 2.22 to 3.46.

7.13 (a) 1.48; (b) 0.49; (c) 1.43; (d) 1.42; (e) 0.477 and 0.467.

D.6 CHAPTER 8

8.2 1/3 for all firms.

8.4 (a) $\bar{Y}_{ht} = 11$, $P_{ht} = 1/3$; (b) $\bar{Y}_{ht} = 10/3$, $P_{ht} = 1/3$.

8.7 (a) $\bar{Y}_{ht} = 46.22$, $P_{ht} = 0.44$.

8.13 (a) $\bar{Y}_{ht} = 134.51$, $P_{ht} = 0.059$; (b) $T_{ht} = 155,762$, $T'_{ht} = 68$; (c) $S_{ht}^2 = 561.149$, $S_{ht}'^2 = 0.0049$, $\widehat{Var}(\bar{Y}_{ht}) = 139.923$, $\widehat{Var}(P_{ht}) = 0.0012$. (d) 115.05 to 153.97 (for μ), 0.001 to 0.117 (for π), 133,229 to 178,295 (for τ), 2 to 135 (for τ').

8.18 (a) $\bar{Y}_{hh} = 46.667$, $P_{hh} = 0.555$; (b) $\widehat{Var}(\bar{Y}_{hh}) = 337.037$, $\widehat{Var}(P_{hh}) = 0.0864$. (c) 16.47 to 76.87 (for μ), 0.07 to 1.04 (for π), 1646.68 to 7,686.65 (for τ), 7 to 104 (for τ').

D.7 CHAPTER 9

9.4 $b^* = 0.1501$, $T^* = 27,925$, $T^*/N = 24.11$.

9.7 (a) $b^* = 0.7431$, $T^* = 138,342$, $T^*/N = 119.38$; (b) $-21,944$ to 298,429 (for τ), -18.95 to 257.71 (for μ).

9.9 (b) $b^* = 0.4419$, $T^* = 316,851$, $T^*/N = 273.62$. (c) 301,807 to 331,895 (for τ), 260.63 to 286.61 (for μ).

9.11 (a) $\hat{Y} = 164.98 + 0.1778X_1 - 0.1347X_2$, $T^* = 196,377$, $T^*/N = 169.58$.

BIBLIOGRAPHY

[1] Arens, A. A., and J. K. Loebbecke. *Applications of Statistical Sampling to Auditing*. Englewood Cliffs, NJ: Prentice Hall, 1981.

[2] Arkin, H. *Sampling Methods for the Auditor: An Advanced Treatment*. New York: McGraw-Hill, 1982.

[3] Arkin, H. *Handbook of Sampling for Auditing and Accounting*. New York: McGraw-Hill, 1993.

[4] Barnett, V. *Sample Survey Principles and Methods*. 2nd ed. London: Oxford University Press, 1991.

[5] Buzzard, K. *Electronic Media Ratings: Turning Audiences into Dollars and Sense*. Stoneham, MA: Focal, 1992.

[6] Chaudhuri, A. *Survey Sampling: Theory and Methods*. New York: Marcel Dekker, 1992.

[7] Chaudhuri, A., and R. Mukerjee. *Randomized Response: Theory and Techniques*. New York: Marcel Dekker, 1987.

[8] Chaudhuri, A. B., and J. W. Vos. *Unified Theory and Strategies of Survey Sampling*. New York: Elsevier Science Publishing, 1988.

[9] Cochran, W. *Sampling Techniques*. 3rd ed. New York: Wiley, 1977.

[10] Deming, W. E. *Some Theory of Sampling*. 1950. Reprint, New York: Dover, 1984.

[11] Deming, W. E. *Sample Design in Business Research*. New York: Wiley, 1960.

[12] Foreman, E. K. *Survey Sampling Principles*. New York: Marcel Dekker, 1991.

[13] Green, R. H. *Sampling Design and Statistical Methods for Environmental Biologists*. New York: Wiley 1979.

[14] Guenther, W. C. *Sampling Inspection in Statistical Quality Control*. London: Oxford University Press, 1987.

[15] Guy, D. M., D. R.Carmichael, and O.R. Whittington. *Audit Sampling: An Introduction*. 3rd ed. New York: Wiley, 1994.

[16] Gwilliam, D. R. *A Survey of Auditing Research*. London: Prentice-Hall International, 1987.

[17] Hajek, J. *Sampling from a Finite Population*. New York: Marcel Dekker, 1981.

[18] Hansen, M. H., W. N. Hurwitz, and W. G. Madow. *Sample Survey Methods and Theory*. Vols. I and II. New York: Wiley, 1953.

[19] Hedayat, A. S. and B. K. Sinha. *Design and Inference in Finite Population Sampling*. New York: Wiley, 1991.

[20] Henry, G. T. *Practical Sampling*. Newbury Park, CA: Sage, 1990.

[21] Kalton, G. *Introduction to Survey Sampling*. Newbury Park, CA: Sage, 1983.

[22] Kasprzyk, D. *et al.*, eds. *Panel Surveys*. New York: Wiley, 1989.

[23] Kish, J. L. *Survey Sampling*. New York: Wiley, 1965.

[24] Krishnaiah, P. R. and C. R. Rao, eds. *Sampling*. Vol. 6 in *Handbook of Statistics*. Amsterdam: North Holland Elsevier Science, 1988.

[25] Lavrakas, P. J. *Telephone Survey Methods: Sampling, Selection, and Supervision*. 2nd ed. Newbury Park, CA: Sage, 1993.

[26] Leslie, D.A., A. D. Teitlebaum, and R. J. Anderson. *Dollar-Unit Sampling: A Practical Guide for Auditors*. Toronto: Copp Clark Pitman, 1979.

[27] Levy, P. S., and S. Lemeshow. *Sampling of Populations: Methods and Applications*. New York: Wiley, 1991.

[28] Oppenheim, A. N. *Questionnaire Design, Interviewing and Attitude Measurement*. New ed. London: Pinter Publishers, 1992.

[29] Rosander, A. C. *Case Studies in Sample Design*. New York: M. Dekker, 1977.

[30] Rossi, P. H., J. D. Wright, and A. B. Anderson. *Handbook of Survey Research*. New York: Academic Press, 1985.

[31] Sarndal, C. E., B. Swensson, and J. H. Wretman, *Model Assisted Survey Sampling*. New York: Springer-Verlag, 1991.

[32] Scheaffer, R. L., W. Mendenhall, and L. Ott. *Elementary Survey Sampling*. 4th ed. Boston: PWS-Kent, 1990.

[33] Schilling, E. G. *Acceptance Sampling in Quality Control: Statistics*. Vol. 42. New York: Marcel Dekker, 1982.

[34] Sonquist, J. A., and W. C. Dunkelberg. *Survey and Opinion Research: Procedures for Processing and Analysis*. Englewood Cliffs, NJ: Prentice-Hall, 1977.

[35] Stephan, F. F., and P. J. McCarthy. *Sampling Opinions: An Analysis of Survey Procedure*. New York: Wiley, 1958.

[36] Stuart, A. *The Ideas of Sampling*. 3rd. ed. London: Oxford University Press, 1987.

[37] Sudman, S. *Applied Sampling*. New York: Academic Press, 1976.

[38] Sukhatme, P. V., S. Sukhatme, and C. Asok. *Sampling Theory of Surveys with Applications*. 3rd ed. Ames: Iowa State University Press, 1984.

[39] Taylor, D., and G. W. Glezen. *Auditing: Integrated Concepts and Procedures*. 5th ed. New York: Wiley, 1991.

[40] Thompson, S. K. *Sampling*. New York: Wiley, 1992.

[41] Warwick, D. P., and C. A. Lininger. *The Sample Survey: Theory and Practice*. New York: McGraw-Hill, 1975.

[42] Webster, J. G., and L. W. Lichty. *Ratings Analysis: Theory and Practice*. Hillsdale, NJ: Erlbaum, 1991.

[43] Wetherill, G. B. *Sampling Inspection and Quality Control*. 2nd ed. London: Chapman and Hall, 1977.

[44] Wilburn, A. J. *Practical Statistical Sampling for Auditors*. New York: Marcel Dekker, 1984.

[45] Williams, B. *A Sampler on Sampling*. New York: Wiley, 1978.

[46] Yates, F. *Sampling Methods for Censuses and Surveys*. 4th ed. London: Oxford University Press, 1987.

Index